Ordination methods for biologists: a non-mathematical introduction using R

Bill Shipley

Bill Shipley
Université de Sherbrooke
Département de biologie
Sherbrooke (Québec)
CANADA J1K 2R1
Bill.Shipley@USherbrooke.ca

ISBN 978-1-7778448-1-3 ISBN 978-1-7778448-0-6 (eBook)

Cover designed by Lyne Labrecque

À Clara, Maëva et la suite.
Vous serez capable de lire ce livre
beaucoup trop vite…

About the author

John William (Bill) Shipley is a full professor at the Université de Sherbrooke, specializing in plant ecology and statistical analysis. He received his honours BSc. from Bishop's University in 1983 and his PhD. in ecology from the University of Ottawa in 1987. He is the author of 3 other books and over 150 peer-reviewed scientific publications.

Other books

Shipley, B. 2000. Cause and correlation in biology: A user's guide to path analysis, structural equations and causal inference. Cambridge University Press, Cambridge UK. ISBN 0-521-79153 7

Shipley, B. 2010. From plant traits to vegetation structure: Chance and selection in the assembly of ecological communities. . Cambridge University Press, Cambridge UK. ISBN 978-0-521-11747-0

Shipley, B. 2016. Cause and correlation in biology: A user's guide to path analysis, structural equations and causal inference with R. Cambridge University Press, Cambridge UK. ISBN 978-1-107-44259-7

Table of Contents

Chapter 1 Introduction 10

1.1) Putting things in order 10

1.2) Who should read this book? 16

1.3) The **R** statistical language and program 16

Chapter 2 Principal components analysis 20

2.1 The goals of PCA 20

2.2) Visualizing how PCA works 26

2.3) Preparing your data 39

2.4) Dealing with proportions in PCA 46

2.5) A covariance matrix or a correlation matrix? 48

2.6) Conducting a PCA and interpreting the output 56

2.7) The "importance" of each principal component 64

2.8) From original variables to principal components 66

2.9) Putting observations on principal component axes 72

2.10) How to choose the principal components 75

2.11) Biplots for visualizing a PCA 81

2.12) Dealing with categorical variables 97

Chapter 3 Correspondence analysis and detrending 109

3.1) siteXspecies data in community ecology 111

3.2) Direct gradient analysis....113
3.3) Correspondence analysis or "reciprocal averaging"....129
3.4) Performing correspondence analysis in R....143
3.5) Detrended correspondence analysis....146
3.6) An empirical example of DCA....159
3.7) Interpreting site scores....164
Chapter 4 Principal coordinates and multidimensional scaling....187
4.1) Distance and (dis)similarity measures....188
4.2) Measures for categorical nominal variables....195
4.3) Measures for binary variables....198
4.4) Measures for categorical multistate ordinal variables....204
4.5) Measures for quantitative variables....206
4.6) Gower's distance....215
4.7) Multidimensional scaling....222
4.8) Principal coordinates analysis....226
4.9) (Non-metric) multidimensional scaling....237
4.10) Procrustes analysis....262
4.11) The Wisconsin understory plant communities....266
Chapter 5 Constrained ordination techniques....281
5.1) Exploratory vs. inferential statistical methods....281

5.2)	Setting the scene	285
5.3)	Redundancy Analysis (RDA)	292
5.4)	Obtaining output from rda()	299
5.5)	Hypothesis testing with rda()	307
5.6)	Graphical output of rda() via triplots	320
5.7)	An empirical example of rda()	329
5.8)	Partial Redundancy Analysis	342
5.9)	Canonical Correspondence Analysis (CCA)	348
5.10)	Partial canonical correspondence analysis	367
5.11)	Hypothesis testing with CCA	370
5.12)	Distance-based redundancy analysis (db-RDA)	377
Chapter 6	Gradient analysis using siteXspecies data	382
6.1)	What is an environmental gradient?	382
6.2)	Simulating environmental gradients	387
6.3)	Using simulations to compare ordination methods	396
6.4)	The "horseshoe" or "arch" effect	399
6.5)	Flexible shortest path adjustments	410
6.6)	Recommendations for siteXspecies ordinations	422
6.7)	Implications for constrained ordination methods	430
6.8)	Putting it all together	432

Chapter 7 Cheat sheet of major R functions....................................444

Bibliography 486

Chapter 1 Introduction

1.1) Putting things in order

The words “order” and “ordination” are closely related. The Latin word “ordinare” means “to put in order”, from which comes the Latin word “ordinatio”. At its simplest, an ordination of ecological data means to arrange these data is some specific order. An ordination method is a mathematical method that produces such a specific order but which specific order to we choose? Well, an order that reveals some ecological property that is not evident simply by looking at the data themselves. All of the complications of the different ordination methods that we will study arise because different hypothesized ecological properties require different ways of ordering our observations in order to reveal patterns that are not immediately obvious.

Let’s start with a very simple example: Imagine that I have randomly placed five $1m^2$ quadrats in a field and, in each quadrat, I have counted the number of rooted stems of a single species (S_1) that I observe. Here are the results:

Quadrat		S_1

Q1		8
Q2		4
Q3		11
Q4		10
Q5		7

Although we don't see any clear pattern in this table, an obvious way of re-arranging it would be to sort the quadrats in terms of the number of stems that each contains.

Quadrat	S_1
Q2	4
Q5	7
Q1	8

Q4	10
Q3	11

By doing this we have *ordinated* these quadrats from the one having the fewest individuals of species S_1 (quadrat Q2) to the one having the most individuals of species S_1 (quadrat Q3). The resulting order is Q2<Q5<Q1<Q4<Q3. This ordination reveals a property of our quadrats that we can call "increasing abundance of species S_1". Actually, since we are doing this using only the information in this one data set, we call this an indirect ordination. I will explain this more fully in Chapter 3. Now that we know the proper order, we could ask what other variable, associated with these quadrats, might explain this increase in abundance? We might expect that such a variable would also systematically increase or decrease as the abundance of species S_1 increases. If so, then the value of this other variable will also increase (or decrease) when we arrange its values following the same ordering our quadrats Q2<Q5<Q1<Q4<Q3. Perhaps this sequence of quadrats also represents an increasing amount of some limiting soil nutrient or a decreasing amount of herbivory.

I forgot to tell you something: When I was observing my quadrats, I also counted the number of rooted stems of a second species (S_2). Here is the full data set:

Quadrat	S_1	S_2
Q1	8	6
Q2	4	4
Q3	11	9
Q4	10	10
Q5	7	8

How should we arrange these data is some specific order so that we again reveal some ecological property that is not evident simply by looking at the data themselves? If, as we did before, we order the quadrats as Q2<Q5<Q1<Q4<Q3 then we are revealing the increasing abundance of S_1. If we order the quadrats as Q2<Q1<Q5<Q3<Q4 then we are revealing the increasing abundance of S_2. Even with only two species, the number of possibilities don't end there. We can combine these abundances in

different ways to reflect different properties of this two-species community. For instance, summing the abundances of the two species per quadrat will give the total abundance of all stems in each quadrat. Taking the absolute differences in the abundances of the two species per quadrat will give the difference in abundances of the two species in each quadrat. In everything that we have done so far, we have implicitly treated a stem of the first species as equal to a stem of the second species. Maybe the first species is 1.5 times bigger than the second species and we want to reflect this in our ordering of the quadrats? If so, then we would multiply each value of S_1 by 1.5 before doing our ordination.

I said earlier that an ordination method is simply a series of mathematical operations on data that produces such a specific order. How does this relate to the different ordinations that I described above? Each of the different possibilities described above represent different mathematical operations on my data. My data table (call it "**X**") is a matrix having five rows (quadrats 1 to 5) and two columns (species 1 and 2). Each cell of **X** holds one unique observation. I will call the value in the cell found in row i and column j as x_{ij}. For instance, x_{32} is the number of stems in row 3 (quadrat Q1) and column 2 (species S_2) and its value is nine. Now, if I want to ordinate my data table to reflect only the increasing abundance of species S_1 then, for each row of my data table (i.e. for each quadrat), I would first calculate the value of the function $f_i=1x_{i1}+0x_{12}$ and then sort the values of f_i. If I want my ordination to reflect only the increasing abundance of species S_2 then I would first calculate the value of the function $f_i=0x_{i1}+1x_{12}$ and then sort the values of f_i. If I want my ordination to reflect the increasing abundance of all stems, and also reflect the fact that species S_1 is 1.5 times bigger than species S_2, then I would first calculate the value of the function $f_i=1.5x_{i1}+1x_{12}$ and then sort the values of f_i.

Of course, any real ecological data set would contain many more than just two species. It is not unusual for real data sets of this type to contain several hundred species and several hundred quadrats. The patterns of association (for instance, their changes in abundance) of these species across these quadrats in such high dimensional data sets is impossible to see by just looking at the data. The purpose of the unconstrained ordination methods in this book (Chapters 2 to 4) is to reveal these patterns of association. This is a type of exploratory statistical analysis. The purpose of constrained ordination methods (Chapter 5) is to test hypotheses about what might be generating these patterns of association. This is a type of inferential statistical analysis.

I began this introduction by talking about data from community ecology: how different species change abundance across the landscape. I dedicate all of Chapter 6 to this type of data because it is one of the most common ecological applications of ordination but any type of data that can be arranged in a matrix of rows and columns can be ordinated and so I have also made sure to include examples from other branches of biology as well.

1.2) Who should read this book?

Of course, I hope that anyone with an interest in ordination methods will find this book useful but I have written it explicitly for biology students who want to understand how these methods work and and how to carry out actual analyses, but who don't have a background in mathematics. Therefore, I explain these methods without delving into the underlying mathematics while still providing you with a good intuitive understanding of what the mathematical operations are doing. I have assumed, when writing this book, that you have already mastered the basics of statistical analysis up to, and including, linear multiple regression and ANOVA. Since this book uses the **R** statistical language and environment, I also assume that you know the basics of using **R**. In summary, this book is aimed at upper level undergraduate or graduate students in biology who want to understand and apply ordination methods to their data.

1.3) The R statistical language and program

R (R-Core-Team 2020) is explicitly[1] called an "environment for statistical computing and graphics". It is an "environment: because it is both a computer language and an integrated suite of software facilities. It has become the most common programming "environment" used by academic statisticians and scientists for at least three reasons. First, it is a sophisticated language using object-oriented programming that is specifically adapted for statistical analysis. Second, being open-source, it is freely available to anyone for download from the central CRAN site at www.r-project.org. Third, this "environment" is organized by "packages", which are groups of **R** functions and associated documentation that are dedicated to specific tasks. Often, whenever a new statistical method is developed and published, a dedicated package is also produced and uploaded onto the main CRAN site. You can then download and install ti so that you can then use it within **R**. This means that the **R** environment is constantly changing as new packages, or new versions of existing packages, are added. This book uses **R** version 4.0.3. **R** is essentially a collection of functions bundled into packages. Details of how to use each function, including the arguments of these functions, can be found by typing `help("X")`, where X is the name of the function and you should get used to using these help files. Most of the dedicated editor programs for **R**, like RStudio, also have facilities for obtaining these help pages.

[1] www.r-project.org

R is a computer language that evolved from the pre-existing S language[2]. The S language was developed by John Chambers and colleagues at Bell Laboratories (formerly AT&T, now Lucent Technologies). The S language was an open source language. A modified open source version of S, now called (appropriately) **R**, was produced when S was converted into the commercial S-PLUS program. Many users use special editors when using **R** that facilitate and organize the **R** code and have many other advantages; this book uses the open-source version of RStudio[3]. I will always show **R** code and **R** objects using `11 point Courier New font` and will show the output of any **R** functions using `10 point Lucida Console font`.

Since the entire **R** project, including the hundreds of packages available for download, are provided free, these come without any guarantee. On the other hand, since most packages are based on peer-reviewed academic papers, the packages are largely free of bugs or other errors. Furthermore, since the more commonly used packages are used by researchers around the world, existing bugs or errors are usually quickly reported and corrected. You will also find a few new **R** functions in this book. I believe that these functions do not contain bugs, and have tested them as best I can, but again… no guarantees.

[2] There is a Wikipedia page that describes the history of the S language.
[3] www.rstudio.com

The production of complete **R** packages takes a lot of work and skill. Rather than receiving monetary payment for this work, the authors of these packages expect to be paid in the universal currency of academics: citations. To pay my dues while avoiding having to cite these functions every time that they are mentioned, I cite them once when they are first mentioned. I have also added a cheat sheet (Chapter 7) that explains each major function, its arguments, and its main output.

Chapter 2 Principal components analysis

2.1 The goals of PCA

Principal components analysis, often referred to by its acronym "PCA", is one of the oldest of the multivariate methods that we will discuss in this book. It is also one of the most commonly used of these multivariate methods. It was first developed by one of the pioneers of modern statistics, Karl Pearson (Pearson 1901) but was independently re-discovered in 1933 by Harold Hotelling who was another important person in modern statistics (Hotelling 1933). There are three main reasons why people use PCA:

1. To visualize the relationships between many variables in just a few dimensions ("dimensional reduction").
2. To obtain equations describing the relationships between many variables in just a few dimensions for communication or for extrapolation to new data.
3. To suggest hypotheses about the underlying causes of this covariation.

Please pay careful attention to the way that I have worded the third reason. PCA cannot "test" hypotheses about the underlying causes of patterns of covariation; for this you must use a different set of statistical methods called "structural equation modelling" (Shipley 2016). However, once we have seen the patterns of covariation (the first objective) and quantified these patterns of covariation in the form of equations that can be used to predict new observations (the second objective), this will sometimes suggest interesting explanations (hypotheses) concerning *why* these patterns might exist (Økland 1996).

An example of DNA variation in European people illustrates each of these goals (Novembre et al. 2008). The data used by these authors consisted of 1387 rows (observational units) and over half a million columns (variables). Each row represented a single person living in Europe and whose ancestors (at least, their grandparents) had all lived in the same geographical region as that person. Each column was the nucleotide of a variable region of human DNA (a SNP, or Single Nucleotide Polymorphism). For instance, different people might have either a cytosine (C) or an adenine (A) nucleotide at a given locus. As the chromosomes of the ancestors of a person mix and combine over the generations, SNPs that are physically close together on a chromosome, or those located on a Y-chromosome, tend to stay together while those that are more distantly located on a chromosome are more likely to be recombined during meiosis. SNPs that are located on different chromosomes will be mixed at each generation. As a result, the patterns of correlation between different SNPs in a person's genome will be determined by his or her genetic history Although the SNPs that people have in their DNA will be correlated, the strength of these correlations are typically very weak between any single pair of loci. In other words, if we plotted the presence/absence of a particular SNP at a given locus over many people, those people who share more ancestors, and those people whose ancestors are shared more recently, will have a slightly greater chance of also sharing the same SNP. This slight positive correlation, when looking at a single SNP, will be very weak. However, since these correlations are spread over millions of loci, the overall pattern of covariation between SNPs could be quite informative even though the amount of information is very low between any single pair. If only we could look at a scatterplot involving a half million axes (one for each SNP) then we could "see" this pattern of covariation. This important insight ignores a substantial problem: our universe only has three physical dimensions and so no one can make such a scatterplot. If only we could somehow reduce this half million axes into just two axes in such a way as to capture this pattern of correlation! This is what PCA attempts to do. Novembre et al (2008) did a PCA of this huge data set. Their results are summarized in Figure 2.1.

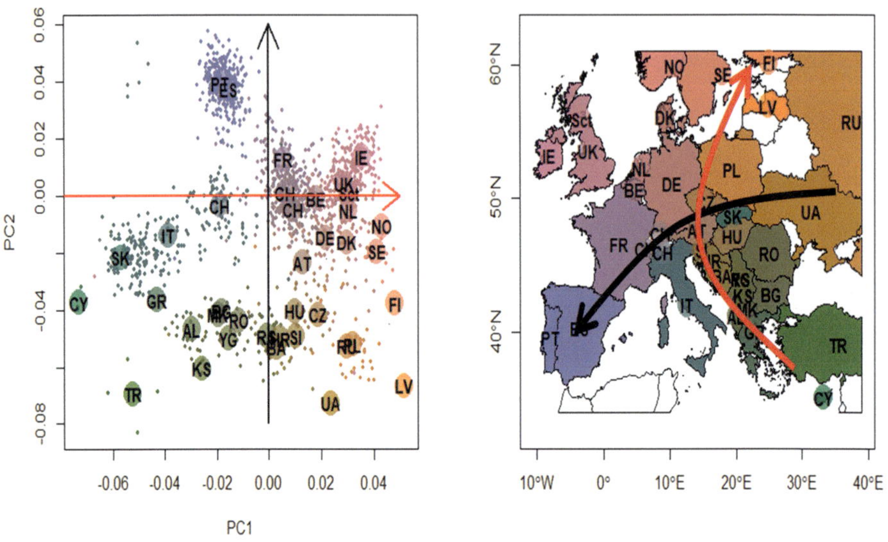

Figure 2.1 The first (red) and second (black) axes of a principal component analysis (left) involving over half a million Single Nucleotide Polymorphisms (SNPs) measured on 1387 individuals of European ancestry. The European country of origin of each of each individual is coded by colour following the map of Europe on the right. The first (red) and second (black) principal components (left) are approximately mapped onto the physical map of Europe (right).

Each of the 1387 small points in Figure 2.1 (left) represents one of the individual people in the data set. The two axes, labelled "PC1" and "PC2" are the two "principal axes" of the PCA. I will explain later what "principal component axes" are but, for the moment, think of them as two independent variables that have captured the most important patterns of correlation among the half million SNPs. The colours of each point show from where in Europe each individual lives. The larger circles indicate the value of the "average" individual from a given European country on the two principal component axes (i.e. mean value of the PC1 and PC2 scores for individuals of that country). This illustrates the first goal of a PCA: to visualize the relationships between many variables (i.e. the SNPs) in just a few dimensions ("dimensional reduction"). In this case, the relationships that exists in half a million dimensions (the original number of SNPs) has been reduced to two dimensions (the two PCA axes).

As you will see, each axis is a linear combination of all half million SNP values. Each axis is a linear function with over half a million variables (i.e. the presence/absence of each SNP). Using these linear functions, it is possible to assign a position of any new individual whose SNP values have been measured on these two axes. This accomplishes the second goal of PCA: To express the main patterns of covariation in the form of equations for communication or for extrapolation to new data. The inset of Figure 2.1 shows a map of modern Europe using the same colour scheme. Amazingly, the placement of individuals on these first two PC axes reproduces the physical map of modern Europe very closely although we have to rotate the actual map to see this concordance[4]. In fact, it is possible to predict the origin of an individual to within a few hundred kilometres simply by knowing the values of their position on these two PC axes (their "loadings"). If you have ever submitted a DNA sample to one of several

[4] You have to rotate the map of Europe clockwise approximately 90 degrees.

companies that tell you from where your ancestors probably lived, then these predictions are likely based on a similar statistical method.

Notice that the first principal component axis (PC1, in red) has one extreme in individuals living in southeastern Europe (Turkey, Cyprus and Greece in the lower right of the map of Europe) and the other extreme in individuals living in northern Europe (the British Isles and Scandinavia in the upper left of the map of Europe). The second principal component axis (PC2, in black) has one extreme in individuals living in the region around the Ukraine and the other extreme in individuals living in the Iberian Peninsula. One possible explanation for this pattern (Lazaridis et al. 2014, Racimo et al. 2020) is that these two PCA axes reflect the population movements of two prehistoric groups into Europe as they spread and intermixed with a preexisting hunter-gatherer population. PC1 is associated with the movement of the first agriculturalists from the Fertile Crescent into Europe during the Neolithic. PC2 is associated with the movements of Eurasian people (identified with the Yamnaya culture) from the Pontic steppe region north of the Black Sea who introduced the Indo-European languages that dominate modern Europe. This illustrates the third goal of PCA: to suggest hypotheses concerning the underlying causes of these complicated patterns of association between variables. In this case, we can hypothesize that the patterns of correlation between these variable regions of DNA are reflecting the mixing of different population groups over thousands of years.

2.2) Visualizing how PCA works

To understand how to reduce the dimensionality of multivariate data without resorting to the mathematical details of PCA, let's begin by generating some artificial data for which we know the relationships between the variables. By doing so, we can "reduce the dimensionality" of these data in a series of graphs without doing a PCA. We do this so that we can better appreciate the difficulties of trying to reach these objectives with the more complicated data with which you will typically use. Here is the **R** code to simulate our artificial data:

```
set.seed(10)
x<-rnorm(n=100,mean=0, sd=1)
y<-rnorm(n=100,mean=0, sd=1)
z1<-0.7*x-0.7*y+rnorm(n=100,mean=0,sd=sqrt(1-
2*0.7^2))
sim1<-data.frame(x=x,y=y,z=z1)
```

The first line initializes the random number generator in **R** using `set.seed()` and is only needed if we want to obtain the exact same data each time we run the code[5]. The next two lines generate 100 independent random values (`n=100`) for the two variables (`x` and `y`) in which each value (i.e. observation or "observational unit") follows a normal probability density (thus "`rnorm`") with a mean of zero (`mean=0`) and a standard

[5] Since these are "random" number generators, different values will be generated each time you – or anyone else – reruns the code unless you all start at the same place (seed number).

deviation of 1 (`sd=1`). The fourth line then creates a new variable (z) that is a linear mixture of the other two variables (x and y) plus a new normally distributed random variable that has a mean of zero and a standard deviation[6] of $\sqrt{1\text{-}2(0.7^2)}$.

Figure 2.2 plots these data as a "scatterplot matrix" using the `pairs()` function in **R**. It is arranged like a matrix with columns and rows. Each of the scatterplots in the upper diagonal of this "matrix" shows a bivariate graph with the vertical axis being the variable listed in the corresponding row and the vertical axis being the variable listed in the corresponding column. Each of the numbers in the lower diagonal of this "matrix" shows the proportion of variation held in common (i.e. the r^2 value) between the variable listed in the corresponding row and column. The scatterplot in row 1 and column 2 shows the relationship between variable x on the vertical axis (the variable represented in row 1) and variable y on the horizontal axis (the variable represented in column 1).

There are three things to notice about these data. First, we need three dimensions (i.e. three axes) in order to plot them because each of the 100 "observations" has three measured attributes (x, y, z) and each attribute varies between observations (i.e. each are "variables"). Second, two of the

---

[6] You could choose any value for the standard deviation of this new random variable. The value that I have chosen has the advantage that the resulting distribution of z will also follow a normal distribution with a standard deviation of 1 (a "standard" normal distribution).

variables (x and y) are independent of each other, and so knowing the value of one of these two variables will tell you nothing about the likely values of the other. Third, the z variable is almost entirely (i.e. 98%) determined by the combination of the first two variables (x and y) even though each of x and y only account for 49% separately of the variation in z. In fact, the values of z lie almost completely on a 2-dimensional surface on the x-y plane that is tilted upwards along the x-axis and tilted downwards along the y-axis. If we only look at the three possible pairs of two-dimensional graphs using the `pairs()` function then we would be doing a rudimentary "dimensional reduction" because we are reducing a three dimensional pattern into a two dimensional graph (Figure 2.2). A statistician would say that we are "marginalizing" over one of the variables but another way to say this is that we are looking at two-dimensional shadows that are cast by the three dimensional pattern when we look at these shadows from different perspectives.

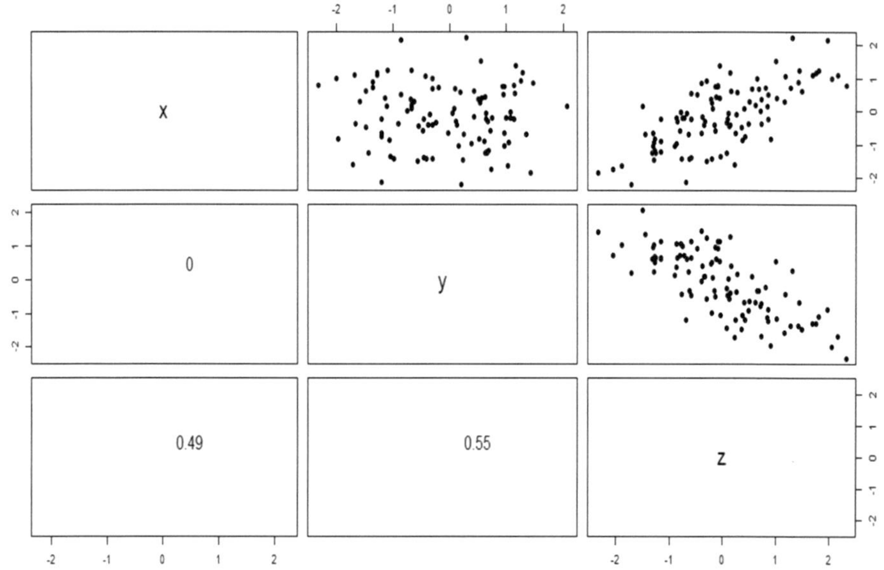

Figure 2.2. A scatterplot matrix of our simulated data involving three variables (x, y, z). Numbers on the subdiagonal are Pearson correlation coefficients.

In Figure 2.2 we can clearly see that the variables x and y are independent of one another (the graph in row 1 and column 2) because the points do not display any systematic tendencies to increase or decrease together. We can also see that each of x and y are partially correlated to z (the other two graphs) because the points do display clear systematic tendencies to increase or decrease together. However, we can't see the fact that the variation in z is almost completely restricted to a 2-dimensional plane that is tilted through this 3-dimensional space.

Since we are only dealing with three dimensions, we can actually make a 3-D « perspective » plot; I did this using the `scatter3D()` function of the "plot3D" package (Soetaert 2019). Figure 2.3 shows these data plotted using three different perspectives, obtained by rotating our perspective plot in different directions. Figure 2.3(b) shows our graph after it has been tilted downwards by 45°. Here, we are looking directly into our 2-D plane and can see that the points are almost arranged on a single line (a single dimension); this is our 2-D plane tilted on its end, which we couldn't see in our scatterplot matrix. In other words, z is very tightly determined by a linear *combination* of x and y even though, from Figure 2.2, we know that z is less tightly determined by each of x or y alone. We know this because we generated z using the **R** code `z1<- 0.7*x -0.7*y + rnorm( n=100, mean=0, sd= sqrt(1-2*0.7^2))`.

Figure 2.3(c) shows our graph after it has been tilted downwards by 180° so that we are looking into it from the top of the z-axis. From this perspective, we don't see the variation in the z-variable at all but we clearly see the points scattered independently on the x-y plane. Keeping Figures 2.3 (a) and (b) in mind, you might now be able see, in Figure 2.3(a), our points arranged very tightly around a 2-D plane of independent values of x and y stretching upwards on a 45° angle along the z-axis.

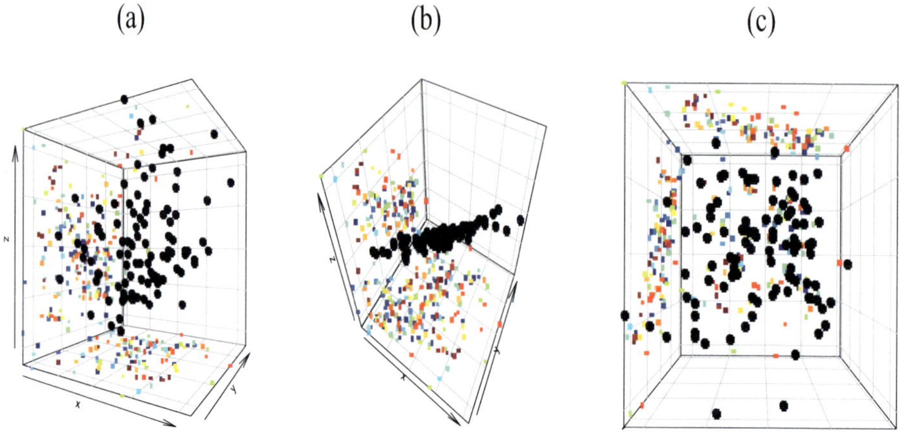

Figure 2.3 (a) A 3-dimensional projection (black points) of our trivariate (x, y, z) set of observations. The origin of this graph is the lower back corner. (b) The same graph after rotating it 45° downwards. (c) The same graph after rotating it downwards. The coloured points in each graph are the 2-dimensional "shadows" that are cast onto each 2D plane.

I have chosen particular rotations of axes in Figure 2.3 so that we can see the patterns of association in fewer dimensions. Keep this image firmly in mind in the next section because this is one way of visualizing what principal component analysis does: it "rotates" the original axes, which are defined by the original variables (here x, y and z) in such a way as to best capture the relationships between the points in fewer dimensions. We will use Figure 2.4, which involves only two dimensions (axes) to illustrate the basic steps by which PCA rotates any number of axes[7].

[7] This formulation views the data as fixed while we rotate the axes. One can equally say that we rotate the data while keeping the axes fixed.

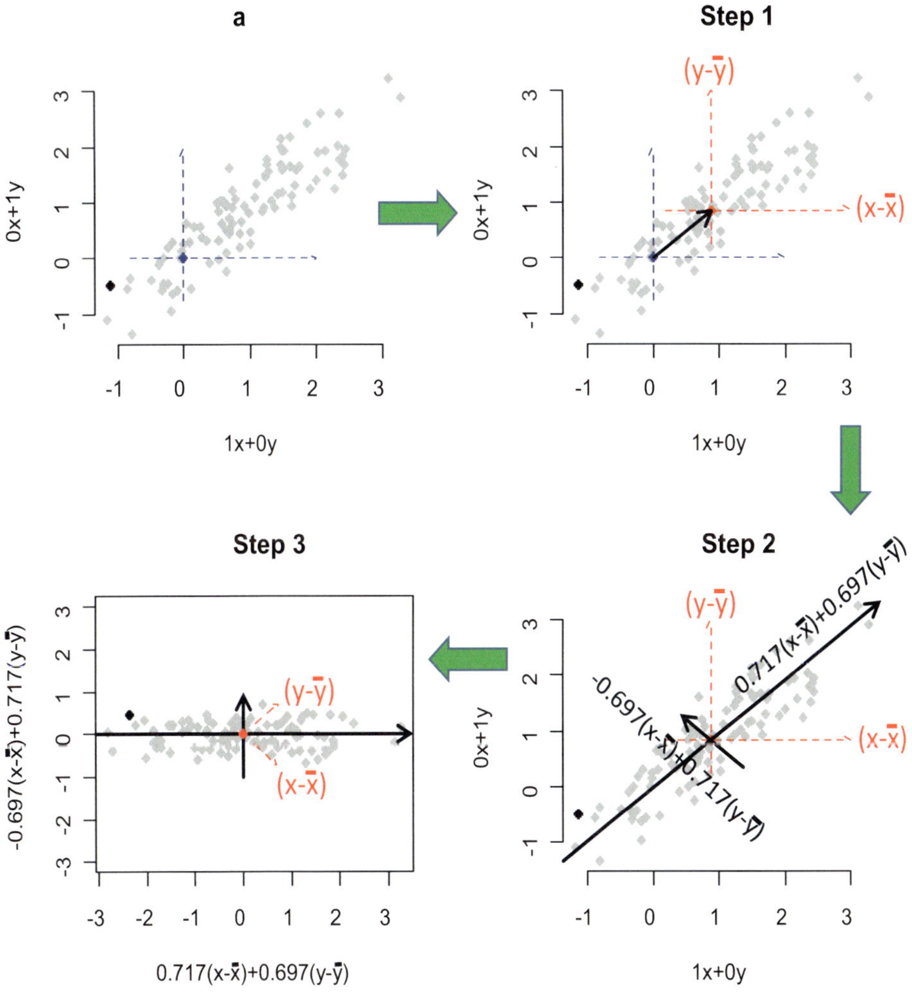

Figure 2.4 A graphical explanation of how a principal component analysis works.

The easiest way to visualize how PCA works is to use an example involving only two variables (x, y), which can be seen in a two-dimensional scatterplot. Figure 2.4(a) shows such a scatterplot with each observation shown by a grey point. I have shown the original horizontal and vertical axes in blue and have emphasized one such observation (the black point). We will follow this black point in each step. The origin of these axes are x=0 and y=0 (the blue dot). When moving along the horizontal axis, a one unit change in the position of a point will result in a one unit change (i.e. 1x) in x and no change (i.e. 0y) in y. This is why I have labelled this axis "1x+0y" rather than the more traditional "x". Similarly, when moving along the vertical axis, a one unit change in the position of a point will result in no change (i.e. 0x) in x and a one unit change (i.e. 1y) change in y. By labelling this axis "0x+1y" rather than the more traditional "y", I make this information more explicit. Notice that whenever we are referring to an axis, we are really referring to a *function* of the variables in the graph (for example, horizontal axis = 1x+0y). Remembering this will make it easier to interpret the axes of a PCA.

STEP 1: Centring the variables

The first step, shown in Figure 2.4 is not necessary but is almost always done except in very special cases (Pielou 1984). It consists of moving the origin from (0, 0) of the graph to the mean values of the variables. The origin of these new axes are therefore $(x-\bar{x})$ and $(y-\bar{y})$.

STEP 2: Rotating the axes

Notice that the variation in the cloud of points (the grey dots) is elongated in one direction. The next step, shown in Figure 2.4, consists of finding the equation of the line that passes through the middle of the cloud of points along this direction of maximum variation. In these data, this line is $0.717(x-\bar{x})+0.697(y-\bar{y})$. Notice what this equation is telling us: if we move one unit along this line then we will change the value of x by 0.717 units and we will also change the value of y by 0.697 units. In other words, the x and y variables do not change independently when moving along this line; instead, they are *positively correlated* as we move along it. We will interchangeably call this equation the first "principal component" of variation, the first "principal component axis", or simply "PC1". In this simple bivariate example, this line is also called the "major axis" and is the line that is produced in a type of regression called major axis regression (Warton et al. 2006) or type II regression. The shorter black line is always perpendicular to this first principal component and is called the second principal component and its equation is $-0.696(x-\bar{x})+0.717(y-\bar{y})$. An equivalent way of saying that the two lines are perpendicular is to say that they are linearly independent of each other. Clearly, the x and y variables do not change independently when moving along this second line either. However, if we move one unit along this second line then we will have increased the y-variable by 0.717 units while decreasing the x-variable by 0.696 units; the x and y variables are *negatively correlated* along this line.

STEP 3: Defining new axes

The last step is to rotate Figure 2.4(c) so that the two principal components become the new "axes", as shown in Figure 2.4(d). If we move by one unit along this new horizontal axis (the first principal component), then we don't change the value along the vertical axis (the second principal component) at all. This means that these two principal component axes are linearly independent of each other. However, if we move along the horizontal axis (i.e. first principal component) by one unit then we will have changed both the values of x and y. Stated equivalently, the x and y variables are now *correlated* in the space defined by these two principal components.

Admittedly, there is not much to be gained by performing these three steps when there are only two variables, as in our example. After all, we can easily see the pattern of association between our two variables without doing these steps. However, the underlying mathematical operations that define the series of rotations and translations of axes that are shown in Figure 2.4 can be performed on any number of axes (i.e. dimensions). We will not study the underlying mathematics in this book but a good reference for these mathematical details, written for biologists with a background in linear algebra, is section 9.1 of Legendre and Legendre (2012). What is easy to observe with only two variables is rather more difficult to observe with three variables (for example, Figure 2.3) and quite impossible to observe with more than three variables. Irrespective of the number of original variables in your data (its *dimensionality*), the result of a PCA is a series of principal component axes, the same number as in your original data. The first axis accounts for the largest proportion of the total variation in your data set. The second axis, which is linearly uncorrelated with the first axis, accounts for the largest proportion of the total residual variation, that is, after removing the variation captured by the first axis. Each successive axis accounts for the largest proportion of the remaining residual variation after removing the variation captured by the preceding axes. Another way of saying the same thing is that principal components are organized based on the amount of variation in the multivariate data set that they capture, from most to least. As a result, the first few principal component axes will capture the largest trends in the data.

The final plot in Figure 2.4 (step 3) is called a (slightly modified) "biplot" of the principal components. We will look more closely at biplots later. The values, or coordinates (the values on the two principal component axes), of each observation on the two axes the final plot in Figure 2.4 are called the "scores" of these observations on each principal component. The values of the two original variables (x, y) in determining these scores are called the "loadings" of these variables on each principal component. The loadings of x and y on PC1 are 0.717 and 0.697, meaning that a one unit change on PC1 results in an increase of 0.717 units of x and 0.697 units of y. This is why the red arrows in Figure 2.4 (step 3), which are the original axes, both point in the same direction as PC1. The loadings of x and y on PC2 are -0.697 and 0.717, meaning that a one unit change on PC2 results in a decrease of 0.697 units of x and an increase of 0.717 units of y. This is why the red arrow labelled "$(x-\bar{x})$" is pointing in the opposite direction as PC2 while the red arrow labelled "$(y-\bar{y})$" is pointing in the same direction as PC2. Finally, the length of each principal component quantifies the standard deviation (i.e. square root of the total joint variance of x and y) that is captured by each principal component. In figure 2.4(step 3) the first principal component accounts for 94.7% of this total variance (i.e. a standard deviation of ~1.29) and the second principal component accounts for 5.3% of this total variance (i.e. a standard deviation of ~0.30) and so PC1 is ~4.3 times as long as PC2. The fact that the first principal component accounts for a large majority of the total variation means that we could ignore the second principal component entirely without introducing too much error. In other words, the relationships between the points would be mostly the same if we simply used the loading of each observation on PC1 (i.e. the value of each observation along this axis) and completely ignored the second principal component. The distance between the points (observations) along this axis would be almost the same as the distances between points in the original graph (Figure 2.4(a)) since the variation in the second dimension (PC2) is very small. We would be reducing a 2-dimensional scatterplot into a 1-dimensional line. We would be *reducing the dimensionality* of the data.

One would almost never do a principal component analysis on a data set involving only two variables[8] since we can easily visualize the complete pattern of covariation simply by looking at a scatterplot like Figure 2.4(a). In Figure 2.3, involving a 3-dimensional data set, we could

[8] Unless you wanted to do a special type of regression called "major axis" regression.

only properly see the entire pattern of covariation by rotating a perspective plot in different directions. As soon as we have four variables in our data then we are in trouble since it is impossible to represent four dimensions on a page (i.e. a 2-D surface).

Imagine what would happen when we analyse data like those from Novembre et al. (2008), which involved over half a million variables (presence/absence of SNPs) and over half a million dimensions! And yet, after a principal component analyses of these data (Figure 2.1) they reduced the dimensionality of their data from over half a million dimensions to only two dimensions, fulfilling the first objective of a PCA. We see that if two individuals have very similar combinations of SNPs, i.e. the two points are very close together in Figure 2.1, then they are overwhelmingly likely to also come from the same geographical location within Europe. In fact, the pattern of points projected onto these first two principal components approximately resembles a map of Europe. If we analyse the DNA of a new individual and obtain their SNPs, then we could then use the equations for these first two principal components to place this new individual onto our graph and determine quite accurately the likely location of his or her ancestors to within a few hundred kilometres. This is the second objective of a PCA: to express the main patterns of association in the data in the form of equations for communication or for extrapolation to new data. Finally, by noting how the two principal components seem to trace two trajectories across Europe that appear to coincide with independent theories about the post-glacial peopling of modern Europe, it suggests hypotheses about how the DNA extracted from ancient individuals should change at different times; the extraction, dating and analysis of "ancient DNA" is currently testing these ideas.

2.3) Preparing your data

Two important assumptions of PCA are that (i) the distributions of all variables approximately follow a normal distribution and (ii) any systematic relationships between the variables are linear. Therefore, PCA requires that each of the variables in your data be numerical values, not factors (i.e. unordered categorical variables). Binary variables (yes/no, presence/absence, male/female etc.) are acceptable but only if you have coded this binary variable as a number (i.e. 0 or 1). These data must be stored as either a matrix or as a data frame. Observations containing missing values must be excluded from the analysis. No observations can contain missing values of any variable (coded as NA) and any such observation will be ignored.

Although the mathematics behind a PCA assumes that all of the variables be normally distributed, this assumption is not critical because we are not testing hypotheses that require exact estimates of probability. It is more important that each variable be symmetrically distributed. In other words, if you look at a histogram of each variable then the distribution of observations around the mean value should be (more or less) mirror images. If this histogram shows that this is not true then you will have to find a transformation that makes the histogram of the transformed values at least approximately symmetric. Two important assumptions of PCA are that (i) the distributions of all variables approximately follow a normal distribution and (ii) any systematic relationships between the variables are linear. Therefore, PCA requires that each of the variables in your data be numerical values, not factors (i.e. unordered categorical variables). Binary variables (yes/no, presence/absence, male/female etc.) are acceptable but only if you have coded this binary variable as a number (i.e. 0 or 1). These data must be stored as either a matrix or as a data frame. Observations containing missing values must be excluded from the analysis. No observations can contain missing values of any variable (coded as NA) and any such observation will be ignored

The Box-Cox transformation (Box and Cox 1964) consists of finding a constant (λ) that converts an original variable (x) into another variable ($y = x^{\lambda}$) such that the new variable (y) is as close as possible to a normal distribution. If λ is close to 1 then you don't need any transformation. If λ is very close to zero, then you should use a logarithmic transformation; i.e. y=ln(x). There are several different **R** functions that find this value of λ and we will use the `boxcox()` function that is found in the MASS package (Venables 2002). I will illustrate this using Figure 2.5(a), which shows a histogram of a numeric vector called "x". I actually generated these values as random values following a chi-squared distribution with 2 degrees of freedom (`x<-rchisq(100,2)`). The vertical red line shows the position of the mean value of x. Clearly, this variable is not normally, or even symmetrically, distributed around the mean. If we used this variable in a PCA without transformation then the few values in the right tail would have an undue influence on the first few principal components. This is because a PCA attempts to capture the patterns of covariation in a data set that maximizes the variance and the variance of x is strongly affected by these few points.

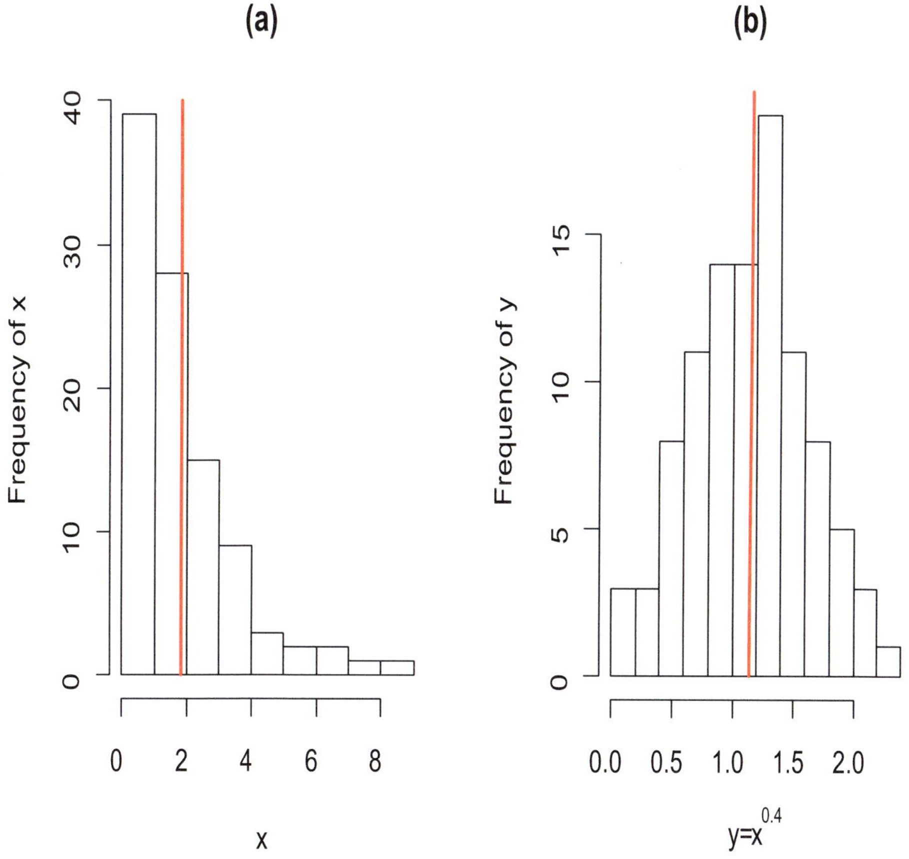

Figure 2.5. A variable that is not normally distributed (left) and the distribution of this same variable after a transformation (right). The vertical red lines show the means of the two distributions.

Here is the **R** code to obtain thc transformation of x to a ncw approximately normally distributed variable, y:

```
library(MASS)
out<-boxcox(x~1,plotit=FALSE)
lambda<-out$x[out$y==max(out$y)]
lambda
y<-x^(lambda)
```

The `boxcox()` function is designed to be used with lm objects, i.e. the output of the "linear model" (lm) function. We trick this function into using our variable alone by specifying, as the first argument, "x~1", which means that we are modelling the variable x without any predictor variables. The second argument (`plotit=F`) is not necessary since it simply supresses the graphical output. The optimal value of our power constant, λ, is 0.4 in this example and so we transform our variable as $y=x^{0.4}$. The second histogram in Figure 2.5 shows the distribution of this transformed variable and it is approximately normally distributed. You should always check the distribution of your variables before beginning a PCA and transform any that deviate strongly from a symmetrical distribution. Remember, however, that transforming a variable also transforms its measurement units, since this will affect your interpretation of this transformed variable.

The second assumption listed above is that any systematic relationships between variables be linear. If your PCA doesn't involve too many variables then this assumption can be checked by looking at scatterplots of the different pairs of variables. A particularly efficient way of doing this is by creating a scatterplot matrix via the `pairs()` function. However, this strategy is only useful if you have a relatively small number of variables (~20) since the resulting graphical output quickly becomes too difficult to view.

What happens if there are nonlinear relationships between variables? The PCA will still capture these relationships – after all, a PCA is simply a rotation of the original axes. However, these nonlinear relationships will not be captured by any single PC axis. Rather, such nonlinear relationships will only be evident when considering several PC axes in combination and this rather defeats the whole purpose of a PCA. To better understand this point, look again at Figure 2.4. Using our original axes, defined by the two variables, we saw a systematic but linear trend in our scattergraph (Figure 2.4(a)). After centering and rotating our two original axes we obtained two new axes, the principal component axes. In this new coordinate system we produced a new scatterplot (Figure 2.4 (step 3)) in which virtually all of this systematic trend was captured by a single PC axis (i.e. PC axis 1). We could pretty much ignore the second PC axis.

In Figure 2.6 I have done the rotation of the axes as done in Figure 2.4 except that in Figure 2.6(a) we see a clear systematic but nonlinear trend even though there is no *linear* relationship between variables x and y (the Pearson correlation coefficient between x and y is 0.01). Even after centering and rotating the axes as we did in Figure 2.4 via a principal component analysis this systematic nonlinear trend in not captured by either PC axis (Figure 2.6b); we still need to look at both PC axes in order to see the relationship. Since one of the main purposes of PCA is to reduce complicated multidimensional patterns of association into patterns of association that are captured by only a few dimensions (axes), the presence of strongly nonlinear patterns of association in many dimensions prevents such dimensional reduction and this defeats the purpose of a PCA. If you see nonlinear trends in a scatterplot of principal components then this is a clue that there exist nonlinear trends in the original data. This point will be particularly important in chapter 6, devoted specifically to the study multispecies assemblages in community ecology, since in such data the relationships between species' distributions is very commonly nonlinear.

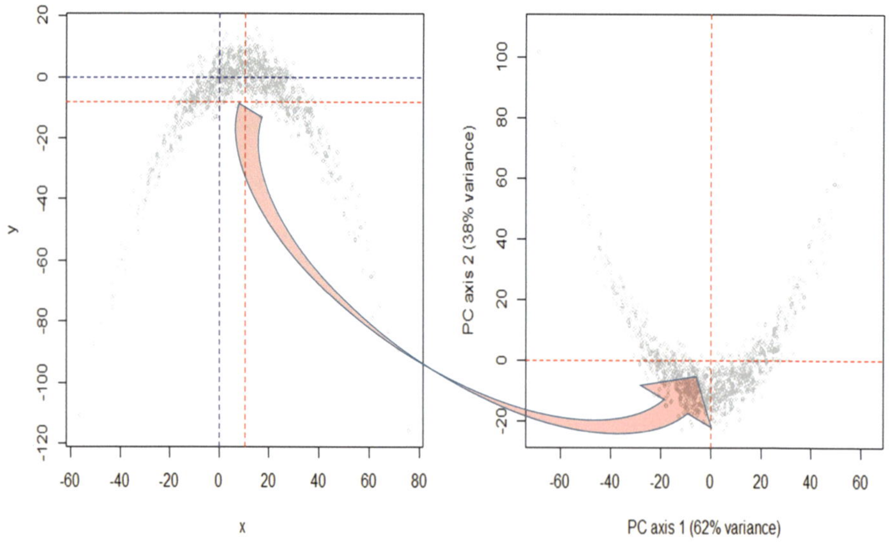

Figure 2.6. (Left) A nonlinear relationship between two variables (x, y) and (right) the resulting principal components.

2.4) Dealing with proportions in PCA

The purpose of PCA is to reveal empirical patterns of correlation in only a few dimensions. By "empirical patterns", I mean patterns of correlation that are imposed by the mechanisms functioning in Nature. This means that we must avoid imposing artificial patterns of correlation that are dictated by purely definitional or logical relationships. One common source of such artificial patterns of correlation is when we express a series of variables as proportions of some whole. For instance, soil scientists often describe the particle size distribution of soils as percentages of sand, silt and clay; i.e. the dry mass of each soil fraction divided by the total mass of the soil. Since these percentages must, by definition, sum to 100%, this creates a dependency between them. To see this, simply answer the following question: if a soil is composed of 60% sand, and 30% silt, what is the percentage of clay? You know the answer without ever observing the soil or even knowing what the words "sand", "silt", "clay" mean. There is a logical, and thus a non-empirical, dependency between these variables in such data. However, PCA doesn't know which patterns are empirical and which are imposed by logic or definitions. From the perspective of the underlying mathematics, there is no difference between empirical and non-empirical patterns but we don't want our analysis to mix together these two very different sources of correlation.

Data having this form are often called "compositional data" and they occur in many different contexts. If you add these three variables in a PCA then this dependency (i.e. correlation) will affect the way the PCA chooses its new axes (Aitchison 1983). Moreover, since one variable is always a linear sum of some combination of the remaining variables, this causes numerical problems that might result in an error message (the covariance matrix is "singular"). Simply dropping one of the proportions, so that the sum doesn't always give a constant value, does not solve this problem. Although there are dedicated packages in **R** that are designed specifically for doing PCA using compositional data (for example, the `pcaCoDaA()` function in the robCompositions package (Templ et al. 2011)), it is quite possible to use compositional data in the two base functions for PCA by first transforming such data. We will learn about these two base functions a bit later. Assuming that you have n variables in the form of proportions (x_1, x_2,…,x_j,… ,x_n) such that these variables sum to unity, three different transformations of compositional data are used before applying such data to a principal component analysis.

The additive log ratio transform consists of (i) choosing one of the proportions as a comparison that you leave out (x_j, it doesn't matter which one) and then (ii) transforming each of the remaining proportions as $y_i = \ln(x_i/x_j) = \ln(x_i) - \ln(x_j)$. The transformed variable, y_i, is then the difference between the ln-transformed values of each remaining proportion and the ln-transformed comparison proportion. This transformation does remove the logical dependence between the remaining proportions, i.e. $y_{i\neq j}$, but it has a weakness: the meanings of the remaining transformed variables (y_i) change, as do the principal components that emerge, depending on how you choose the comparison proportion.

A second transformation, called the centered log ratio transform, is Aitchison's (1983) solution. This transformation consists of (i) calculating the geometric mean of the proportions: $g=\left(\prod_{i=1}^{n} x_i\right)^{\frac{1}{n}}$ then (ii) transforming the proportions, without leaving any out, as $y_i=\ln(x_1/g)=\ln(x_i)-\ln(g)$. Note that $\ln(g)=\frac{1}{n}\sum_i \ln(x_i)$ which is simply the mean of the ln-transformed proportions and so an entirely equivalent formula is $y_i=\ln(x_i)-\overline{\ln(x)}$. Each of the transformed variables, y_i is therefore the difference between the ln-transformed proportion (x_i) and the mean of the ln-transformed proportions.

2.5) A covariance matrix or a correlation matrix?

PCA can use two different measures of linear association between variables: covariances and correlations. This book assumes that you already understand how to calculate and interpret variances, covariances and correlations and that you know what a "standardized" variable is (i.e. $s(x_{ij})$), and so I will simply give the equations here. The notation "x_{1j}" means "the jth observation of variable x_1" while the notation "x_{ij}" means "the jth observation of the ith variable" in a whole set of variables called "x". A "covariance" is simply the "variance" of a variable with itself. A correlation is simply the covariance between two standardized variables and can only take values between -1 and +1. Two variables that are independent of each other will have (on average) a covariance and a correlation of zero.

$$\mathrm{Var}(x_i)=S^2(x_i)=\frac{\sum_{j=1}^{n}\left(x_{ij}-\bar{x}_i\right)\left(x_{ij}-\bar{x}_i\right)}{n-1}=\frac{\sum_{j=1}^{n}\left(x_{ij}-\bar{x}_i\right)^2}{n-1} \qquad (2.1)$$

$$\mathrm{Cov}(x_1,x_2)=S(x_1,x_2)=\frac{\sum_{j=1}^{n}\left(x_{1j}-\bar{x}_1\right)\left(x_{2j}-\bar{x}_2\right)}{n-1} \qquad (2.2)$$

$$s(x_{ij})=\frac{\left(x_{ij}-\bar{x}_i\right)}{\sqrt{S^2(x_{ij})}} \qquad (2.3)$$

$$\mathrm{Cor}(x_1,x_2)=$$
$$r(x_1,x_2)= \qquad (2.4)$$
$$\frac{\sum_{j=1}^{n}\left(x_{1j}-\bar{x}_1\right)\left(x_{2j}-\bar{x}_2\right)}{\sqrt{\sum_{j=1}^{n}\left(x_{1j}-\bar{x}_1\right)^2}\sqrt{\sum_{j=1}^{n}\left(x_{2j}-\bar{x}_2\right)^2}}=\frac{\mathrm{Cov}(x_1,x_2)}{\sqrt{\mathrm{Var}(x_1)}\sqrt{\mathrm{Var}(x_2)}}$$

The above equations apply to pairs of variables (here, x_1 and x_2). We will always deal with *multivariate* data sets in this book, which means that there are always more than two variables measured on each observational unit. Therefore, there will always be more than one pair of variables. We will put all of these pairs of covariances or correlations into a matrix, remembering that the covariance of a variable with itself is a variance and that the correlation between a variable and itself is 1.0. If you have saved your data in a matrix of data frame called `my.dat` then typing `var(my.dat)` will give the covariance matrix of these data and typing `cor(my.dat)` will give the correlation matrix of these data.

There is no absolute rule governing when to choose a covariance matrix (i.e. use variables in their original measurement units) and when to choose a correlation matrix (i.e. use variables that have been standardized). However, the general rule of thumb is to use a correlation matrix when the original variables have different measurement units or when you want to concentrate on the patterns of correlation between the variables without also considering the natural variation in the variables. You would use a covariance matrix when all of the variables have the same measurement units and when the differences in the range of variation between these variables have biological meaning.

Every variable has an associated measurement unit. These measurement units are also carried over to the statistics that are calculated from these variables. For instance, imagine that we measure the wing lengths of a series of birds in centimetres: L=(10.5, 9.6, 12.3, 11.8, 13.1). The mean (11.46 cm) and standard deviation (1.40 cm) of wing length is also be measured in centimetres. The variance (1.97 cm^2) of wing length, being the square of the standard deviation, is measured in square centimetres. If we decide to instead measure wing length in millimetres then this is a transformation (L') of our variable: $L'=10L$. If we now calculate these same statistics on wing length when measured in millimetres, we find that the mean (114.6 mm), the standard deviation (14.0 mm) are ten times larger while the variance (197 mm^2) is 100 times (i.e. 10^2) larger. Of course, the wings have not grown 10 times longer; rather, the measurement unit has increased by 10 times.

Since the standard deviation of a variable has the same measurement units as the variable itself, then we can create a "unitless" variable by dividing the variable by its standard deviation. Actually, it is more correct to say that the units of the original variable have been replaced by new units (standard deviations) since a one unit change in the new variable is equal to one standard deviation of the original variable. This transformation is called a *standardization* and the new variable is called a *standardized* variable. Many authors call a standardized variable one that has undergone a slightly more complicated transformation in which the variable is also centred about its mean value, i.e. $(x-\bar{x})/S_x$, but this is more correctly called a *centred and standardized* variable although the two versions only differ by a constant ($\bar{x}/S_x$). The `scale(x, center =`

`TRUE, scale = TRUE)` function in **R** performs this transformation since it both centres and scales the variable by default. If we standardize a series of variables having different measurement units then we will have removed the original measurement units and replaced them all with the same measurement unit (standard deviations). Therefore, if our original variables have different measurement units, it is common to conduct a PCA using such standardized variables.

The covariance between variables is also affected by the measurement units that we choose and this has important consequences for PCA. Imagine that we have also measured the weight in grams of the same birds from which we measured the wing lengths: W=(3.9, 5.0, 5.6, 4.8, 6.7). The covariance between the lengths (in centimetres) and the weights (in grams) is 1.07 cm·g. If, however, we express the lengths in millimetres, then the covariance becomes 10.7 mm·g. Since a principal component analysis consists of creating new axes that are linear functions of the original variables, the directions and lengths of these new axes are also affected by the measurement units that we choose. This means that transformations of our variables will affect these principal component axes and that the interpretation of our principal component axes is intimately linked to the measurement units that we choose.

We have already discussed the question of data transformations before conducting a principal component analysis. Up to now, we have viewed such transformations as solutions to problems of non-normality or non-linearity. When we have variables in our data that have different measurement units, and especially when these measurement units reflect different physical properties (mass, length, time, etc.) then the principal component axes will reflect, and be strongly affected by, this mix of measurement units. Sometimes we will want our PCA to reflect a particular choice of measurement units but, often, we do not want our view of the patterns of association between the observations in a set of principal component axes to change simply because we have chosen to measure wing length in centimetres rather than millimetres. In this case, the solution is to standardize all of the variables by dividing each one by its standard deviation. Since PCA almost always used variables that are centred about their means, this means transforming each variable to its scaled and centred version.

A variance – covariance matrix is a square symmetric matrix with the same number or rows and columns as there are variables and can be obtained using the `var(X)` function where X is a data frame or matrix. It has variances on the diagonal and covariances everywhere else. The matrix is "symmetric" because the same values are found, as mirror images, above and below the diagonal. Since a "variance" is simply a covariance of a variable with itself, a variance - covariance matrix is also simply called a "covariance" matrix. The elements of this matrix have measurement units. Changing the measurement units will change the values in the matrix. A correlation matrix is also a square symmetric matrix with the same number or rows and columns as there are variables and can be obtained using the `cor(X)` function where X is a data frame or matrix.. Its diagonal elements (i.e. its variances) are all equal to one and its off-diagonal elements (i.e. its covariances) are Pearson correlation coefficients which can only vary between ±1.

What happens if we calculate a covariance matrix using standardized versions of our variables? In this case, the variances of all of the variables are the same: one. The covariance between two standardized variables is equal to the Pearson correlation coefficient between them. In other words, the covariance matrix involving standardized variables is a correlation matrix. What has all this to do with a principal component analysis? I said earlier that you must decide if you want your principal component axes to reflect the choice of measurement units of your variables. By standardizing all of the variables, or equivalently by choosing to use a correlation matrix, then you have removed the differences in the original measurement units. These two equivalent choices, (i) covariance matrix vs. correlation matrix or (ii) original variables vs. standardized variables, are reflected in the two numerical methods that underlie a PCA.

I said earlier that one would normally choose a correlation matrix when conducting a PCA involving variables having different measurement units. When all of the variables possess the same measurement units then the choice between a correlation matrix and a covariance matrix is more nuanced. Choosing a correlation matrix not only forces the variables to have the same units (i.e. standard deviations from the mean), it also forces them to have the same variance (i.e. one). When choosing a covariance matrix involving variables that all have the same measurement units, the differences in the variances between such variables will not be due to different measurement units but they will still reflect the natural variation between them. Depending on the purpose of the analysis, these differences in the natural variation between the variables can be important or simply a distraction. If the differences in the natural variation between the variables are biologically important then you would choose a covariance matrix. If the differences in the natural variation between the variables are not biologically important, you would choose a correlation matrix. As an example, imagine that you have measured lengths of different body parts of different species birds using the same measurement units (centimetres): the lengths of the wings, the legs, the tail and the beak. If your data set includes birds of very different sizes (hummingbirds to ostriches) then the lengths of all of these body parts will strongly covary. Because of this, you can be sure that the first principal component will consist of all of these variables being positively related to each other and therefore capturing the underlying property of "size". The first principal component is revealing the fact that the body parts of bigger birds… are generally bigger. This pattern of covariation is obvious without conducting a PCA. In this case, then it might be appropriate to conduct the PCA using a correlation matrix since doing so will reveal how different body parts change relative to each other after removing this obvious pattern. Another strategy, explained later, is to use a covariance matrix but to ignore the first principal component.

2.6) Conducting a PCA and interpreting the output

There are many statistical programs that can conduct a principal component analysis. This wealth of choices arises because PCA was one of the first such methods to be developed and also because it is still commonly used. All of these different programs use one of two different (but almost identical) general methods to perform a PCA: (i) an *eigenvalue decomposition*, which examines the covariances or correlations between the variables or (ii) a *singular value decomposition*, which examines the covariances or correlations between the observations[9]. In all but a few, very special and rare situations, these two methods give essentially identical results. These special situations occur when some variables are almost perfectly correlated with some combination of other variables[10]. When this very rare situation occurs then a singular value decomposition has more stable numerical properties. As you might expect, there are also many different **R** packages that perform PCA. We will use two functions from the stats package that is automatically available when you start **R**;

---

[9] Actually, a singular value decomposition can do both, depending on whether one uses its **U** or its **V** matrix.

[10] If there are more variables than observational units then a singular value decomposition is the only method that can be used.

these two functions are called `prcomp()`, which uses a singular value decomposition, and `princomp()`, which uses an eigenvalue decomposition.

It is time to leave behind the simple examples. We will use a subset the empirical data that established the "worldwide leaf economic spectrum" (Wright et al. 2004); these data are available in the electronic appendix that accompanies that publication. The authors extracted, from previously published studies, data on a set of leaf traits from 2548 species of plants. Several of the leaf variables are expressed relative to both leaf mass or to leaf projected area (amount of nitrogen, net photosynthetic rate, respiration rate). For instance, net photosynthetic rate is reported in two different ways: the net amount of carbon dioxide fixed per second per leaf area and the net amount of carbon dioxide fixed per second per leaf mass. Why not express these variables per leaf rather than per unit leaf mass or per unit leaf area? Because the leaves of different species vary enormously in size. A leaf of *Wolffia globosa* (Roxb.) Hartog & Plas is about 0.2 mm in diameter while *Gunnera manicata* Linden has leaves up to 3m in diameter. Comparing the amount of nitrogen or the photosynthetic rate per leaf essentially reduces to comparing differences in leaf size. Since these mass-based and area-based traits will necessarily have an association generated by such standardizations, we will only use the mass-based versions. Each observational unit (i.e. line in the data set) therefore reports the mean value of each trait for a single species. However, not every published paper reported values on all of the leaf traits and so there are many lines with missing values. We will use the four leaf traits (Table 2.1) for which there are the most complete data giving a total of 492 lines (each line representing one species) for which values. These data are saved in a data frame called `GLOPNET2`. The data available in the online appendix for these four leaf traits are given as $\log_{10}$ transformed values. I used the `boxcox()` function to verify that a logarithmic transformation is appropriate to normalize these data. This is the case.

Do we use a correlation matrix of a covariance matrix to conduct our PCA? The original measurement units of each of these variables are different and this argues in favour of using a correlation matrix. On the other hand, these variables have all been transformed to their base10 logarithms and this transformation replaces the original units of each variable by the same (log10) measurement units. These new logarithmic units are no longer sensitive to changes in the original measurement units. For instance, our previous example of lengths of bird wings, in centimetres, was L=(10.5, 9.6, 12.3, 11.8, 13.1). The standard deviation of the log10-transformed values is 0.054. If we express these wing lengths in either millimetres or in metres then the standard deviation of the log10 transformed values is still 0.054. This means that the log10-transformed values of the variables in Table 2.1 are no longer sensitive to changes in the original measurement units and so we could reasonably use either a correlation matrix or a covariance matrix. Using a correlation matrix would erase the natural variability between the log-transformed variables since the correlation of a variable with itself (i.e. the covariance of a standardized variable with itself) is always unity. Using a covariance matrix based on these log-transformed variables would still take into account the natural variability between the variables without introducing effects that are sensitive to the original measurement units. For example, log10(Nmass), the proportion of the leaf mass that is nitrogen, has a standard deviation of 0.22 while log10(LMA), the leaf dry mass divided by the leaf projected area, has a standard deviation of 2.02. It is presumably because the original authors did not want to include such differences in this natural variability in their "leaf economic spectrum" that Wright et al. (2004) chose to use a correlation matrix. Choosing between a covariance matrix and a correlation matrix when using PCA has to be based on the underlying biological question, not only on statistical criteria. We will first conduct a PCA of these data using a covariance matrix and later we will do the same thing using the correlation matrix.

Leaf trait (name and column number in GLOPNET2)	Original measurement units	Description	Mean ±SD using log10 units
Leaf lifespan (logLL, 10)	Months	The average lifespan of a leaf	0.95 ±0.43
Leaf nitrogen content (logNmass, 11)	$100\times g\,g^{-1}$	The percentage of the leaf dry mass that is nitrogen	0.22±0.19
Leaf net maximum	$\mu mol\,g^{-1}\,s^{-1}$	Net amount of CO_2 fixed per gram leaf dry	1.97±0.31

photosynthetic rate (logAmass, 12)		mass per second in a leaf exposed to full sunlight	
Leaf mass per area (logLMA, 16)	g m^{-2}	Leaf dry mass per projected surface area	2.02±0.26

Table 2.1. A summary of the GLOPNET2 data set used in the subsequent principal component analysis. The GLOPNET2 data is a subset of the original GLOPNET data of Wright et al. (2004).

Here are the two equivalent versions of the `princomp()` function of the stats library of **R** using our `GLOPNET2` data set.

princomp()

```
princomp(~logLL+logNmass+logAmass+logLMA, data = GLOPNET2, subset, cor = FALSE, scores = TRUE, covmat = NULL, na.action, fix.sign = TRUE, ...)
```

```
princomp(GLOPNET2[,c(10,11,12,16)], cor = FALSE, scores = TRUE, covmat = NULL, subset, fix_sign = TRUE, ...)
```

The first version is the "formula" version. This allows you to explicitly use and name a subset of the variables in a data frame (`GLOPNET2`). The second version uses all of the variables in the data frame or matrix unless you explicitly select only certain columns. We have requested that the analysis be based on a covariance matrix by assigning `cor=FALSE;` changing this to `cor=TRUE` would instead request that the analysis be based on a correlation matrix. Both versions use all lines in the data set by default but you can select a subset of lines using the `subset=`

argument. You could also first calculate a covariance or correlation matrix, via `var()` or `cor()`, and then input this matrix into the `princomp()` function using the `covmat=` argument, rather than inputting the full data set. The `fix_sign=` argument reminds us that the directions of the principal components are arbitrary in any PCA which is why you might find the same PCA will have axes that are mirror images of one another when done in different statistical programs. The underlying numerical algorithm used in the `princomp()` function is called an *eigen analysis* of the covariance or correlation matrix. This numerical algorithm requires that there be more observational units (lines in the data set) than variables (columns of variables in the data set). This is rarely a limitation with biological data but it can sometimes occur. For instance, the data on SNPs of Europeans that we discussed previously could not be analysed using the `princomp()` function because there were many more variables (SNPs) than observational units (individual people) in that data set.

Here are the two equivalent versions of the `prcomp()` function of the stats package of **R** using our `GLOPNET2` data set.

```
prcomp(~logLL+logNmass+logAmass+logLMA,      data      =
GLOPNET2, subset, na.action, retx = TRUE, center =
TRUE, scale. = FALSE,...)
prcomp(GLOPNET2[,c(10,11,12,16)], retx = TRUE, center
= TRUE, scale. = FALSE, tol = NULL, rank. = NULL, ...)
```

Unlike the `princomp()` function, the `prcomp()` function cannot use a covariance or correlation matrix as input. This is because, unlike `princomp()`, the `prcomp()` function does not perform an eigen analysis of the covariance or correlation matrix. Instead, `prcomp()` uses a method called a *singular value decomposition* of the data. This means that `prcomp()` can only use a data frame of the actual data set. In this sense, it `prcomp()` is less versatile. However, a singular value decomposition allows one to obtain principal components in cases when you have more variables than observations, as was the case in the European SNP data. In this sense, `prcomp()` is more versatile[11]. In the vase majority of cases, the `princomp()` and `prcomp()` functions will give the same result. If `prcomp()` does not actually use a correlation or covariance matrix, then how does one choose between the two? Remembering that a covariance matrix involving standardized variables is a correlation matrix, you would instead choose between using standardized variables (`scale.= FALSE`, i.e. a covariance matrix) and unstandardized variables (`scale.= TRUE`, i.e. a correlation matrix).

prcomp()

In each case, the basic output of a PCA consists of three pieces of information and this output is in the form of a list. The first piece of information is the standard deviation of each principal component axis and

[11] A singular value decomposition also provides more numerical stability that is advantageous in some very rare cases.

is called "`sdev`" in the output list of each function. The proportion of the total variation in the data that is captured by each principal component is the square of these values. Note that some other PCA programs might report "eigenvalues" because the square root of such eigenvalues are the standard deviations of the principal components. The second piece of information is the way that each of the original variables have been linearly combined to form each principal component axis. This is called the "`rotation`" in the output of the `prcomp()` function and is called "`loadings`" of the `princomp()` function. The third piece of information is the position of each observation on each principal component axis. This is simply called "`x`" in the `prcomp()` function and is called "`scores`" in the `princomp()` function. Table 2.2 summarises the elements of the output list for each function.

`prcomp()` name	`princomp()` name	Explanation
`sdev`	`sdev`	These are the standard deviations of each principal component axis. These are also the "lengths" of each principal component axis. The square of these values (called *eigenvalues*) give the variances of each principal component. The sum of these squared values give the total variance. The square of these values divided by the sum of these squared values gives the proportion of the total variance captured by each principal component.

`rotation`	`loadings`	A square matrix, with the number of rows and columns equal to the number of variables in the data set. Each column gives the values by which each variable are multiplied in order to define each principal component.
`center`	`center`	The variable means. If you chose to centre the values (the default option) then these are the values which are subtracted from each variable before rotating original axes.
`scale`	`scale`	The standard deviations of each variable. If you chose to analyse the correlation matrix then these are the values by which each variable was divided in order to convert the observations to standardized values.
`x`	`scores`	The coordinates of the observations on each of the principal components. This is a matrix with as many rows as there are observations and as many columns as there are principal components (and variables). Each row gives the coordinates of a given observation.

Table 2.2. The elements of the output list of the two principal component analysis functions described in this chapter.

2.7) The "importance" of each principal component

PCA converts the original axes into an equal number of new axes (principal components). The first axis accounts for the most variance. The second axis accounts for the most variance after removing the variance accounted for by the first axis. Successive axes each account for the most variance after removing the variance accounted for by all of the preceding axes. Since one goal of PCA is to allow us to see patterns in the full data while using only a few axes (i.e. dimensional reduction), it is important to know how much of the total variance in the data have been captured by each principal component axis. If the `prcomp()` or `princomp()` output has been saved in an object called `GLOPNET.PCA1`, then you get this information by typing `summary(GLOPNET.PCA1)`. The output is:

```
Importance of components:
                          Comp.1     Comp.2    Comp.3     C
omp.4
Standard deviation     0.5727325 0.17839657 0.1301287 0.120
14222
Proportion of Variance 0.8384702 0.08134988 0.0432843 0.036
89566
Cumulative Proportion  0.8384702 0.91982004 0.9631043 1.000
00000
```

There are four principal components in this output because there were four variables in the analysis. The first line lists the standard deviation of each principal component axis (these are also the square roots of the eigenvalues). If you type `GLOPNET.PCA1$sdev` then you will also get the first line of the above output. The square of the standard deviation of a principal component axis is the amount of variance captured by it. This is the information given in the second line ("Proportion of Variance"). Therefore, proportion of the total variance that is captured by the first axis

("Comp.1") is $0.5737325^2 \sim 0.838$ log10 units, or ~83.8%. Stated another way, if you only used the positions of each observation along the first principal component, and ignored the other three principal components, then you would still capture 83.8% of the original pattern of covariation. The second axis ("Comp.2") captures an additional 8.1% of the total variance. If you only used the positions of each observation along the first two principal components, and ignored the other two principal components, then you would still capture 83.8 + 8.1 = 91.9% of the original pattern of covariation. The total variance captured by all four principal components is the sum of these squared values (here, 0.3912155). The third line of the output lists the total proportion amount of the variance accounted by all of the principal component axes up to and including the listed principal component.

2.8) From original variables to principal components

How have the original four axes (i.e. the original four variables) been combined to make these four principal component axes? This information is given by the loadings object of the `princomp()` function (`princomp(…)$loadings`) or the rotation object of the `prcomp()` function (`prcomp(…)$rotation`). Here are the loadings for our GLOPNET analysis:

```
GLOPNET.PCA1$loadings

Loadings:
         Comp.1 Comp.2 Comp.3 Comp.4
logLL     0.722  0.630  0.283       
logLMA    0.385 -0.691  0.516  0.329
logNmass -0.263  0.266         0.926
logAmass -0.511  0.233  0.808 -0.180
```

Each number is the "loading" of a given variable on a particular principal component axis. Entries listed as "…" mean that the value is less than 0.001. We say that the *loading* of logLL on the first principal component ("Comp.1") is 0.722. The name "rotation" used in the `prcomp()` function refers to the fact that this number is also the amount by which the original logLL axis has been rotated in order to produce this first principal component. Each column of the output gives us the equation by which we can calculate the position of any observation on each principal component. Remembering that each variable has been centred about its mean, the equation for the first principal component, using the first column of `GLOPNET.PCA1$loadings`, is:

$$\text{axis1}=0.722\left(\text{logLL-}\overline{\text{logLL}}\right)+0.385\left(\text{logLMA-}\overline{\text{logLMA}}\right)-$$
$$0.263\left(\text{logNmass-}\overline{\text{logNmass}}\right)\text{-}0.511\left(\text{logAmass-}\overline{\text{logAmass}}\right)$$

A similar equation exists for each of the subsequent three principal component axes. The second objective of PCA, given at the beginning of this chapter, was to express the main patterns of covariation in the form of equations for communication or for extrapolation to new data. These equations fulfil this second objective. If we obtain values of the four variables for the leaves of a new species, not included in the analysis, then we can predict its value on each principal component by using these equations. Alternatively, you can obtain such predictions by using `predict(object, newdata, ...)` where "`object`" is a saved object from the `prcomp()` or `princomp()` functions and "`newdata`" is a data frame containing the new observations.

The numerical interpretation of the first principal component is that an increase along it by one unit is the same thing as simultaneously increasing logLL by 0.722 units, increasing logLMA by 0.385 units, decreasing logNmass by 0.263 units and decreasing logAmass by 0.511 units. Qualitatively, species that occur higher on this first principal component have longer lifespans (logLL) and higher leaf mass per area (logLMA) – in other words, leaves that are denser and/or thicker – while also having less leaf nitrogen per mass (logNmass) and lower maximum net photosynthetic rates per mass (logAmass).

The third objective of PCA is to suggest hypotheses about the underlying causes of this covariation. This is done by looking at how the variables load onto a principal component axis. For instance, we note that the variables logLMA and logLL load positively on the first axis while logAmass and logNmass load negatively on this same axis. This means that LogLMA and logLL are positively correlated with each other, logAmass and logNmass are also positively correlated with each other (since they have the same sign for their loadings) but that the set of (logLMA, logLL) is negatively correlated with the set of (logAmass, logNmass). The biological hypothesis proposed by Wright et al. (2004) was that this first principal component represents a very general trade-off between two opposing abilities of leaves: the ability to rapidly acquire nutrients and carbon and the ability to maintain these nutrients for a long time. Leaves with lots of nitrogen (primarily photosynthetic enzymes) allows the plant to more quickly acquire nutrients (high logNmass) and convert CO_2 into sugars (high logAmax). These variables lead to a high metabolic rate and a rapid rate of resource acquisition. However, this is associated with a short lifespan (small logLL) and leaves that are not structurally tough (low logLMA). This leads to a rapid turnover of acquired resources. Importantly, such an interpretation of a PCA does not come from the statistics; it is a biological interpretation – a working hypothesis. PCA can suggest hypotheses but does not test them.

You can judge the importance of each variable in defining a principal component axis by comparing their loadings on that axis. It is not correct to compare the numerical values of loadings on different principal components (Legendre and Legendre 2012). Notice that the loading of logLL on the first principal component is 0.722 while the loading of logNmass on this same principal component is -0.263. This means that logLL is $|0.722/-0.263| \sim 2.7$ times more important than logNmass in defining this first PCA axis. The first principal component is therefore most strongly determined by the trade-off between logLL (loading = 0.722) and logAmass (loading = -0.511), and less strongly by the trade-off between logLMA (loading = 0.385) and logNmass (-0.263).

How should we interpret the second principal component? It is a mathematical property of PCA that each principal component axis is linearly uncorrelated with all of the others. This means that the second principal component represents the patterns of covariation[12] between the four original variables after removing, or *holding constant*, the variation in the first principal component. The same interpretation applies to each of the successive principal components. The second principal component of our analysis therefore describes the relationships between the four original variables for leaves who have the same values on the first principal component. This second principal component is capturing a relationship between the *residuals* of the four leaf traits after removing the main pattern given by the first principal component. After accounting for the general the pattern of covariation captured on the first principal component, some species have a lower than expected value of logLMA (thus, a loading of -0.691) while simultaneously having larger than expected values of the other three leaf traits, especially logLL (having a loading of 0.630). This may, or may not, lead to additional biological hypotheses concerning secondary tradeoffs.

[12] Patterns of covariation, not patterns of correlation, because we chose to use the covariance matrix.

We can plot the loadings on the principal components. Since these loadings represent the rotations of the axes (i.e. the original variables) in the original coordinate system that were used to produce the new principal component axes, we represent these loadings as arrows[13]. The angle between each arrow and the principal component axes represent the covariance between the original axes (variables) and the new principal component axes. Using a correlation matrix, these covariances are also correlations. Figure 2.7 shows such a plot using the first two principal components.

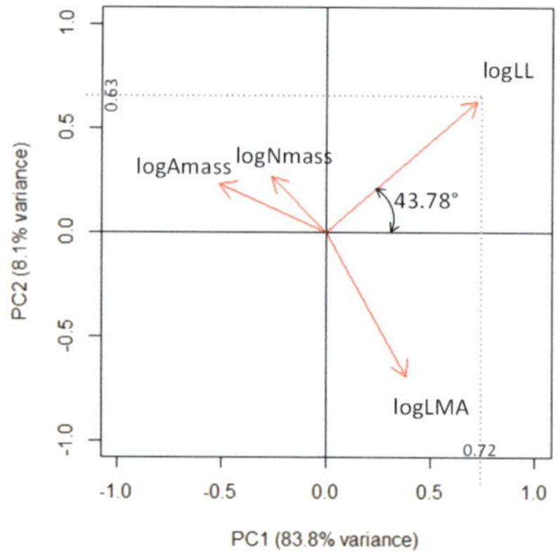

Figure 2.7. The loadings of the four measured variables on the first two principal component axes using the GLOPNET data. The loadings for logLL on these two axes are 0.72 and 0.63.

[13] Some computer programs show this using lines with tick marks to make clear that these are the original axes of the variables.

2.9) Putting observations on principal component axes

People are so used to positioning observations onto the horizontal and vertical axes of a scatterplot that they often forget what constitutes an "axis". An axis is actually a mathematical function that converts our observations into the values of this function. A set of these functions defines a *coordinate system*. I have emphasized this in Figure 2.4(a) by labelling the horizontal and vertical axes by these mathematical functions. In the original coordinate system of these data, the horizontal axis (the "x axis") is defined as "axis 1 = 1x+0y" and the vertical axis (the "y axis") is defined as "axis 2 = 0x+1y". We traditionally call these two axes the "x" and "y" axes because the horizontal axis is defined only by x (i.e. 1x+0y) and the vertical axis is defined only by y. These two axes are orthogonal (oriented at 90° from each other) and are also linearly independent of each other (cos(90°)=0).

As you now know, a PCA consists of rotating the original axes (i.e. the original coordinate system) and (optionally) moving (or *translating*) the origin of these axes. The result of these rotations and translations give us the principal components, whose equations we have already seen. Inputting the values of each of the original variables for each sampling unit into these equations gives the coordinates of these observational units on our new principal component axes. For instance, we already know that the first principal component axis in our analysis is:

$$\text{axis1}=0.722\left(\text{logLL-}\overline{\text{logLL}}\right)+0.385\left(\text{logLMA-}\overline{\text{logLMA}}\right)-0.263\left(\text{logNmass-}\overline{\text{logNmass}}\right)-0.511\left(\text{logAmass-}\overline{\text{logAmass}}\right)$$

The mean values of the four variables are given in Table 2.1. If we input these mean values plus the values of each of the four variables for each of the 492 species (i.e. observational units) in the GLOPNET2 data set into this equation in **R** then this would give us the coordinates of these 492 species on this first principal component axis:

```
0.722*(GLOPNET2$logLL-mean(GLOPNET2$logLL))
+0.385*(GLOPNET2$logLMA-mean(GLOPNET2$logLMA))
-0.263*(GLOPNET2$logNmass-mean(GLOPNET2$logNmass))
-0.511*(GLOPNET2$logAmass-mean(GLOPNET2$logAmass))
```

The values of each of the 492 species when input into this equation are called the scores of the observations on this first PCA axis. Happily, we don't need to do this calculation because the scores of each observation on each principal component axis are given as a matrix called `GLOPNET.PCA1$scores`. There are always as many rows in this matrix as there are observational units (here, 492) and as many columns as there are principal components and variables (here 4). We can also plot the scores of each species on these first two principal component axes (Figure

2.8) where each point represents one species.

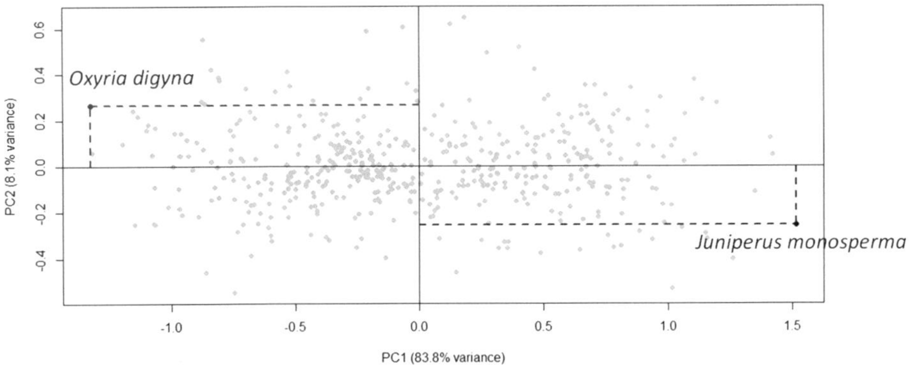

Figure 2.8. The scores of each of the 492 species on the first two principal components using the covariance matrix calculated from the GLOPNET data set. The scores of *Juniperus monosperma* are (1.5174844, -0.2560386) and the scores of *Oxyria digyna* are (-1.3248177, 0.2670361).

This plot is the best representation of the relationships between the leaves of these species in only two dimensions. Two points that are close together in this plot are species whose leaves are very similar based on the four measured traits. Two points that are far apart in this plot are species whose leaves are very different. The two species that are identified in the plot (*Juniperus monosperma* and *Oxyria digyna*) therefore have very different leaf traits because the straight line between them is long. Furthermore, by looking at Figure 2.7, we also know how these species differ. *Juniperus monosperma*, because it has the largest score on the first principal component, must have very dense, thick leaves (high LMA) that are long-lived (LL) but with little leaf nitrogen (Nmass) and with a low net photosynthetic rate (Amass). *Oxyria digyna*, because it has the smallest score on the first principal component, must have the opposite combination of trait values.

2.10) How to choose the principal components

The main purpose of PCA is to reduce the dimensionality of a multivariate data to just a few principal components but in a PCA there are always as many principal component axes as there are variables. This means that we must discard some (often most) principal components. Which ones should we keep? There is no single answer to this question. We expect, since we have randomly sampled our observations, that some of the multivariate variation in our data is simply random sampling variation. We want to discard this variation. Since we can only physically draw graphs in two or three dimensions, we also rarely want to keep more than two or three principal components. Since each successive principal component captures less and less of the total variation, we usually want to use the first two or three principal components but this is not always the case.

The loadings on each principal component axis (i.e. the linear combination of the original variables that define an axis) allow us to interpret these axes. Sometimes an axis can account for a large proportion of the total variation but it reflects a property for which we have no interest. If so, then we are perfectly justified in ignoring such an axis. For example, imagine that we have measured various length measurements of different body parts in many different species of birds: the length and width of the wings, the length and width of the beak, the total length of the body and the length of the legs. A PCA of such data (either based on a covariance or on a correlation matrix) would show that the first PC axis has positive loadings on all of these variables. In other words, if you typed `princomp(...)$loadings` then the values output in the first column (the loadings on the first principal component) would all be positive. This pattern of loadings means that, when comparing across bird species, larger lengths for one body part tend to be associated with larger lengths of all of the other body parts. We would interpret this first principal component as representing differences in "size" since bigger bird species tend to have longer lengths of all of their body parts. The fact that this is the first principal component means that this size effect is the dominant trend in our data. However, we don't need to conduct a principal component analysis to see this obvious allometric pattern and so this first principal component is not revealing anything that we didn't already know. The subsequent principal components capture the patterns of covariation between the body parts that are independent of the first principal component, i.e. after removing this general size effect. If we are actually interested in such size-independent patterns, then we are perfectly justified in ignoring this first principal component.

How do you know which principal components to keep and which principal components to ignore? The answer to this question must remain somewhat vague: keep those axes that (i) capture variation beyond purely random sampling variation, (ii) that measure some composite property that is of biological interest, and (iii) that are as few in number as possible while respecting (i) and (ii). Since the goal of PCA is to reduce the dimensionality of the original data, most users limit themselves to two – only occasionally three – principal components. There are statistical procedures to help you with point (i) but not with point (ii) because it is a biological, not a statistical, goal.

The first, somewhat subjective and graphical, method of identifying those axes that capture important variation is called a "scree" plot. Figure 2.9 shows a scree plot for the GLOPNET2 data. A scree plot consists of plotting the variance captured by each principal component (the eigenvalues) associated with each axis as a function of the order (fist, second etc.) of each axis. These values are the squares of the `$std` objects in the output (Table 2.2). Such plots typically show an initial rapid drop followed by a more linear decrease[14]. There are three rules of thumb associated with interpreting scree plots. The first rule of thumb is that the point where the initial rapid drop changes to a more linear decrease is an indication of the number of "important" axes. In Figure 2.9 this occurs

[14] A "scree" is a collection of broken rocks ay the base of a cliff or mountain. This visual image (a sharp drop down the cliff and then a more gradual drop once you reach the broken rocks) evokes the pattern that you look for in a scree plot.

right after the first principal component. A second rule of thumb is to concentrate on the axes whose eigenvalues are greater than the mean of the eigenvalues or, equivalently, whose standardized eigenvalues are greater than unity. A standardized eigenvalue is the eigenvalue associated with a principal component divided by the mean of all of the eigenvalues. If your PCA is done on a correlation matrix (or, equivalently, using standardized variables) then the mean eigenvalue is always unity. The mean of the eigenvalues is obtained by `mean(GLOPNET.PCA3$sdev^2)` and I have added this as a horizontal line in Figure 2.9. Only the first principal component would be kept based on this second rule of thumb described above. The third rule of thumb is to keep those principal components whose eigenvalue is greater that the value expected from a "broken stick" distribution. The red line in Figure 2.9 shows these values. A broken stick distribution gives the expected values obtained by randomly choosing a value between 0 (no variance associated with PCA axis 1) and 1 (all of the variance associated with PCA axis 1) and then successively doing the same thing for each of the remaining axes. Only the first principal component would be kept based on this third rule of thumb. All three rules of thumb tell us to concentrate only on the first principal component. Unfortunately, these three rules of thumb don't always completely agree.

Although there is a `screeplot()` function in the base library of **R** it is rather rudimentary and does not plot our rules of thumb. The `screeplot()` function in the vegan package is better. The full name of this function, and

screeplot()

the name that you must specify when searching in the help files, is `screeplot.cca` but this is (confusingly) not the name that you use to call the function. I have also added the `bstick=TRUE` argument. This argument adds a "broken stick" distribution which represents the distribution of the proportion of variance that would be accounted in mutually independent normal data. Eigenvalues whose values are greater than that of the broken stick distribution indicate axes that are (probably) accounting for real patterns in the data. Here is the code for producing Figure 2.9:

```
library(vegan)
screeplot(GLOPNET.PCA3,bstick=TRUE,type="barplot",npc
s=4,main="PCA of GLOPNET2, using covariance
matrix",ylab="Proportion of variance (eigenvalues)")
abline(mean(GLOPNET.PCA3$sdev^2),0,lwd=2)
```

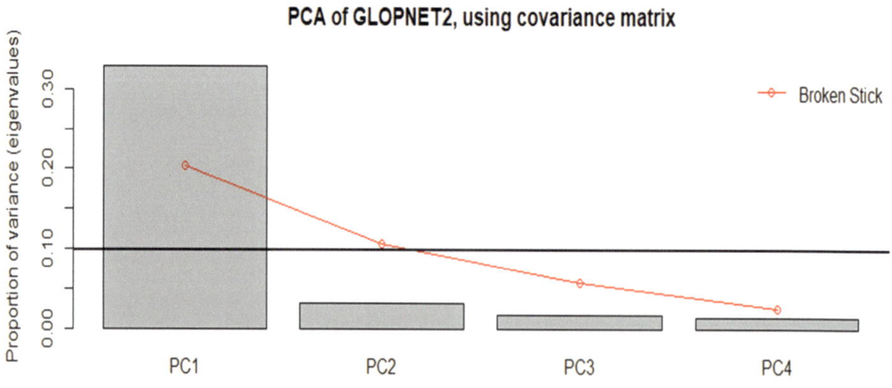

Figure 2.9. A scree plot of the eigenvalues associated with the PCA of the GLOPNET2 data using a covariance matrix. The solid horizontal line shows the

mean eigenvalue and the red line shows the values of a broken stick model with four breaks.

There is a second, more formal, method of identifying those axes that capture variation beyond purely random sampling variation. This consists of conducting an inferential test with respect to each successive axis by assuming, given the null hypothesis, that the axis in question only captures random sampling variation. You would only consider those principal components for which this null hypothesis is rejected. Several of such tests are reviewed in Forkman et al. (2019) who also gives **R** code for two versions of a parametric bootstrap method. I have reproduced this code, slightly modified, in Boxes 2.1 and 2.2. The code in Box 2.1 is used for analysing a covariance matrix on unstandardized data and the code in Box 2.2 is used for analysing a correlation matrix; i.e. a covariance matrix using standardized data. The null hypothesis that the axis in question only captures random sampling variation and it is assumes that your data are normally distributed. Therefore, a significant null probability (i.e. a value smaller than your chosen significance level) is evidence that the axis does *not* represent only random sampling variation and therefore that it is capturing real multivariate associations. Because each successive principal component accounts for successively smaller proportions of the total variation in your data, once the test identifies an axis that has a non-significant null probability it stops testing. Since our PCA of the GLOPNET2 data used a covariance matrix, we would use the code in Box

2.1 (`Forkman.test.unstandardized.PCA`). Here is how to call this function with our data and its output; I have increased the default number of bootstrap runs (the "N.Boot" argument) to 10,000:

```
Forkman.test.unstandardized.PCA(GLOPNET2[,c(10,11,12,
16)],N.Boot=10000)
  axis.rank p.value
1         1  0.0000
2         2  0.0000
3         3  0.2143
```

The result tells us that the first two principal component axes represent systematic patterns of covariation but the third (and therefore the fourth) principal component do not.

2.11) Biplots for visualizing a PCA

The most common, and the most useful, graphical tool for visualizing the results of a PCA is called a "biplot". As the name suggests, a biplot is a superposition of two different plots onto one figure, namely the scores and the loadings plots that we already studied. However, since the scores and loadings plots will generally have different scales, it is necessary to scale either (or both) in order to place them together on the same plot. The scaling is represented by a constant c, which can vary continuously between 0 and 1. Two common types of scaling are used (c = 0 or 1) with c=0.5 sometimes used as a compromise between the other two.

Each scaling emphasizes certain geometric relationships between the observational units (the distances between them) or between the original variables. A further unfortunate complication is that different authors use different names for these choices. I will call these “variable preserving” (c=1) and “distance preserving” (c=0) biplots. Table 2.3 presents the different choices, names, and properties of each type of biplot with respect to principal components conducted using either a covariance or a correlation matrix. The values of the scaling coefficient (c) in this table reflects the values used in the `biplot()` function and the similar, but more powerful, `ggbiplot()` function of the ggplot2 library (Wickham 2016) of R.

Constant (c)	Names	Properties
PCA using covariance matrix		
c=1	Variable preserving biplot (Scaling 2 biplot, Correlation biplot, GH biplot)	A projection of the tip of the arrowhead onto a principal component axis at 90° indicates the importance of the variable in defining that principal component.
		The Euclidean distance between two points in the biplot is an estimate of the Mahalanobis distance between the points in the data.
		The angle between each arrow and a principal component axis is equal to the covariance between the variable and the principal component axis.
		The angle between each pair of arrows is equal to the covariance between the two variables.

c=0	Distance preserving biplot (Scaling 1 biplot, JK biplot)	A projection of a point onto a principal component axis at 90° indicates the loading of that sampling unit on that principal component.
		The Euclidean distance between two points in the biplot is an estimate of Euclidean distance between the points in the data.
		The angle between each arrow and a principal component axis is equal to the covariance between the variable and the principal component axis.
PCA using a correlation matrix		
c=1	Variable preserving biplot (Scaling 2 biplot, Correlation biplot, GH biplot)	A projection of the tip of the arrowhead onto a principal component axis at 90° indicates the importance of the standardized variable in defining that principal component.
		The Euclidean distance between two points in the biplot is an estimate of the Mahalanobis distance between the points in the data.
		The angle between each arrow and a principal component axis is equal to the correlation between the variable and the principal component axis.
		The angle between each pair of arrows is equal to the correlation between the two variables.
c=0	Distance preserving biplot (Scaling 1 biplot,	A projection of a point onto a principal component axis at 90° indicates the loading of that sampling unit on that principal component.
		The Euclidean distance between two standardized points in the biplot is an

	JK biplot)	estimate of Euclidean distance between the standardized points in the data.
		The angle between each arrow and a principal component axis is equal to the correlation between the variable and the principal component axis.

Table 2.3. The two most common types of biplots based on the scaling constant (c).

A basic biplot is produced using the `biplot()` function but there are arguments available to change many aspects of these plots. For instance, the biplot in Figure 2.10 was produced using these lines of code:

biplot()

```
biplot(GLOPNET.PCA1,scale=0,col=c("black","red"),
xlabs=rep(".",sum(full.lines)),cex=c(2,1),
main="Distance preserving",xlab="PC1 (83.8%
variance)",ylab="PC2 (8.1% variance")
abline(h=0)
abline(v=0)
```

The first argument (`GLOPNET.PCA1`) is an object saved from a call to `princomp()` or to `prcomp()`; in this case, from our PCA of the GLOPNET2 data based on a covariance matrix. The second argument (`choices=`) specifies the two principal component axes to plot; in this case the first and second axes. These are the default values for this argument. The third argument (`scale=`) specifies which type of scaling is desired; this is the value of the scaling constant (c) as explained in Table 2.3. The other arguments control different graphical aspects of the plot.

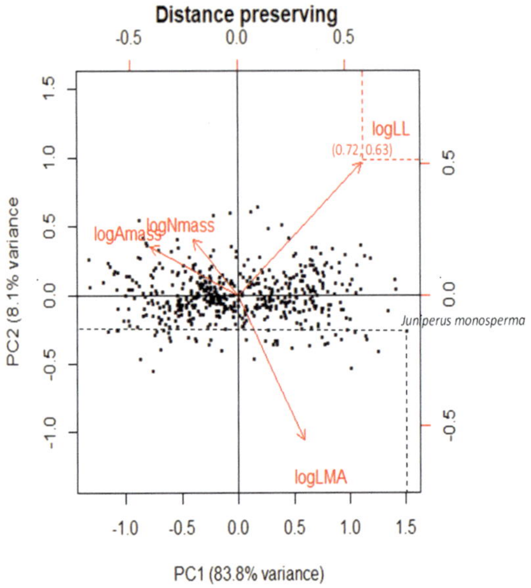

Figure 2.10. A distance preserving biplot based on principal component analyses done using the covariance matrix of the GLOPNET2 data.

Since a biplot is a superposition of a scores plot and a loadings plot, it requires two sets of axes. Figure 2.10 shows a distance preserving biplot (i.e. `scale = 0`) of the PCA based on the covariance matrix. The left vertical and the lower horizontal axes are for the scores (the points) of the observational units (here, species) on the chosen principal components. The right vertical and upper horizontal axes are for the loadings of the variables, which are the arrows in the biplot.

The loadings for any variable with respect to the chosen PCA axes can be obtained by projecting the tip of the arrow for that variable onto the two PCA axes. To obtain the loading of the variable on the first PCA axis, you draw a vertical line from the arrow tip until in meets the upper horizontal axis of the biplot at a 90° angle. To obtain the loading of that variable on the second PCA axis, you draw a horizontal line from its arrow tip until the line meets the right vertical axis (Figure 2.10, dotted red lines). These dotted red lines are not part of the `biplot()` output but I have added them here to illustrate the point.

Because the loadings and the scores are in different units but you are placing both on the same graph, you must scale either the scores or the loadings. You specify your choice using the `scale=` argument of the biplot function. If you chose `scale=0`, as done for Figure 2.10, then the distances between the points (the scores) on the biplot will approximate the Euclidean distances of the points in the actual data and these distances will be measured in the original units of the data. In our example, these distances will be in log10 units since all of our variables were log10 transformed. I said that this choice of scaling "approximates" the actual Euclidean distances because we are still ignoring the other principal components. The quality of the approximation depends on the proportion of the total variance that is captured by the two chosen principal components in this type of biplot. The more the total variance that is captured by the two chosen PCA axes, the better the approximation. We will call this type of biplot a "distance preserving" biplot but be aware that different authors have used different names. For instance (Legendre and

Legendre 2012) also call this type of biplot a "scale 1" biplot while the SAS statistical program calls it a "JK" biplot.

To obtain the score of a point on the first PCA axis (here, PCA1), you simply draw a vertical line from the point until it meets the lower horizontal axis of the biplot at a 90° angle. To obtain the score of a point on the second PCA axis (here, PCA2), you simply draw a horizontal line from the point until it meets the left vertical axis of the biplot at a 90° angle. For instance, the scores for the species *Juniperus monosperma* on these first two axes in Figure 2.10 are shown by the dotted black lines although these lines are not actually produced by `biplot()`. By comparing the biplot in Figure 2.10 to the plots shown in Figures 2.7 and 2.8, you will notice that a distance preserving biplot is simply the combination of the two. If you chose a distance preserving biplot then you will loose some information concerning the relationships between the loadings (i.e. the variables).

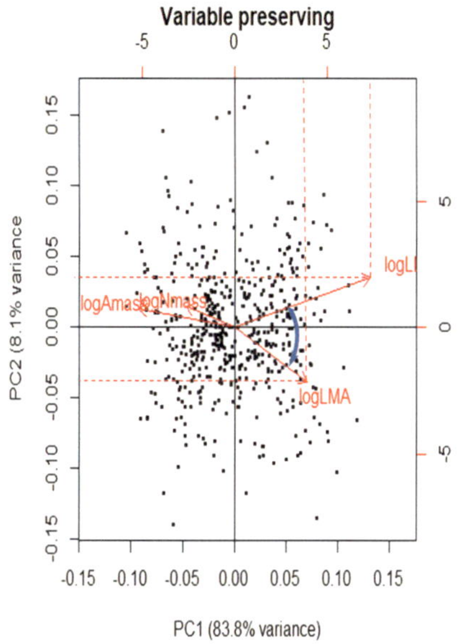

Figure 2.11 A variable preserving biplot of the GLOPNET data using a covariance matrix

The other common choice of scaling for biplots is `scale=1`. If you choose this type of scaling then the resulting biplot conveys more information about the relationships between the variables (thus, loadings). In particular, the angle (a) between two arrows approximates the Pearson linear correlation (r) between the two variables; the exact relationship is $r=\cos(a)$. An angle of 0° (i.e. two arrows superposed in the same direction) means that the two variables are perfectly positively correlated. As the angle between the arrows increases the correlation between the variables decreases until, at an angle of 90°, the correlation between the two variables is 0[15]. Angles of more than 90° represent increasingly negative correlations. For example, the arrows for logAmass and logNmass in Figure 2.11 have a very small angle between them, we know that these two variables are tightly and positively correlated. Since the angle between the arrows for logLMA and logLL is small, but slightly larger than for the other two arrows, we know that logLMA and logLL are also positively correlated but somewhat less tightly than for the other two variables. Finally, since the angles between the two pairs of variables (logLMA and LogLL versus LogAmass and LogNmass) are large (almost 180°), we know that the correlations between these two sets of variables is strong but negative.

A projection of the tip of the arrowhead onto a principal component axis at 90° indicates the importance of the variable in defining that principal component. For example, the projection of logLL is about twice as important as logLMA in determining the first principal component axis since it has about twice the projected distance. Looking at the second principal component axis, we see that logLMA and logLL have about equal importance and are about twice as important as logAmass and logNmass for this second axis.

[15] Note that cos(0°)=1 (the two arrows are superimposed and the correlation is 1) while cos(90°)=0 (the two arrows are at right angles and the correlation is 0).

We will call this type of biplot a "variable preserving" biplot because it provides the best two-dimensional representation of the relationships between the variables. However, a variable preserving biplot rescales the scores. The distances between points in this variable preserving biplot are not approximations of the actual Euclidean distances between the points in the data; rather, the distances between points in this type of biplot are approximations of the points when the variables have been standardized to unit standard deviations. This type of distance is called a Mahalanobis distance. In order to better understand the differences between "distance preserving" and "variable preserving" biplots, we must go on a slight detour to discuss the relationship between two (among many) types of distance measures: Euclidean and Mahalanobis distances.

A "Euclidean" distance is a distance between two points based on Euclid's famous formula to calculate the hypotenuse of a triangle. Imagine that we have two variables X and Y and we have measured the values of X and Y for two observations (Figure 2.12). We want to calculate the Euclidean distance, d(1,2), between these two points. The solution is shown in Figure 2.12. This Euclidean distance is affected by the measurement units of the variables; if Y was measured in (say) metres then (Y_2-Y_1) would be 2 metres but if we change to centimetres, then (Y_2-Y_1) would be 200 centimetres. This would profoundly change the value of the

Euclidean distance between the two points. This is important point to remember when we discuss the Mahalanobis distance.

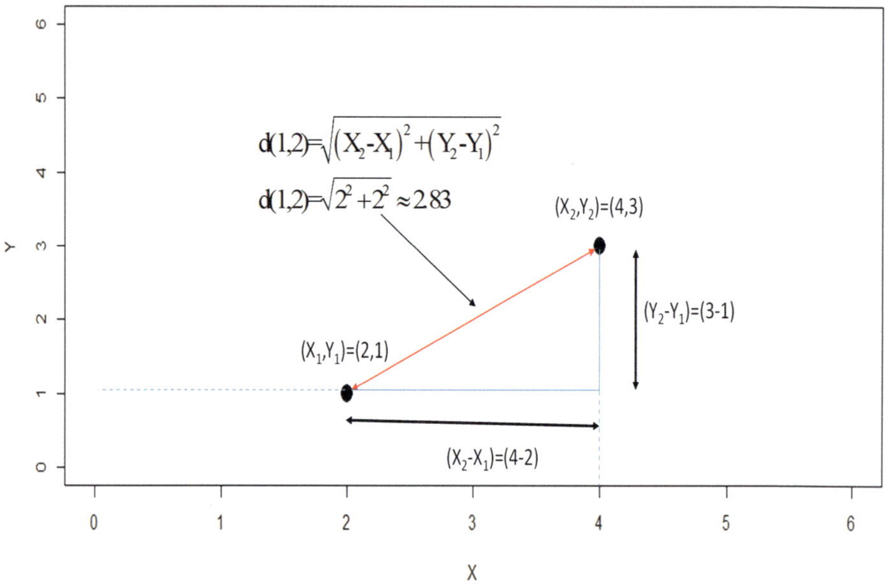

Figure 2.12. Calculating the distance between two points in a 2-dimensional space using Euclid's formula.

When we have more than two axes, either using the original variables in our data set or using the new principal component axes, the distance between any two points (i, j) can be calculated using a generalization of Euclid's formula. If a_{ik} and a_{jk} are the values of points i and j for axis (or variable) k, and if there are a total of D axes (or dimensions), then the distance between these two points in this D-dimensional space is $d(i,j)=\sqrt{\sum_{k=1}^{D}\left(a_{ik}-a_{jk}\right)}$. An important property of PCA is that the Euclidean distance between points using the original axes (variables) is maintained when these points are plotted using the entire set of principal component axes. This is because (as you now know) a PCA is just a rotation of the original axes. This makes intuitive sense if look at Figure 2.4, since the points remained stationary while we rotated and translated the axes; in other words, the distances between these points didn't change. This property holds whether you conduct your PCA using the original unstandardized variables (using a covariance matrix) or using standardized variables (using a correlation matrix).

A Mahalanobis distance is named after the Indian statistician who first introduced it, Prasanta Chandra Mahalanobis (Mahalanobis 1936). A graphical explanation of the relationship between Euclidean distances and the associated Mahalanobis distances is shown in Figure 2.13. This figure plots the scores of the GLOPNET data on the first two principal components; you can compare it to Figure 2.10. One species, shown by the red dot, occurs at the centre with a score of zero on both of the two principal components. We can calculate the Euclidean distances of three species, shown by the black, blue and orange dots, to the species (red dot) that occurs right at the center. These Euclidean distances are the lengths of the arrows. The black and blue dots have almost equal Euclidean distances to the red dot, 0.33 and 0.34 respectively. The orange dot has a Euclidean distance to the red dot that is three times longer (0.99).

Remember that these distances are dependent on the measurement units of the two principal components and the principal component scores are dependent on the measurement units of the original variables; here, the variables are measured in log10 units.

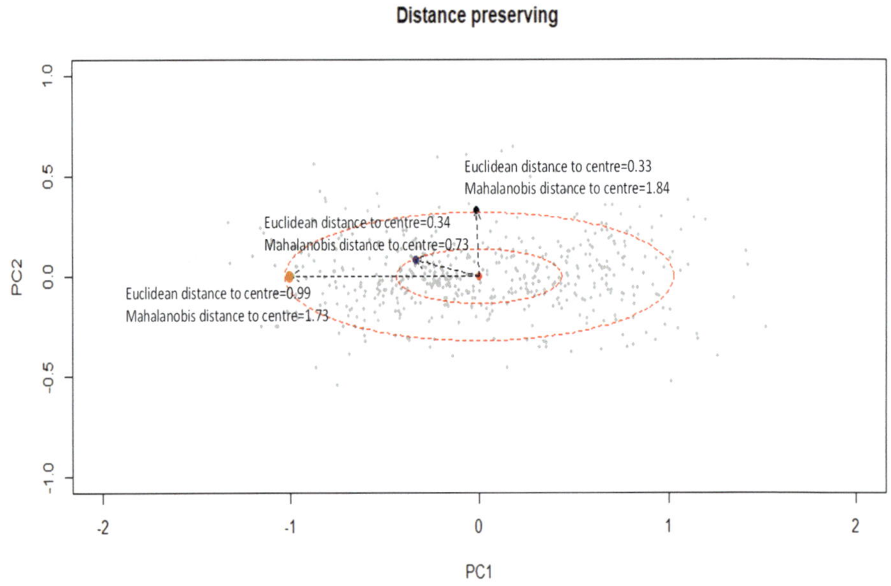

Figure 2.13. A score plot of the GLOPNET data using the first two principal components and with a "distance preserving" scaling. The red dot is at the grand mean of the scores of the two principal component axes and the red ellipses show points having the same Mahalanobis distances (0.73 and 1.84 standard deviations). The black arrows show the Euclidean distance between three points (black, blue and orange points) and the red dot.

A single normally distributed variable can be measured in either its original units or in standard deviations. Measuring this variable in standard deviations is equivalent to standardizing it. Imagine that we have a normally distributed variable whose mean is 0 and whose standard deviation is 5. If we observe a value of 15 then we can equivalently say that it has a Euclidean distance from its mean of 15 units (i.e. 15-0) or that it has a Euclidean distance from its mean of 3 standard deviations. What happens if we have more than one measured variable on each observational unit, each of which are normally distributed? In this case, we have a multivariate normal distribution. Principal components analysis assumes that the variables follow a multivariate normal distribution and so the scores in Figure 2.13 would also follow a (bivariate) normal distribution. However, we can also measure each of these variables in units of standard deviations. The red point in Figure 2.13 is at the exact centre of the distribution, i.e. at the means of the scores along both axes. If we start at the red point and draw a horizontal line parallel to the horizontal axis then the spread of the PCA scores along this horizontal line is the variation along the first principal component when PC2 is fixed at zero. The distance of each point along this horizontal line can be measured in standard deviations from the mean. The standard deviation of the scores of the first principal component (PC1) in Figure 1 is ~0.57, which is the first value given if we type `GLOPNET.PCA1$sdev`. In other words, a point along our horizontal line at PC1=0.57 and PC2=0 is one that is located at one standard deviation from the red dot.

If we start at the red point and draw a vertical line parallel to the vertical axis then the spread of the PCA scores along this vertical line is the variation along the second principal component when PC1 is fixed at zero.

The standard deviation of the scores of the second principal component (PC2) in Figure 1 is ~0.18, which is the second value given if we type `GLOPNET.PCA1$sdev`. A point along our vertical line at PC1=0 and PC2=0.18 is also located at one standard deviation from the red dot. If we start at the red dot and draw a line at 45° with respect to PC1 then one standard deviation will be intermediate (~0.3) between 0.57 and 0.18. If we start at the red dot and sweep around it by 360°, while keeping our line equal to one standard deviation, then we will obtain an ellipse. Such an ellipse would mark equal distances from the grand mean when measured in units of standard deviations. This is a Mahalanobis distance. A Mahalanobis distance is a Euclidean distance when the variables are measured in units of standard deviations. In general, the covariation between the axes is also taken into account but, for PCA, the axes are always uncorrelated. The two red ellipses in Figure 2.13 show such ellipses marking distances of 0.73 and 1.84 standard deviations. Therefore, points on the same ellipse have the same distance from the centre in terms of standard deviations. The orange and black dots in Figure 2.13 have almost the same Mahalanobis distances, 1.84 and 1.73 respectively, since they are both close to the outer ellipse even though their Euclidean distances, as indicated by the lengths of the arrows, are very different. The black and blue dots have almost the same Euclidean distances on the two principal components (0.33 and 0.34) but very different Mahalanobis distances (1.84 and 0.73 standard deviations from the mean). The black and orange dots have very different Euclidean distances on the two principal components (0.33 and 0.99) but almost the same Mahalanobis distances (1.84 and 1.73).

When you plot your points using only two principal component axes in a scores plot then the distances between the points in this plot will be only approximations of the actual distances calculated using all of the principal component axes. The accuracy of the approximation is determined by the percentage of the total variance captured by these two principal component axes; a higher percentage of the total variance results in a better approximation. The distance between two points indicates the degree of similarity of these two points in terms of the underlying variables. Two points that are close together are very similar in terms of their values on the underlying variables. Two points that are far apart are very different in terms of their values on the underlying variables.

The effect of rescaling the measurement units of the scores to Mahalanobis distances also changes the numerical values of the scores; i.e. the way the original variables are combined to determine the principal components. Using Mahalanobis scaling, the projection of the tip of the arrowhead onto a principal component axis at 90° indicates the importance of the variable in defining that principal component. The angle between each arrow and a principal component axis is still an approximation of the correlation between the variable and the principal component axis but the angle between each pair of arrows is now also an approximation of the correlation between the two variables. This is why this type of biplot is called a "variable preserving" biplot.

Each of the two choices that you make – using a covariance or correlation matrix and using a distance preserving or variable preserving scaling – will affect the biplot. We have already discussed when you would chose a covariance or correlation matrix. You would choose a distance preserving biplot when your primary interest is on the relationships between the observational units (i.e. the principal component scores) and the units of measurement of the variables are important. You would choose a variable preserving biplot if your primary interest is on the relationships between the underlying variables (i.e. the principal component loadings) or when the units of measurement of measurement of the underlying variables are not important.

2.12) Dealing with categorical variables

You cannot use categorical variables in a principal component analysis. However, studying how the observations (i.e. the principal component scores) or the variables (i.e. the principal component loadings) relate to categorical variables is an important part of principal components analysis. Look again at Figure 2.1(left) which plots the scores of the first two principal components of the SNP profiles of Europeans. People whose ancestors lived in different countries (i.e. different categorical groups) are plotted using different colours. By categorizing the country from which each individual lives and differentiating these categories by colours, the most important message from that study becomes immediately obvious: people whose ancestors came from the same geographical locations had very similar scores and the further apart people live in Europe, the more different the scores. This strategy of comparing the scores of a PCA on different axes based on some external grouping is a powerful way of aiding in the interpretation of a PCA.

The GLOPNET data also contains categorical variables relating to different categorical attributes of each species. One variable, "GF", in the GLOPNET data set classifies each observation (species) into one of five growth forms: S=shrub, T=tree, H=herbaceous, G=grass and V=vine.

It is relatively straightforward to plot the scores of each observation using different symbols, colours or sizes. Of course, you could produce increasingly sophisticated graphs but details of graphing in **R** is not the topic of this book. For instance, the `dataEllipse()` function of the car package (Fox and Weisberg 2019) can produce such graphs along with confidence ellipses. Here is some simple code to plot the scores of the species on the first two principal component axes using different colours. The first line extracts the names of the different growth forms that are stored in the GLOPNET data set. The second line sets up the plot of the scores of the species on the first two principal components

(`GLOPNET.PCA1$scores`) but doesn't add the points (`type= "n"`). The for loop then successively adds the points of each group using a different colour for each group. The final line adds a legend to the plot.

```
growthforms<-unique(GLOPNET2$GF)
plot(GLOPNET.PCA1$scores[,2]~GLOPNET.PCA1$scores[,1],
type="n",xlab="PC 1",ylab="PC 2")
for(i in 1:length(growthforms)){
  sel<-GLOPNET2$GF==growthforms[i]
points(x=GLOPNET.PCA1$scores[sel,1],y=GLOPNET.PCA1$sc
ores[sel,2],col=i,pch=16)
}
legend(x="topleft",pc=16,col=1:5,legend=growthforms)
```

The basic `biplot()` function does not have an argument allowing us to plot different colours and symbols for the different groups. Instead, we will install[16] and load the `ggbiplot()` function (Vu 2011), which allows for more flexibility:

```
library(devtools)
install_github("vqv/ggbiplot")
library(ggbiplot)
```

The use of the ggplot2 package is beyond the scope of this book but some useful introductions to it can be found in Wickham (2016). Here is how to create our distance-preserving biplot based on a covariance matrix using the ggplot2 library:

```
ggbiplot(GLOPNET.PCA1,groups=GLOPNET2$GF,scale=0,elli
pse=T,ellipse.prob = 0.75)
```

[16] from: https://www.datacamp.com/community/tutorials/pca-analysis-r

The first argument is an object from either `princomp()` or `prcomp()`. The second argument specifies that we want to differentiate the points based on the categorical variable "GF" in the GLOPNET2 data frame. It is possible to choose your own combination of point types and colours but we will use the default choices here. The third argument specifies that we want a distance-reserving scaling. The fourth and fifth arguments specify that we want to add ellipses around each group enclosing 75% of the points in each group. Now, we will calculate the average scores of each of the five growth forms on the first two axes:

```
unique.GF<-unique(GLOPNET2$GF)
group.means<-
data.frame(x=rep(NA,5),y=rep(NA,5),GF=unique.GF)
for(i in 1:5){
  sel<-GLOPNET2$GF==unique.GF[i]
  group.means$x[i]<-mean(GLOPNET.PCA1$scores[sel,1])
  group.means$y[i]<-mean(GLOPNET.PCA1$scores[sel,2])
  group.means$GF[i]=unique.GF[i]
}
```

This is how to add these group means to the biplot:

```
ggbiplot(GLOPNET.PCA1,groups=GLOPNET2$GF,scale=0,elli
pse=T,ellipse.prob=0.75)+ggtitle("Distance
preserving")+geom_point(mapping=aes(x=x,y=y,colour=GF
,size=2.5),group.means)
```

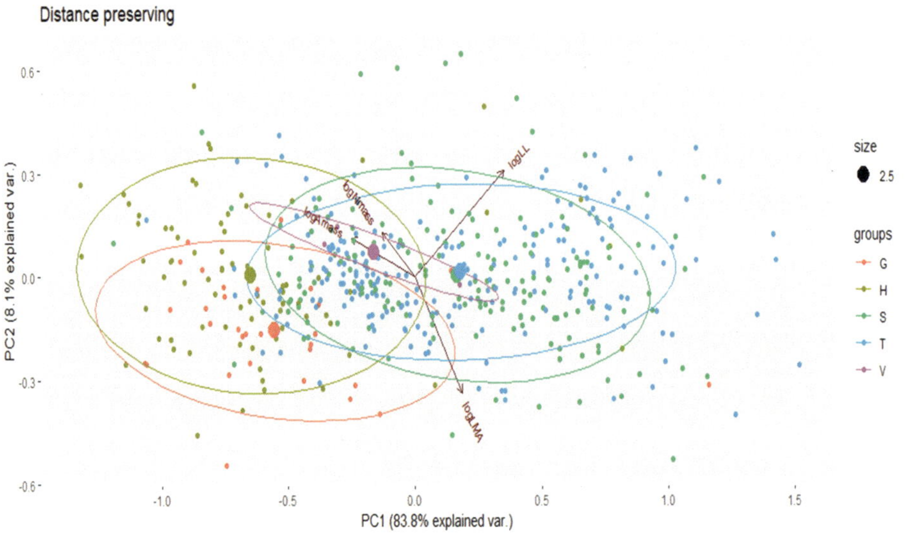

Figure 2.14. A distance-preserving biplot of the GLOPNET data with different colours indicating each of the five growth forms (G=grasses, H=herbs, S=shrubs, T=trees, V=vines). The mean scores of each group are shown in the larger points. The ellipses indicate the 75% confidence intervals for each group.

The resulting biplot is shown in Figure 2.14. We can clearly see that the non-woody species (grasses and herbs) are located towards the left of the first principal component, that the woody species (shrubs and trees) are located towards the right of the first principal component, and that the three vine species are between the woody and non-woody species. In other words, the non-woody species have leaves that are less dense (lower logLMA) and are shorter-lived (lower logLL) but have more leaf nitrogen (logNmass) and a higher maximum net photosynthetic rate (logAmass). Furthermore, the average scores of the shrubs and trees are almost indistinguishable while the herbs and grasses seem to be distinguished along the second principal component and so the grasses tend to have somewhat denser leaves than the herbs (logLMA) and somewhat less leaf nitrogen (logNmass).

The above description describes the patterns between the growth forms in the sample of species included in the GLOPNET data. If we want to extrapolate beyond this sample, and say something about the differences between growth forms in general, then we must determine the probability that such differences might occur by chance. In order to do this, we will conduct an analysis of variance on these scores. Since, by assumption, the data are normally distributed, the scores are also normally distributed. This means that you could test for differences in the scores along a given principal component using an Analysis of Variance (ANOVA). For instance, we can test if the scores (`GLOPNET.PCA1$scores`) along each of the first two axes of our PCA differ significantly between groups (`GLOPNET2$G`) as follows:

```
summary(aov(GLOPNET.PCA1$scores[,1]~GLOPNET2$GF))
```

```
              Df Sum Sq Mean Sq F value Pr(>F)
GLOPNET2$GF    4  53.44  13.360   60.27 <2e-16 ***
Residuals    487 107.95   0.222
```

```
summary(aov(GLOPNET.PCA1$scores[,2]~GLOPNET2$GF))

              Df Sum Sq Mean Sq F value   Pr(>F)
GLOPNET2$GF    4  0.795 0.19882   6.515 4.11e-05 ***
Residuals    487 14.863 0.03052
---
```

As expected, there are highly significant differences between the scores of the five growth forms along both the first and second axes. However, it is possible for the differences between groups to display no significant differences along any single axis, but to be different when both axes are taken into consideration, Skalski et al. (2018) have recently shown how to conduct an omnibus test, which they call an Analysis of Distance[17] involving combinations of axes. This test is a simple modification of a standard ANOVA and so assumes normality of the PCA scores. The **R** code is available in Appendix S4 of that publication and I have re-written it (Box 2.3) as an **R** function. This function, `Skalski.adonis()`, has three arguments. The first argument is the input data frame and this data frame must have separate named numerical columns containing the principal component scores on each axis of interest as well as a named factor column containing the codes for each group. Here is how to test for significant differences between the five growth forms on the first two principal

Skalski.adonis()

[17] WARNING! There is another function called adonis in the vegan library of R that we will use in other chapters. This other function is a non-parametric ANOVA that is appropriate for any type of dissimilarity measure and that is based on a permutation method that is also appropriate for non-normal distances/dissimilarities. It is more general that the method described here and is also appropriate for PCA, but requires notions of dissimilarity that will be discussed in other chapters.

components based on the covariance matrix. First, we create a new data frame called `GLOPNET.PCA1.ANODIS`:

```
GLOPNET.PCA1.ANODIS<-data.frame(GF=GLOPNET2$GF,
PC1=GLOPNET.PCA1$scores[,1],PC2=GLOPNET.PCA1$scores[,
2])
```

Next, we call the `Skalski.adonis()` function. You give the columns of the data frame (here, `GLOPNET.PCA1.ANODIS`) containing the axis scores using the `PC.axes=` argument (here columns 2 and 3):

```
Skalski.adonis(dat=GLOPNET.PCA1.ANODIS,PC.axes=c(2,3)
,Groups=1)
```

The output reports a Fisher's F value of $F_{8,974}$=60.27 given 8 and 974 degrees of freedom with a null probability of less than 0.001, which is expected since these groups also differed significantly along each axis separately. Are there significant differences between only the trees and shrubs? We can answer this question by conducting the ANODIS using only the lines involving either trees or shrubs:

```
sel<-
GLOPNET.PCA1.ANODIS$GF=="S"|GLOPNET.PCA1.ANODIS$GF  ==
"T"
skalski.adonis(dat=GLOPNET.PCA1.ANODIS[,sel],PC.axes=
c(2,3),Groups=1)
```

There is no convincing evidence that the scores on the first two axes differ significantly when looking only between the shrubs and the trees since the resulting F-statistic is $F_{2,760}$=0.074 with a null probability of 0.93. Are there significant differences between only the herbs and the grasses? We can answer this question using only the lines involving either herbs or grasses:

```
sel<-GLOPNET.PCA1.ANODIS$GF=="G" |
GLOPNET.PCA1.ANODIS$GF=="H"
skalski.adonis(dat=GLOPNET.PCA1.ANODIS[sel,],PC.axes=
c(2,3),Groups=1)
```

There is no convincing evidence that the scores on the first two axes differ significantly when looking only between the grasses and the herbs since the resulting F-statistic is $F_{2,210}$=1.077 with a null probability of 0.34. Since we conducted these last two tests after looking at our data, they are called *post hoc* tests. In order to maintain the overall significance level (α) when conducting n separate tests, you could use a Bonferonni correction by adjusting the alpha significance level for each separate test to α/n. The adjusted alpha level in our case would be 0.05/2=0.025.

BOX 2.1. The following is **R** code to conduct a parametric bootstrap test of significance of principal component axes obtained from a covariance matrix on centered but unstandardized (unscaled) variables.

```
Forkman.test.unstandardized.PCA<-
function(dat,alpha=0.05,N.Boot=1000){
  n <- nrow(dat) ; n
  p <- ncol(dat) ; p
  M <- min(n - 1, p)
# Variables are centred, but not scaled
  Y <- scale(dat, center = TRUE, scale = FALSE)
# Singular value decomposition
  svd.Y <- svd(Y)
  lambda <- svd.Y$d
# Test statistic
  T <- {lambda^2/rev(cumsum(rev(lambda^2)))} [1:(M - 1)]
# Bootstrap Sampling
  T.boot <- matrix(NA, nrow = N.Boot, ncol = M - 1)
  for(Component in 1:(M - 1)){
    K <- Component - 1
    for(Boot in 1:N.Boot){
      E.boot <- matrix(rnorm((n - 1 - K)*(p - K)), nrow = n
- 1 - K, ncol = p - K)
      lambda.boot <- svd(E.boot)$d
      T.boot[Boot,   Component]   <-   lambda.boot[1]^2   /
sum(lambda.boot^2)
    }
    if   (mean(T.boot[,   Component]   >   T[Component])   >
alpha){break}
  }
 p.values<-colMeans(T.boot > matrix(rep(T, N.Boot), nrow =
N.Boot, byrow = TRUE)) # p-values
 n.tested<-sum(!is.na(p.values))

data.frame(axis.rank=1:n.tested,p.value=p.values[1:n.tested
])
}
```

BOX 2.2. The following is **R** code to conduct a parametric bootstrap test of significance of principal component axes obtained from a correlation matrix on centered and standardized (scaled) variables.

```
Forkman.test.standardized.PCA<-
function(dat,alpha=0.05,N.Boot=1000){
  n <- nrow(dat) ; n
  p <- ncol(dat) ; p
  M <- min(n - 1, p)
  # Variables are centred, and divided by standard
deviations
  Y <- scale(dat, center = TRUE, scale = TRUE)
  # Singular value decomposition
  svd.Y <- svd(Y)
  lambda <- svd.Y$d
  # Test statistic
  T <- {lambda^2/rev(cumsum(rev(lambda^2)))} [1:(M - 1)]
  # Sampling
  T.boot <- matrix(NA, nrow = N.Boot, ncol = M - 1)
  Theta.hat <- matrix(NA, ncol = 1, nrow = M - 1)
  Matrix.0 <- matrix(0, nrow = n, ncol = p)
  Matrix.1 <- matrix(1, nrow = n, ncol = p)
  for(Component in 1:(M - 1)){
    K <- Component - 1
    Matrix.Component <- matrix(Component, nrow = n, ncol =
p)
    Theta.hat <- ifelse(Matrix.Component == Matrix.1,
Matrix.0, svd.Y$u[, 1:K]%*%diag(svd.Y$d[1:K], nrow = K,
ncol = K)%*%t(svd.Y$v[, 1:K]))
    Variance.hat <-
t(svd.Y$d[Component:M])%*%svd.Y$d[Component:M]/((n - 1 -
K)*(p - K))
    for(Boot in 1:N.Boot){
      Y0.boot <- matrix(rnorm(n*p, as.vector(Theta.hat),
sd = sqrt(Variance.hat)), nrow = n, ncol = p)
      Y.boot <- scale(Y0.boot, center = TRUE, scale = TRUE)
      lambda.boot <- svd(Y.boot)$d
      T.boot[Boot, Component] <- lambda.boot[Component]^2
/t(lambda.boot[Component:M])%*%lambda.boot[Component:M]
    }
    if (mean(T.boot[, Component] > T[Component]) >
alpha){break}
```

```
}
  p.values<-colMeans(T.boot > matrix(rep(T, N.Boot), nrow
= N.Boot, byrow = TRUE))
  n.tested<-sum(!is.na(p.values))

data.frame(axis.rank=1:n.tested,p.value=p.values[1:n.test
ed])
}
```

BOX 2.3. The following is **R** code to conduct a parametric ANOVA of PCA scores on more than one axis as a function of factor variables

```
Skalski.adonis<-function(dat,PC.axes,Groups){
  #dat is the data frame containing the PC scores in
columns
  #given in PC.axes and groups given in columns given in
Groups
  n.axes<-length(PC.axes)
  ssTreat.x<-ssError.x<-dfTreat.x<-dfError.x<-
rep(NA,n.axes)
  for(i in 1:n.axes){
    my.mod<-aov(dat[,PC.axes[i]]~dat[,Groups],data=dat)
    test<-summary(my.mod)
    ssTreat.x[i] = test[[1]][1,2]
    ssError.x[i] = test[[1]][2,2]
    dfTreat.x[i] = test[[1]][1,1]
    dfError.x[i] = test[[1]][2,1]
  }
  # Add Sums of Square for overall test
  sumSS.trtmnt<-sum(ssTreat.x)/sum(dfTreat.x)
  sumSS.error<-sum(ssError.x)/sum(dfError.x)
  # F-test
  F.all =sumSS.trtmnt/sumSS.error
  F.df1 = sum(dfTreat.x)
  F.df2 = sum(dfError.x)
  #null probability
  null.p<-1-pf(F.all,F.df1,F.df2)
  list(Group.SS=sumSS.trtmnt,Residual.SS=sumSS.error,
       F.stat=F.all,df1=F.df1,df2=F.df2,null.prob=null.p)
}
```

Chapter 3 Correspondence analysis and detrending

Correspondence analysis (CA) was originally developed outside of biology in the 1930s by Hirschfeld (1935) and later by Benzécri (1973) in order to aide in the visual interpretation of contingency tables but is used in community ecology. What is the relationship between contingency tables and community ecology? A very common type of data structure in community ecology is the siteXspecies data table, in which the rows represent different sampling locations (sites) and the columns represent each of the possible species that could be found in each site. The entries in such siteXspecies tables are sometimes the number of individuals of a given species that were observed in a given site; i.e. the abundance of that species. This general structure (counts in each combination of rows and columns) is a contingency table. However, correspondence analysis can be used even if the entries are positive real numbers (like biomass or percent cover) and not just whole numbers (counts). It was first popularized in ecology by Hill (1973) as an exploratory ordination method designed specifically for studying the species composition of ecological communities and for gaining insight into the underlying causal environmental variables that might control the changes in species composition between sites. This general objective is called "indirect gradient analysis" by community ecologists. It is "indirect" because the underlying environmental variables controlling the ecological community (i.e. the gradient) are not actually included in the analysis; rather, the environmental variables are only inferred after the correspondence analysis is complete and based on the ordering of the sampling units and of the species.

Hill called his method "Reciprocal Averaging" for reasons that will become clear later in this chapter. Reciprocal averaging, also called the "power algorithm", is actually a general numerical technique for obtaining eigenvectors in matrix algebra (Gourlay and Watson 1973) and is not limited to Hill's method. For instance, reciprocal averaging and can also be used (using a different weighting scheme) to produce a principal component analysis without using the numerical methods of eigen analysis that are at the heart of the **R** functions that we studied in Chapter 2. However, correspondence analysis of data on species abundances across sites suffers from some important weaknesses that are called the "compression" and the "arch" effects. In response, Hill and Gauch (1980) proposed a modification of correspondence analysis to address these weaknesses that they called "detrended" correspondence analysis, or DCA. An advantage of both CA and DCA for gradient analysis in community ecology is that the reciprocal averaging algorithm maps very nicely onto the theory of realized ecological niches and onto ecological expectations of how such realized niches should change across habitats. However, a number of ecologists (Minchin 1987, Wartenberg et al. 1987, Legendre and Legendre 2012) have also criticized DCA because of the somewhat *ad hoc* way in which it tries to solve the two weaknesses of CA. We will look more closely at the ability of different ordination methods to perform gradient analysis in community ecology in Chapter 6.

3.1) siteXspecies data in community ecology

A community ecologist typically collects data by recording the abundance or presence/absence of each of S different species occurring in each of Q different sample units. Such data structures are called siteXspecies data. Abundance is commonly measured in terms of the number of individuals of each species, the biomass of each species, or associated measures of either. In this context, a sampling unit is the area of physical space in which each species is recorded. For instance, a plant ecologist might record the plant species found in different 1m^2 quadrats in a grassland or different 10m diameter areas in a forest. For such studies, a sample unit is an individual quadrat or 10m diameter circle. A limnologist studying the phytoplankton of a single lake might record the species of plankton found in different water samples of lake water placed onto microscope slides. For such studies, each water sample placed on a microscope slide is a sampling unit. An insect ecologist might record the number of individuals per species found in different pit traps. For such studies, each pit trap is a sampling unit. The result of all such studies is a two-dimensional contingency table. This table consists of rows, representing each sampling unit, and columns, representing each species that has been recorded in the study. The cells of the table contain the abundance, or presence/absence, of each of the possible species found in that sampling unit. In this sense, you can think of each species as a variable attribute that is measured on each sampling unit.

We have seen this structure of data before because principal component analysis (Chapter 2) uses the same type of data. For instance, the GLOPNET data of Chapter 2 consisted of rows representing species and columns representing leaf traits. In each case, the rows represent the sample units and the columns represent the attributes recorded in each sampling unit. PCA produces new axes, also called unobserved or *latent* variables, which explain the major patterns of covariation among the observed attributes. Correspondence analysis also produces new axes, typically interpreted in terms of unobserved or latent environmental variables, which explain the patterns of covariation in abundance among the observed variable attributes (species). However, for reasons that will be explained in Chapter 6, PCA is often not very well suited for siteXspecies data. Before explaining correspondence analysis (CA) and its detrended modification (DCA), we must review some basic ecological theory and how this theory relates to correspondence analysis.

3.2) Direct gradient analysis

The theoretical justification for using correspondence analysis (or "reciprocal averaging") when analyzing the changes in community composition as a function of unidentified environmental variables was extensively developed by ter Braak and colleagues in a series of publications (Ter Braak 1985, Ter Braak and Barendregt 1986, ter Braak and Looman 1986, Ter Braak and Gremmen 1987, Ter Braak and Prentice 1988). To understand this justification, we must first take a detour and consider how the abundances of species typically change along known environmental gradients. This is a detour because we are now analyzing the changes in community composition as a function of known and measured (not unidentified and inferred) environmental variables. In other words, we are detouring to talk about *direct* gradient analysis, not *indirect* gradient analysis.

We can characterize the environment at any point in space by the values of a set of physical properties that affect the fitness of organisms. Every organism (and, by extension, every species) can only survive, grow and reproduce within a characteristic range of those environmental variables that affect its evolutionary fitness. Certain values of these environmental variables, in interaction with other species, will allow the optimum combination of growth, survival and reproduction (i.e. evolutionary fitness) of that species. If we set aside history and demographic stochasticity, then a species will be most abundant in sites having this optimum combination of environmental conditions. This optimum combination of environmental conditions defines a single point (μ) in this "environmental space". Here, I am using the word "space" in its mathematical sense not in its physical or geographical sense. Other combinations of values of these environmental variables, in interaction with other species, will result in a somewhat reduced ecological performance and so the species will be less abundant. Finally, there will be combinations of environmental variables that do not allow the species to maintain a viable population. The species will be absent or very rare in such combinations of environmental variables.

This is Hutchinson's concept of a multidimensional realized niche (Hutchinson 1957). Let's consider only one environmental variable for simplicity even though realized niches will involve many more environmental variables. In this simplified case, the concept of a realized niche would look like Figure 3.1. The environment to which the species is optimally adapted is the environmental value at point μ. The degree to which its performance decreases as the environment deviates from this optimum is quantified by its "tolerance" (t). ter Braak's (1988) link to correspondence analysis assumes that the mathematical function describing how its abundance changes over this environmental variable follows a Gaussian[18] response function (Figure 3.1). If we had more than one environmental variable then our function would be a multivariate version of the Gaussian function.

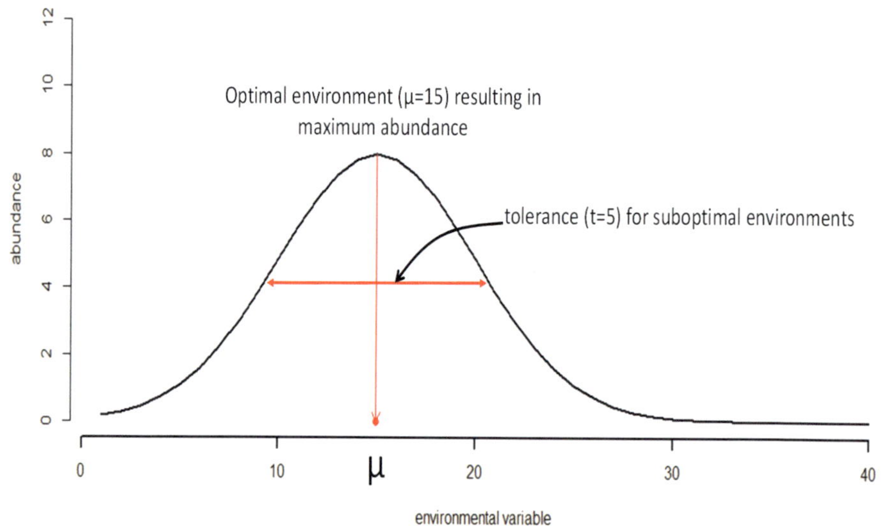

Figure 3.1. A typical response curve of a species at different values of some environmental variable. Typically, a species will have some maximum performance (μ, abundance, fitness) in the environment to which it is best adapted and will decrease away from this optimum. The performance of this species will decrease as the environment deviates from this optimum with a certain level of tolerance (t) for suboptimal environmental conditions.

[18] A Gaussian response function, which uses the same mathematical equation describing a normal probability distribution, is named after the mathematician Carl Friedrich Gauss who is credited with this equation.

The assumption that changes in abundance as a response function of an environmental variable can be described by a Gaussian equation is not as limiting as it might first appear. For instance, this assumption doesn't mean that changes in abundance must follow such a function in the units and scaling that we might choose to measure this environmental variable. In fact, as you will see, correspondence analysis in community ecology doesn't even use the environmental variables; rather, they are only inferred after the correspondence analysis is complete. Rather, ter Braak's (1988) assumption is that there exists some transformation of the underlying environmental variable for which the performance of the species over this transformed environmental variable is Gaussian. It is also important not to confuse the niche axes, representing values of environmental values devoid of any spatial or temporal context, and the actual values of these environmental values as you might encounter them in nature as you travel across the landscape (Figure 3.2).

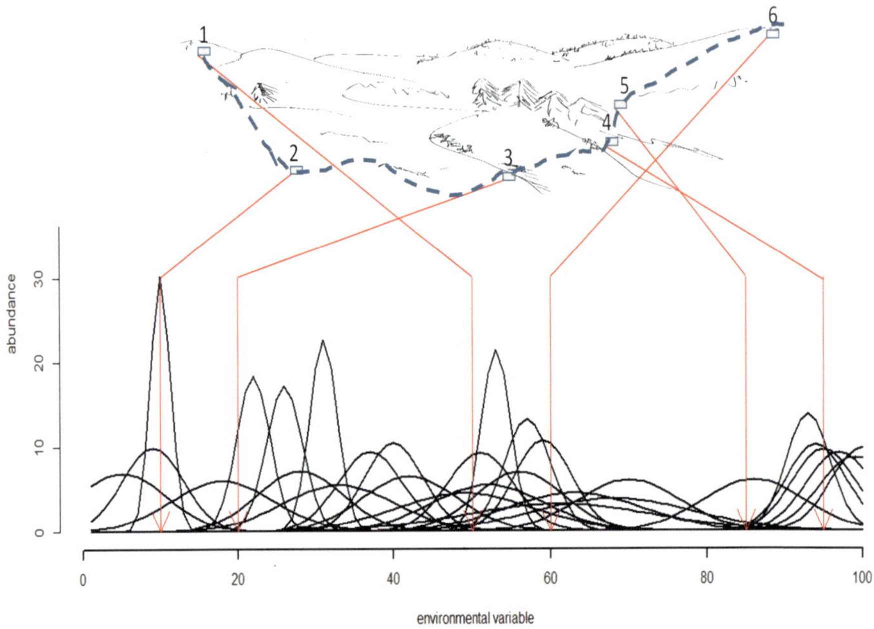

Figure 3.2. Imagine walking a transect in the landscape shown at the top. You sample the community composition at six sites, starting your walk at point 1 and ending at point 6,. There are 30 species that exist in this area and their realized niches are determined by a single environmental variable. As you walk along your transect you will notice that the identity and abundance of the species will change in seemingly random ways since the environment does not change linearly in space.

This potential confusion comes in two parts. First, the "underlying environmental variable" that is captured by an ordination axis is not necessarily (actually, almost never) a single environmental variable as we might define it. Species are almost always simultaneously affected by several different environmental variables. If these different environmental variables are correlated across the landscape then the species will experience, and respond to, this correlated combination of environmental variables is a coordinated way. The underlying environmental "variable" that is captured by the ordination axis reflects this correlated combination of environmental variables. In other words, a single ordination axis can (and usually does) capture a correlated combination of several underlying physical variables related to the environment.

The second potential confusion comes from the fact that it is quite rare for such environmental variables to change across the landscape in the same order as they are represented in niche space. Imagine that you have walked across the landscape shown in Figure 3.2 and then record the abundance of each species that you find in a series of six sites. The values of the environmental variable at each site change as shown in Figure 3.2. The environmental variable has a value of 50 at the first site, a value of 10 at the second site, and so on. Importantly, the physical distance between sites has little in common with the amount of difference in the underlying environmental values that are controlling the distribution and abundance of the species. For instance, sites 1 and 6 in Figure 3.2 are the farthest apart on the landscape but are the most similar in terms of their environments. Each time the environmental variable changes, different groups of species will be better adapted to those environmental conditions. The ecological communities will differ at each site as the environment "moves" through niche space. This is the phenomenon of community assembly by species filtering (Grime 1979, Shipley 2010).

I said above that we have to "set aside history and demographic stochasticity". What does this mean? Norway Maples (*Acer platanoides* L.) are native to Europe and to western Asia. They did not occur in North America until the late 1700s. The lack of Norway Maples in North America until the late 1700s was not caused by a lack of available habitats within the realized niche of this species. We know this because Norway Maples were introduced into North America in the 1700s by European settlers. Norway Maples grow so well in my region (Quebec, Canada) that the species is now considered invasive. The reason why Norway Maples were never found in North America before the 1700s has nothing to do with a lack of suitable habitats but rather because Norway Maples evolved in Europe after Europe and North America became separated by an important dispersal barrier: the Atlantic Ocean. The reason is history, not niches. On a more local level, it is common for a species to be absent from sites in which it could happily grow and reproduce if only its seeds could reach such sites. Dispersal between sites has an important element of chance, especially if the dispersing population is small and so the fact that a species is absent from a site does not always mean that the site is outside of its realized niche. This is a component of demographic stochasticity. Both the past history and the current environment of sites will affect the composition of the component species and both can be captured by the axes of correspondence analysis. These points must be kept in mind when linking species' abundances, niche theory and ter Braak's (1988) argument for ordination by correspondence analysis and its detrended version.

In the imaginary walk shown in Figure 3.2, the ordering of the changes in the environment did not mirror the ordering of this environmental variable in niche space. Such a disconnect between the changes of environmental variables in physical space and those in a theoretical niche space are the rule, not the exception, and this severely complicates our ability to identify which environmental variables are actually filtering our species and therefore controlling the structure and composition of ecological communities. Indeed, the main purpose of indirect ordination methods is to help the ecologist to identify these underlying environmental variables. However, in some special habitats, the ordering of environmental variables in physical space does closely corresponds to their ordering in niche space. These special habitats allow us to more easily see the link between correspondence analysis, which does not explicitly use environmental variables, and direct gradient analysis that does use environmental variables.

Figure 3.3. The shoreline of Axe Lake when the water level is low. At the high water phase, the entire foreground is inundated. The yellow flowers of *Lysimachia terrestris* are higher on the shoreline. Photo by Paul Keddy.

The phenomenon of plant zonation at different water depths along the shorelines of lakes and rivers in northern countries provides a nice empirical example. Areas higher on the shoreline are inundated with water when the snow melts during the spring or during years with unusually high water levels. During the summer, the water level of the lake decreases and the soil at higher elevations on the shoreline is exposed while the soil is only rarely exposed to the air at the lowest elevations (Figure 3.3). Of course, "elevation" on a shoreline is not an environmental variable having any direct causal connection to plant performance but it shoreline elevation is strongly correlated with many other variables (soil oxygen levels, soil humidity, nutrient availability etc.) that do directly control plant performance. As a result, a physical transect going from high on the shoreline and moving downslope towards the bed of the lake or river will approximate the changes in the important environmental variables controlling the species composition of shoreline vegetation. In this special habitat, walking downslope along the shoreline is like walking along a niche axis for the shoreline plants. Because of this special, and unusual, concordance between a change in physical space (shoreline height) and a change in environmental space (a niche axis), Keddy (1983) established 25 transects perpendicular to the water line of Axe Lake in Ontario (Canada). For each transect, and for each interval of 5 centimeters above or below the water level at a fixed date, he recorded the presence or absence of each species. Figure 3.4 plots the distribution of three of these species as a function of the height of the shoreline. We see that *Lysimachia terrestris* is most abundant at a height of ~30 cm above the water line and is entirely absent at heights below 0 cm. *Lobelia dortmanna* and *Eriocaulon septangulare* are most abundant at ~18 cm and ~20 cm below the water line but *Eriocaulon septangulare* is rather more tolerant of different positions along this gradient of depth.

Axe Lake, at the time in which Keddy was conducting his research, was minimally affected by humans. Sadly, this is no longer the case. It is increasingly difficult to find data from relatively undisturbed shoreline wetlands. Students who are collecting new data should be aware of the likelihood that they are now describing vegetation that has already has significant human impacts, impacts that tend to obscure the search for basic causal factors as opposed to anthropogenic effects. On shorelines, the impacts of cottages, hydroelectric dams and all-terrain vehicles may over-ride natural gradients; equally, in forests, impacts of logging, introduced tree-diseases, and over-grazing by deer may now obscure natural gradients. This emphasizes the need for statistical and experimental methods of quantifying and identifying the myriad underlying environmental gradients (natural and anthropogenic) that structure natural communities.

In the case of Axe Lake, there is another factor at play: history. Axe Lake has many rare species that likely arrived near the end of the last ice age, when a large glacial lake called Lake Algonquin covered the area. You can read more and see a map of the Lake Algonquin shoreline in (Keddy 1981). The presence of these species reminds us that some important causal factors in plant communities may lie buried in the past: it is unlikely that some of the species in this data set would occur if Axe Lake had not been part of a glacial lake that formed about 11,000 years ago, after the end of the last ice age, and then disappeared. The importance of post-glacial events after the ice age is important everywhere in the northern hemisphere. For Quebec, Marcel Raymond's classic guide (Raymond 1950) describes many of these events, including the post glacial Champlain Sea, as they likely influenced the distribution of plant species in the province.

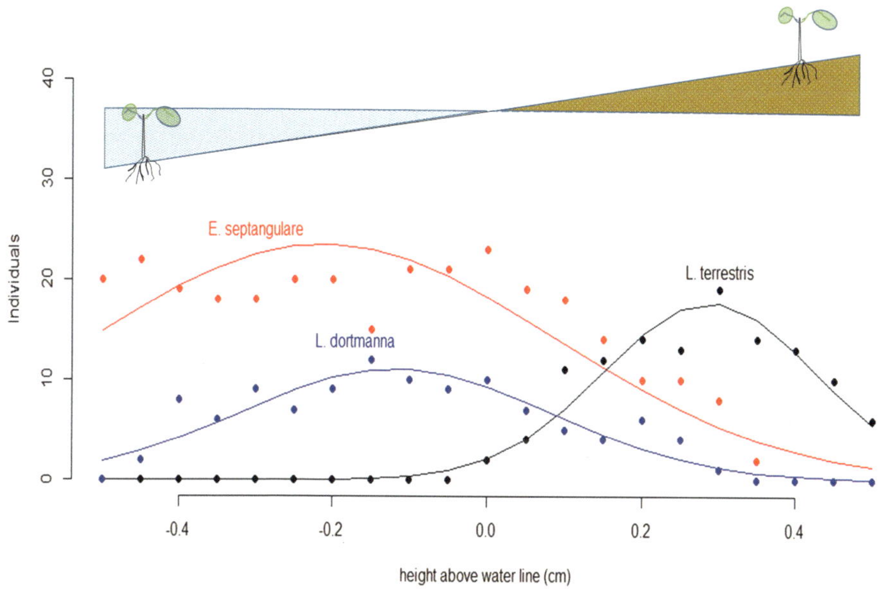

Figure 3.4. The distribution of three species on the shoreline of Axe Lake as a function of the height (cm) above or below the waterline on a fixed date. Each point is a count of the number of individuals of a species. The solid lines are fitted values of a generalized linear quadratic regression, which models a Gaussian response curve.

We can fit a statistical model to describe the way in which the abundance of each species will change along our environmental gradient, as seen in Figure 3.4. This is called a "Gaussian" or unimodal regression by ter Braak and Looman (1986) because the form of the equation is the same as the one describing a Gaussian, or Normal, probability distribution. However, the equation has nothing to do with a normal probability density and so, to avoid confusion, I will call it a Gaussian *response function*. Imagine that we construct a data table in which each row represents a sampling unit (for example, a quadrat on the shoreline of Axe Lake) and each column represents a species. The entries of this data table, y_{ij}, will be the abundance in column j (j=1, S) of the jth species found in the quadrat of row i (i=1, Q). For instance, y_{23} would be the abundance of species #3 found in quadrat #2. The value of the environmental variable (x) in sampling unit i is indicated by x_i. The variation in the abundance of species j in sampling unit i that is not due to the environmental variable be "e_{ij}".

$$y_{ij}=A_j e^{\frac{-(x_i-\mu_j)^2}{t_j}+e_{ij}}$$

$$y_{ij}=A_j e^{\left(\frac{1}{t_j}\right)x_i^2+\left(\frac{2u_j}{t_i}\right)x_i-\left(\frac{u_j^2}{t_j}\right)+e_{ij}}$$

$$y_{ij}=A_j e^{b_2x_i^2+b_1x_i+b_0+e_{ij}} \quad (3.1)$$

$$\ln(y_{ij})=\ln(A_j)+b_2x_i^2+b_1x_i+b_0+e_{ij} \quad (3.2)$$

$$\mu_j=\frac{-b_1}{2b_2} \quad (3.3)$$

$$\mu_j = \frac{\sum_{i=1}^{Q} y_{ij} x_i}{\sum_{i=1}^{Q} y_{ij}} = \sum_{i=1}^{Q} \left(\frac{y_{ij}}{\sum_{i=1}^{Q} y_{ij}} \right) x_i = \sum_{i=1}^{Q} p_{ij} x_i \qquad (3.4)$$

The equation describing a Gaussian response function is given in Equation 3.1. You might recognize this equation as being similar to that of a normal probability density. Taking the natural logarithm of y_{ij} gives a quadratic linear equation (Equation 3.2). The parameters of this equation can be obtained from a multiple regression in which the ln(abundance) is regressed on the environmental variable (x) and its square (x^2). The point on the environmental gradient (μ_j) at which the abundance of the species j is maximum is when the first derivative of Equation 3.2 is zero, which is given in Equation 3.3. There is only one such point for a given species, which is why ter Braak and Looman (1986) also call this a *unimodal* model. Obtaining the environmental value that maximizes the abundance (μ_j) of a species by Equation 3.3 requires that we can fit the multiple regression and this requires that we have actually measured the environment. This type of gradient analysis is called a "direct" gradient analysis because we have directly included the environmental variable in the analysis.

It is also possible, in certain situations, to obtain the environmental value that maximizes the abundance (μ_j) without actually fitting a quadratic regression like in Equation 3.2. This second method involves a *weighted average* of the environmental values experienced by species j. A "weighted" average means that the environmental values (x_i) experienced by species j in each site i are multiplied by the proportion (p_{ij}) of all individuals of species j in site i that occur over all possible sites. Equation 3.4 explains how to calculate this weighted average. Notice that μ_j is also the value of the environmental gradient (x_i) at which the *average* individual of species j is found (if abundance is measured as the number of individuals) or the value at which the average unit of biomass is found (if abundance is measured in terms of biomass). This is the first of two "averages" in the algorithm of reciprocal averaging.

The use of a weighted average to describe the point along an environmental gradient at which a species reaches it maximum abundance goes back to the pioneers of vegetation analysis (Curtis 1955, Whittaker 1956). Using a weighted average as an estimator of the environmental variable that maximizes the species' abundance, rather than obtaining this from a regression, is key to the theoretical link between correspondence analysis (a method of indirect gradient analysis) and Gaussian response regression using observed environmental variables (a method in direct gradient analysis). ter Braak and colleagues (Ter Braak and Barendregt 1986, ter Braak and Looman 1986, Ter Braak and Prentice 1988) showed that Equation 3.4 is an unbiased estimator of the point on the environmental gradient (μ_j) at which the abundance of the species j is maximum given certain important assumptions that will be given later.

The environmental variable (x_i) in our Axe Lake example is the measured height of the shoreline at point i. Since the abundance (y_{ij}) of each of the three species in our shoreline example was estimated as the number of transects at a given height interval in which a species was observed (i.e. counts), it is reasonable to assume that such counts follow a Poisson sampling distribution. Given this statistical assumption, we can estimate the parameters of Equation 3.1 by first fitting a quadratic generalized linear regression model with a Poisson link to Equation 3.2 and then back-transforming these predicted values from their natural logarithms. The topic of generalized linear regressions is outside the scope of this book but you can find a gentle introduction in Crawley (2013, Chapter 13). We can do this in **R** using the following **R** pseudocode:

```
fit<-glm(y~x+I(x^2),family="poisson",data=…).
```

For instance, the resulting regression equation for *Lobelia dortmanna* in Figure 3.2 is ln(individuals)=2.23-2.92height-12.39height^2. Each of the solid lines in Figure 3.2 is the predicted abundance of the species from such a Gaussian response regression when back-transformed from natural logarithms. You might be wondering what this has to do with correspondence analysis since the whole point of correspondence analysis is to ordinate our species and sample units when we haven't measured, and might not even know, which environmental variables are controlling our community structure? The answer can be found by looking at how "reciprocal averaging", a.k.a. correspondence analysis, works.

3.3) Correspondence analysis or "reciprocal averaging"

In the Axe Lake example, we could do a direct gradient analysis, using a Gaussian response regression, because we already knew the important underlying environmental gradient that controlled the vegetation: height on the shoreline. We now leave behind direct gradient analysis to consider the more challenging (but more common) situation in which we don't even know which underlying variable, or combination of variables, controls the distribution of our species. Correspondence analysis is a type of indirect gradient analysis in community ecology because it attempts to order individual sampling units and species in such a way as to reveal underlying, but unmeasured, environmental variables that are structuring these ecological sampling units. Correspondence analysis, like PCA in the previous chapter, is also a type of eigen[19] analysis. However, the best way to understand how correspondence analysis works in community ecology is to look at the algorithm of "reciprocal averaging" since this algorithm produces the same result as the eigen analysis while linking to direct gradient analysis in the form of Gaussian response regression.

Reciprocal averaging consists of estimating two types of weighted averages that are iterated many times, each one feeding into the other, until

---

[19] We don't discuss eigen analysis or matrix algebra in this book but it is at the heart of many of the ordination methods described in this book.

these estimates remain constant from one iteration to the next; thus, the name of "reciprocal averaging". Ter Braak and Prentice (1988) call these two basic steps the *regression* and *calibration* steps. In what follows, we are working with a siteXspecies table of values in which each row represents one of the Q sampling units (quadrats, sites, etc.) and each column represents one of the S species in the data set. We do not know which underlying environmental gradients are controlling the distributions of our species or even how many independent underlying environmental gradients might exist. An important objective of correspondence analysis and its detrended version is to identify such unknown underlying environmental gradients.

For illustration and simplicity, imagine that Dr. Keddy had placed quadrats at heights of 0.4, 0.2, 0, -0.2 and -0.4 m on the shoreline of Axe Lake and recorded the abundances of the three species given in Figure 3.4. He sends us the siteXspecies data (Table 3.1) but he has forgotten to tell us that these quadrats came from a lakeshore habitat and that these five quadrats were placed at different heights along the shoreline. Imagine that we also have no knowledge of the biology of these three species. In other words, the nature and values of the underlying environmental gradient are unknown to us. Furthermore, he has rearranged the order of the quadrats so that the order no longer follows the changes in shoreline height.

Sampling unit	*Lysimachia terrestris*	*Eriocaulon septangulare*	*Lobelia dortmanna*	Row sum $y_{i+} = \sum_{j=1}^{S} y_{ij}$
Q1	2	23	10	35
Q2	13	0	0	13
Q3	0	20	9	29
Q4	0	19	8	27
Q5	14	10	6	30
Column sum $y_{+j} = \sum_{i=1}^{Q} y_{ij}$	29	72	33	Total abundance $y_{++} = \sum_{i=1}^{Q} \sum_{j=1}^{S} y_{ij} = 134$

Table 3.1 The abundances of three species recorded in each of five quadrats. The last row gives the total abundance of all species in a given quadrat. The last column gives the total abundance of a particular species over all quadrats.

After initializing, the basic algorithm of reciprocal averaging consists of repeatedly applying a "regression" and a "calibration" step until the predicted position of each sampling unit and each species stabilizes as follows:

Initialization step: The reciprocal averaging algorithm starts by assigning arbitrary values to the unknown environmental gradient for each of the i=1,Q sampling units; i.e. the values of x_i in Equation 3.3. We will therefore simply assign the numbers 1, 2, 3, 4 and 5 as our initial and arbitrary guess of the values of x_j (the unknown underlying environmental gradient) for each of the five quadrats in Table.3.1. We could choose whatever numbers we want but, since these numbers represent the values of the unknown environmental gradient for each sampling unit, and since we don't know the nature of this environmental gradient, the absolute values that we choose are meaningless.

Regression step: In the regression step, we pretend that we know the values of the underlying gradient (i.e. x_j in Equation 3.4) but we don't know the values of the environmental optima of the species (i.e. μ_j in Equation 3.4). If the distribution of the jth species along an environmental gradient follows a Gaussian response function (or, more generally, a function that is symmetric about its maximum value), then its maximum abundance (μ_j) will occur at the point on the environmental gradient given by the weighted average given in Equation 3.4. The regression step of reciprocal averaging therefore consists of estimating the environmental optimum of each species based on our current estimate of the values of the unknown environmental gradient (x_j) using Equation 3.4. For *Lysimachia terrestris* (species 1), and given our initial and arbitrary guess of the values of the underlying environmental gradient, this would give $\mu_1 = (2(1) + 13(2) + 0(3) + 0(4) + 14(5))/29 = 3.38$. The values for the other two species would be $\mu_2 = 2.902$ and $\mu_3 = 3.00$.

Calibration step: In the calibration step, we reverse our assumptions made in the regression step. Now, we pretend that we know the values of the environmental optima (μ_j) for each species but we don't know the value of the underlying environmental gradient for each sampling unit (i.e. x_i). Given our understanding of the nature of realized niches, the most abundant species in a sampling unit will have their environmental optima close to the value of the underlying environmental gradient of that sampling unit. This is because these will be the species that are best adapted to the local environment. Species whose environmental optima are farther from the value of the underlying environmental gradient in the sampling unit will be less abundant. Therefore, we can estimate the value of the unknown environmental value of our sampling unit by calculating the environmental optimum of each species in the sampling unit weighted by its relative abundance in that sampling unit. This is done by multiplying the relative abundance[20] of each species in the quadrat by its estimated environmental optimum (Equation 3.5). Notice that Equation 3.5, just like Equation 3.4, is a weighted average. In fact, x_j is the estimated environmental optimum of an average individual in sampling unit i.

$$x_i^* = \frac{\sum_{j=1}^{S} y_{ij}\mu_j}{\sum_{j=1}^{S} y_{ij}} = \sum_{j=1}^{S} p_{ij}\mu_j \qquad (3.5)$$

[20] "Relative" because the abundance of each species is expressed relative to the total abundance of all species in the quadrat or site.

Again, since we don't actually know the nature of the underlying environmental gradient, the absolute values of x^*_j are meaningless. We therefore rescale these estimates of x^*_i in units of standard deviations of the species' optima to give the new values of x_j. Box. 3.1 gives the equations to do this. Of course, the **R** functions described later do all of these calculations for you.

Box 3.1

To do this, we first calculate the weighted average species optimum (Equation 3.5) over all sampling units $\left(\overline{x^*}\right)$.

$$\hat{x} = \frac{\sum_{i=1}^{Q} x_i^* \sum_{j=1}^{S} y_{ij}}{\sum_{i=1}^{Q}\sum_{j=1}^{S} y_{ij}} = p_{+j} x_i^* \quad (3.6)$$

Next, we calculate the standard deviation of x_i^* (Equation 3.6).

$$S_x = \sqrt{\frac{\sum_{i=1}^{Q}\left(\left(x_i^* - \overline{x^*}\right)^2 \sum_{j=1}^{S} y_{ij}\right)}{\sum_{i=1}^{Q} y_{ij}}} = \sqrt{\sum_{i=1}^{Q} p_{i+}\left(x_i^* - \overline{x_i^*}\right)^2} \quad (3.7)$$

Finally, we rescale our current estimates of the position of each sampling unit along the unknown environmental gradient using Equation 3.7.

$$x_i = \frac{\left(x_i^* - \overline{x_i^*}\right)}{S_x} \quad (3.8)$$

Putting all of these steps together, here is the basic reciprocal averaging algorithm for correspondence analysis; we will modify this basic algorithm later:

Initialization step. Choose initial estimates (x_i) of the unknown environmental gradient.

1. Do the next four steps.
   A. Given x_i, calculate new species scores (μ_j) by weighted averaging using Equation 3.4
   B. Given the new μ_j, calculate new estimates of the environmental gradient x_i^* (i.e. the sampling unit scores) using Equation 3.5.
   C. Rescale the estimates of the environmental gradient to give x_i using Equations 3.6 to 3.8.
   D. Calculate the difference between the previous estimates of the environmental gradient with these new estimates.
2. If the difference between the previous and current estimates of the environmental gradient are greater than some small amount then repeat steps 1A to 1D; otherwise stop.

Imagine that we have been given the data set of vegetation observed in Axe Lake, but not the environmental gradient, i.e. the height of each quadrat on the shoreline. The values of the underlying vegetational gradient for our five quadrats (i.e. the *site scores*, x_i) and the values for the species optima for our three species (i.e. the *species scores*, μ_j), after applying the above algorithm to the vegetation data, are given in Table 3.2.

Reciprocal averaging result	Gradient value (site or species score)
Quadrat 1 (x_1)	-0.68
Quadrat 2 (x_2)	2.48
Quadrat 3 (x_3)	-0.50
Quadrat 4 (x_4)	-0.69
Quadrat 5 (x_5)	0.79
Lysimachia terrestris (μ_1)	1.46
Eriocaulon septangulare (μ_2)	-0.42
Lobelia dortmanna (μ_3)	-0.36

Table 3.2. The predicted positions, using reciprocal averaging of each of the five quadrats and each of the three species. The values of x_i are the gradient positions of each quadrat ("site scores") and the values of μ_j are the gradient positions of the optimum abundance of each of the three species ("species scores").

The values in Table 3.2 tell us that the ordering of our quadrats that best approximates the unknown environmental gradient is Q4, Q1, Q3, Q5, Q2. The values of this unknown gradient range from -0.69 standard deviations to 2.48 standard deviations. Quadrats 4, 3 and 1 are quite close to each other along this unknown environmental gradient while quadrats 5 and (especially) 2 are much further way from the first three quadrats. *Eriocaulon septangulare* has maximum abundance at -0.42, *Lobelia dortmanna* at -0.36 and *Lysimachia terrestris* at 1.46 standard deviations from the mean.

After obtaining these results, imagine that Dr. Keddy then sends us the actual heights on the shoreline for our five quadrats (although you can get these values directly from Figure 3.4). Figure 3.5 plots the positions of our quadrats and species along the hypothetical gradient that our algorithm has identified and the actual positions of the quadrats and species along the physical gradient (height on the shoreline).

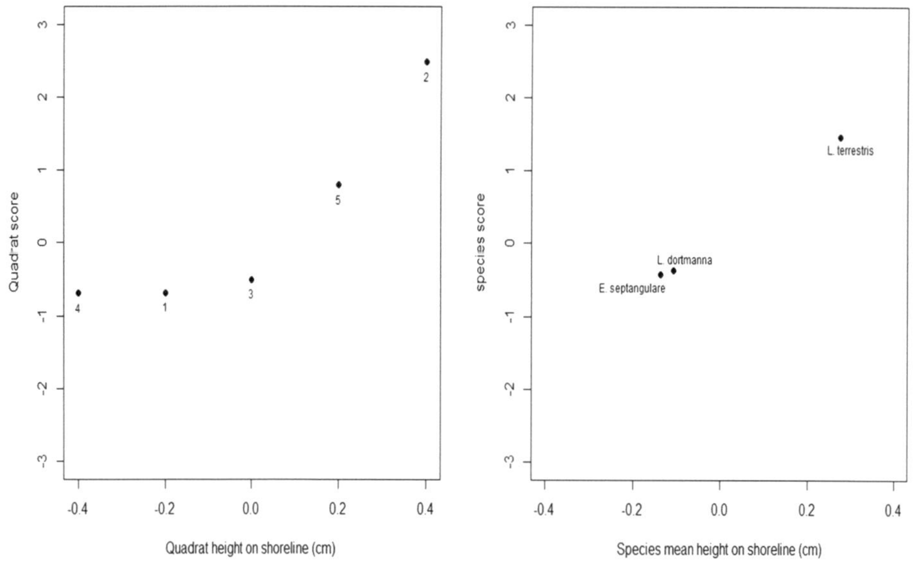

Figure 3.4. The values on the y-axis show the positions of each quadrat (left) and each species (right) along the latent "vegetational" gradient as identified using reciprocal averaging and the actual positions of each quadrat and species in height (cm) above or below the water level at a fixed day.

Figure 3.5 shows some important information. First, notice that the reciprocal averaging algorithm has successfully ordered our quadrats and species relative to the actual height above or below the water line on the shore. It did this even though neither the reciprocal averaging algorithm nor we knew the environment from which these quadrats came! Second, there is a clear nonlinear relationship between the positions of the quadrats along the unidentified gradient produced by the reciprocal averaging algorithm and the physical heights along the shoreline. This is because the gradient that is revealed by our algorithm is based on how the plants experience the gradient as measured by their changes in abundance. The algorithm does not on how quantify it based on our perception in terms of centimetres of altitude relative to some arbitrary height. Compare Figure 3.5 with Figure 3.4. The scores assigned to the quadrats in Figure 3.5 that were located at heights of -0.4, -0.2 and 0 cm along the shoreline changed very little. The vegetation at heights from -0.4 cm to 0 cm along the shoreline consisted of the dominant *Eriocaulon septangulare,* the subdominant *Lobelia dortmanna*, no *Lysimachia terrestris*, and that the relative abundances of these species changed very little (Figure 3.4). Based on the patterns of abundance of these three species, the environment *that is relevant to the vegetation* in these quadrats changed very little even though the quadrat heights varied from -0.40 to 0 cm. From 0 to 40 cm along the shoreline, the composition of the vegetation in Figure 3.4 changed much more rapidly. It changed from a community dominated by *Eriocaulon septangulare* along with the subdominant *Lobelia dortmanna* to a community dominated by *Lysimachia terrestris*. In fact, at 40 cm, the first two species are completely missing. The environment *that is relevant to the vegetation* in these quadrats changed very rapidly based on the ecological behaviour of the three species and this is what we see in Figure 3.5. In other words, the "gradient" that is detected by the reciprocal averaging algorithm is calibrated in terms of the changes in the vegetation. There is no reason to expect that a scaling as experienced by the vegetation will be linearly related to our chosen environmental variables. Furthermore, you should always consider the possibility that the changes in species composition that define the axis of the correspondence analysis might also reflect the effects of past history rather than only environmental changes causing species filtering. This is unlikely to be the case in our example because all of the sampling units came from a single small lake and the plants can presumably easily disperse within it. It would likely be the case, given the known biogeography of the region, if our data set had included many different lakes that were geographically distant.

My description of the reciprocal averaging algorithm results in a single axis. However, you can also extract additional subordinate axes if you suspect that there is more than one underlying gradient that are sorting your species. When I say "underlying gradient", remember that I am referring to independent combinations of environmental variables as experienced by the organisms, not necessarily to independent combinations of environmental variables as we measure them. To extract subordinate axes, we simply add a new step after step 1B of our algorithm (Ter Braak and Prentice 1988). After we have obtained our final values of the site scores on our first axis (x_{i1}), this additional step consists of applying our reciprocal averaging algorithm again except that we subtract the scores on our first axis from those obtained at each iteration using Equation 3.9. This is repeated for each successive subordinate axis and results in the site and species scores on each new axis being linearly independent of all those obtained before it in the same way as each axis in a principal component analysis is linearly independent of the other axes.

$$f = \sum_{i=1}^{Q} \frac{y_{i+}}{T} x_i^* x_{i1} \qquad (3.9)$$
$$x_i^* = x_i^* - f x_{i1}$$

I said earlier that the species scores of a correspondence analysis are approximations of the species maxima that would be obtained based on a Gaussian response model, such as we did for the Axe Lake data in Figure 3.4. However, this approximation only strictly holds for the first axis of correspondence analysis, not for subsequent axes (Ter Braak 1985). Here are the conditions under which the approximation is tight[21] (Ter Braak and Prentice 1988):

---

[21] Since, by definition, an indirect gradient analysis is conducted when we haven't already identified and measured the environmental variables which compose the environmental gradient, we are not really in a position to "test" these assumptions.

1. The site scores are equally spaced, or at least uniformly distributed, over the whole range of occurrence of the species along the underlying unknown environmental gradient. In practice, this means that some sampling units must not be located close together (or "clumped") along the underlying unknown environmental gradient and then separated by large empty regions along the underlying gradient.
2. The site scores are closely spaced in comparison to the species tolerances. In practice, this means that the sampling units must be located sufficiently closely together along the underlying unknown environmental gradient so that we can observe the full range of the Gaussian response function; i.e. the abundance of each species increasing up to its maximum and then decreasing.
3. The species scores are equally spaced, or at least uniformly distributed, over the whole range of occurrence of the sampling units along the underlying unknown environmental gradient. In practice, this means that some species maxima must not be located close together (or "clumped") along the underlying unknown environmental gradient of site and then separated by large empty regions along the underlying gradient.
4. The species scores are closely spaced in comparison to their tolerances. In practice, this is the same requirement as point 2.

5. The species have equal tolerances and equal maximum abundances. In practice, differences between species in terms of their maximum abundance (the species maxima) and tolerances do not appear to affect the approximation greatly (Ihm and Groenewoud 1975, Kooijman 1977, Ter Braak 1985). This relative insensitivity is good news because this last condition is almost never be true in real data sets.

3.4) Performing correspondence analysis in R

It is relatively easy to program the steps of the reciprocal averaging algorithm into the **R** language. Happily, you don't have to do this because there are at least two functions in **R** that can perform a correspondence analysis much more quickly and efficiently[22].

The first of these functions is the `corresp()` function in the MASS **package** of **R** (Venables and Ripley 1994). The two equivalent versions of this function are:

corresp()

```
corresp(x=, nf= )
corresp(formula= , nf, data= )
```

[22] These functions are based on a different algorithm that uses an eigen analysis, not the reciprocal averaging algorithm. This allows them to be much more rapid.

You can either enter the siteXspecies matrix or data frame (the "`x=`" argument) or you can specify the species names to include from a siteXspecies data frame in the form of a formula argument. Since real ecological data will contain many species, the first version is much more convenient. The "`nf=`" argument specifies the number of axes that we want to extract from your data. If we save our data set (Table 3.2) as a matrix or data frame called `QXS`, then the following **R** command will produce a correspondence analysis with two axes: `CA<-corresp(QXS, nf=2)`. The output gives the site scores (`CA$rscores`), the species scores (`CA$cscores`), and a measure of the correlation between the site and species scores (`CA$cor`). This last element of the output list is called a "canonical correlation" and measures the proportion of the variance in the site scores of a given axis that is shared by the variance in the species scores of that same axis. Here is the output of this function if we ask for a single axis:

```
corresp(QXS[,2:4],nf=1)

First canonical correlation(s): 0.7674974

 Row scores:
         4          1          3          5          2
-0.6847876  2.4779359 -0.5047700 -0.6862327  0.7946969

 Column scores:
  Lysimachia.terrestris Eriocaulon.septangulare        Lobel
ia.dortmanna
              1.9018094              -0.5500727            -
0.4711285
```

The second function, called `decorana()`, is in the vegan package (Oksanen et al. 2010):

`CA<-decorana(veg=QXS[,2:4],ira=1)`. We will use the vegan package repeatedly in subsequent chapters. The argument "ira=1" in `decorana()` is to specify that we want correspondence analysis rather than the detrended version (i.e. ira=0). Unlike `corresp()`, which requires that we choose how many axes to extract, `decorana()` always extracts four axes. There is a `summary()` function associated with `decorana()`. To obtain the information for only the first axis, equivalent to the output from `corresp()`, you would type:

```
out.dca$evals[1]

      RA1
0.5890523

out.dca$rproj[,1]

        4         1          3          5         2
-0.8198598  2.9667007 -0.6043342 -0.8215898  0.9514482

out.dca$cproj[,1]
  Lysimachia.terrestris Eriocaulon.septangulare        Lobel
ia.dortmanna
              2.9667007              -0.8580781            -
0.7349303
```

The first call (`out.dca$evals[1]`) gives the eigenvalue associated with the first axis, the square root of which is the same as the "canonical correlation" of `corresp()`. The second call (`out.dca$rproj[,1]`) gives the site scores on the first axis and the third call (`out.dca$cproj[,1]`) gives the species scores on the first axis. The site and species scores that are output by `decorana()` differ by a constant amount from those output from `corresp()`. I haven't been able to find an explanation for this difference but suspect that the difference lies in how the two functions rescale the axes. The eigenvalues, site and species scores are also output using the `summary()` function.

3.5) Detrended correspondence analysis

Unfortunately, correspondence analysis suffers from two weaknesses: the "compression effect", which is a compression of the site and species scores at the extremes of the latent environmental gradient, and the "arch effect", which exists when extracting more than one axis. These two weaknesses prompted Hill and Gauch (1980) to modify correspondence analysis, a method that they called *detrended* correspondence analysis (DCA) and which is also called DECORANA. We will learn a better way of correcting the arch effect in Chapter 6.

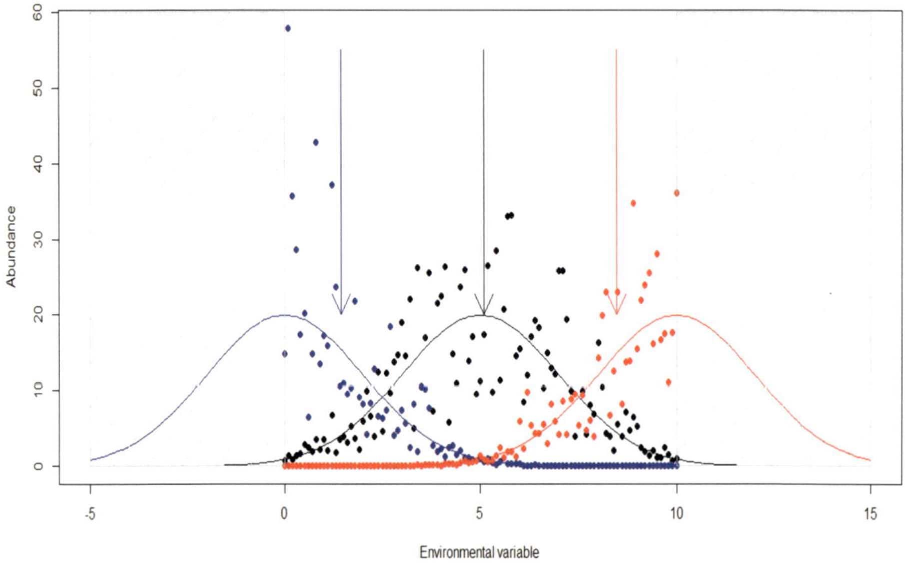

Figure 3.6. The solid lines show the theoretical Gaussian distribution of three species along a single environmental gradient. The shaded regions show the parts of the environmental gradient that were not actually sampled. The points show the sample points. The arrows show the weighted means (species scores) for each species based only on the sample points.

The compression effect occurs when there are species whose abundance distributions along the unknown underlying environmental gradient extend beyond the environmental values actually encountered in the sample data. This is a violation of the second and fourth of the conditions of Ter Braak and Prentice (1988) linking correspondence analysis and direct gradient analysis via Gaussian response regression.

These conditions require that the sampling units be located sufficiently closely together along the underlying unknown environmental gradient so that we can observe the full range of the abundance of each species increasing up to its maximum and then decreasing. In other words, there cannot be species whose distribution extends outside of the environmental conditions contained within the sampled gradient. The distributions of such species are located in the shaded regions in Figure 3.6. A moment's reflection will convince you that this must occur in every real data set.

The measured values of the environmental gradient in Figure 3.6 range from 0 to 10 units but the range of values needed to completely describe the Gaussian distributions of the three species is from -5 to 15. Since the weighted means are unbiased estimates of the optima of these Gaussian curves (Ter Braak and Barendregt 1986, ter Braak and Looman 1986, Ter Braak and Prentice 1988) we should see these arrows pointing at the points along the environmental gradient at which each species reaches maximum abundance. The arrows show the positions of the calculated weighted means of these three species in our simulated data, as given by Equation 3.4. The positions marked by these arrows would be the species' scores in a one-dimensional correspondence analysis. The arrow for the black species is indeed pointing at the point at which it reaches its maximum abundance but this is not the case for the other two species. The weighted mean (and species score) of the blue species is shifted to the right while the weighted mean of the red species is shifted to the left. The weighted mean is a biased estimate of the species optimum for both the blue and red species. This bias arises because we are missing data on half of the distributions of these two species. As a result, the scores of species whose distributions extend beyond the measured environmental gradient at both extremes will be compressed towards the centre. Since the site scores are weighted averages of the species scores, sites that are located towards the extremes of the environmental gradient will also be compressed towards the centre. This is not a problem with Gaussian regression since it is produced by a quadratic regression of ln(abundance) on the measured the environmental variable and so we can calculate the species' optima using the regression coefficients and Equation 3.3. This is a problem with correspondence analysis since the species optima are calculated using weighted means (Equation 3.4). As a result, the compression effect will commonly occur at the extremes of the correspondence analysis axes.

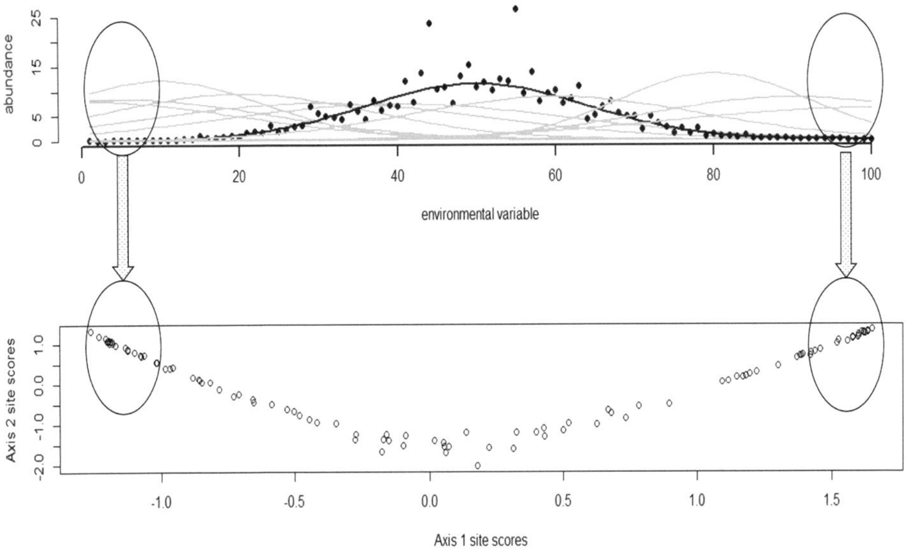

Figure 3.7. The top panel shows the Gaussian response functions (grey lines) of 11 species along a single environmental gradient. The points show the simulated abundances of one of the species; these points are not plotted for the other species for visual clarity. The bottom panel shows the site scores on the first two correspondence analysis axes for these data.

The arch effect[23] of subordinate axes in correspondence analysis is an even more serious weakness. Unlike the compression effect, which is caused by "missing" data (Figure 3.6), the arch effect is simply a mathematical artifact that does not correspond to any real structure in the

[23] The "horseshoe" effect is an extreme version of the arch effect. We will study the horseshoe effect in Chapter 6.

data (Hill and Gauch 1980). Correspondence analysis requires that axes be *linearly* independent of those that precede them, just as in principal components analysis (Chapter 1). However, the subordinate axes can still display *nonlinear*, and sometimes quite strong, dependencies on those that precede them. This means that it is difficult to interpret axes beyond the first one in terms of underlying environmental gradients since these subordinate axes will partly reflect this mathematical artifact.

Figure 3.7 illustrates this problem. The top panel shows simulated data in which eleven species display Gaussian response functions along a single environmental gradient. The lower panel shows the site scores of 100 "sites" on the first two axes of the resulting correspondence analysis. Since there is only one environmental variable generating the patterns in these simulated data, the first axis of the correspondence analysis should correctly arrange the 100 sites along this environmental variable. This is what happens. Since there is only one environmental variable generating the patterns in these simulated data, we would also expect the second axis to reflect only the random variation of the sampled sites around the theoretical Gaussian curve. The scores on the second axis should be independent of those on the first axis and should also be independent of the environmental values of each site. However, this is not what happens. The second axis is linearly independent of the first axis (the Pearson correlation coefficient is close to zero) but there is an obvious nonlinear relationship between the scores on the two axes, forming an "arch". This occurs because the algorithm of correspondence analysis forces the second axis to be linearly independent of the first axis but does not place any constraints on nonlinear relationships. This is a very common result when correspondence analysis is applied to data in community ecology. The site scores inside the circles at the extremes of the first axis in Figure 3.6 also display the compression effect since the distance between the points at the extremes are closer together (i.e. compressed) relative the distance between the points in the centre of the first axis. These compressed site scores are found in the sites at the extremes of the underlying environmental variable in which the species Gaussian response curves are truncated.

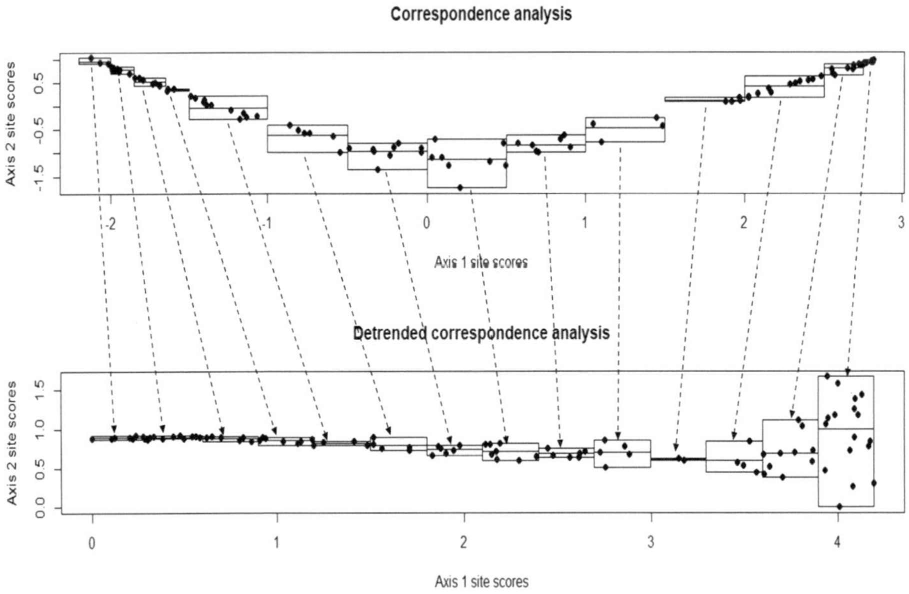

Figure 3.8. An illustration of how detrended correspondence analysis removed the nonlinear relationship between axes in correspondence analysis.

Figure 3.8 illustrates how "detrended" correspondence analysis (DCA) attempts to correct the compression and arch effects, based on two axes. If there are more than two axes then the process is repeated for each successive pair of axes. First, the range of the site scores on the first axis is divided into a series of segments having approximately the same number of points. The default number of segments in 26. This number was chosen based on the personal experiences of early users and changing the number of segments will affect the result. Within each segment, the site scores on the second axis are centred around the mean site score of the second axis within that segment. This produces "residuals" of the site scores on the second axis. Plotting these residuals against the site scores on the first axis removes the arch effect.

Next, the algorithm calculates the variance of the species scores between the sites occurring within each segment. The contraction effect occurs because species whose true optima occur close to the extremes of the underlying environmental gradient (and will therefore be more abundant in sites close to the extremes of the first axis) will have their species scores pulled closer the centre of the underlying environmental gradient. The variation in the species scores, for those species that actually occur in these extreme sites, will be lower than the variation in the species scores for those species that actually occur in sites located close to the middle of the underlying environmental gradient. DCA attempts to correct for the compression effect by a process of nonlinear rescaling. In effect, the species scores within each segment are rescaled; i.e. they are multiplied by a value that is constant within each segment but that changes between segments. This constant is chosen so that the average species tolerance, assuming a Gaussian distribution function, is unity. The site scores are then recalculated as the weighted means of these new species scores. The effect is to stretch or contraction the site scores, as can be seen in Figure 3.8. Figure 3.9 shows how the species scores along the first axis have been modified in DCA and CA relative to the actual species optima. Since the actual species optima in our simulated data vary from 0 to 100 (Figure 3.7), they have all been divided by 25 to fit in Figure 3.9.

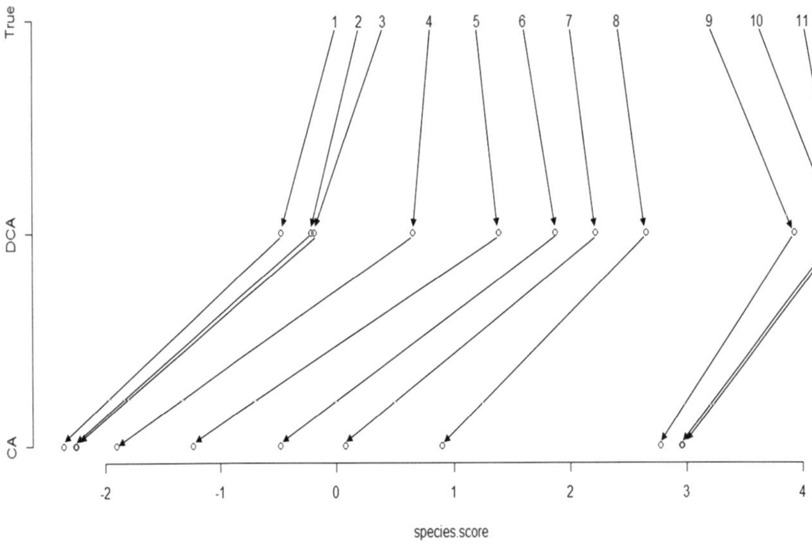

Figure 3.9. A comparison of the true species optima as shown in Figure 3.7 and the species scores produced by correspondence analysis (CA) and detrended correspondence analysis (DCA).

The compression effect along the first axis is clearly seen in the species scores of correspondence analysis since the scores of the first three species and the last three species are much closer together than are the true species means. We can also see the modifications of the species scores that the detrending step has produced although, at least in these simulated data, the modifications don't seem to have made much difference.

The nonlinear rescaling process that Hill and Gauch (1980) proposed was based on an explicit assumption of how species should enter and then leave the community as the environment changes. Their rescaling was designed so that the rate of appearance and disappearance of species should be constant along the gradient. Of course, this necessarily implies that species richness is also constant across the underlying gradient. If so then, excluding species whose distribution is truncated, the variation in the species scores within a sampling unit will the same for all sampling units. Since it is impossible to know the actual value of the constant variance, they set this variance to unity, which they called a "standard scaling". Using this standard scaling, axes are also in standard deviation units and the total length of an axis can be reported in standard deviation (SD) units. Using this standard scaling, it would take about four standard deviations along a DCA axis for a species to appear, rise to its optimum, decrease and then disappear again. A full turnover of the species between a pair of sampling units would occur if the two sampling units are about four standard deviation units distant along a DCA axis. A 50% change in the species composition would occur between sample units that are around 1 standard deviation apart. It is very important to remember that the standard scaling that underlies DCA is dependent on the assumption of a constant rate of turnover of species across sampling units, which, in turn, requires that species richness does not change along the DCA axis. Community ecologists know, based on both empirical observations and on theoretical arguments, that species richness *does* change along many environmental gradients at local, regional and global scales. Therefore, the scaling of environmental gradients in DCA based on standard deviation units should not be taken seriously but such a scaling is still useful for comparing the amount of change in the composition of species across sampling units in different studies.

The original DECORANA program of Hill (1979), which provided the first computer program to perform DCA, contained a coding error that sometimes affected results. This coding error was not identified until much later (Oksanen and Minchin 1997). The first versions of the popular CANOCO program (Ter Braak 1987), as well as early versions of the popular PC-ORD program (McCune and Mefford 1995), both of which included the original FORTRAN subroutine of the DECORANA program, also contained this coding error but this coding error was subsequently corrected in both programs. The `decorana()` function of the vegan package does not have this coding error. It also uses a slightly different method of rescaling to correct the compaction of scores at the extremes of the axes, as explained in the **R** documentation of the function. Let's now look at the `decorana()` function of the vegan library:

```
decorana(veg, iweigh=0, iresc=4, ira=0, mk=26, short=0,
         before=NULL, after=NULL)
```

The first argument (`veg=`) is a matrix or data frame containing the abundance data. These abundance data can be counts of individuals, estimates of biomass, or any other reasonable proxy for population density of each species in each sampling unit. Rows are sampling units (quadrat, site etc.), columns are species, and each value in each cell is the observed abundance of the species in column j observed in the sampling unit in row i. If a given species is not observed in a given sampling unit then its abundance is zero. Missing values are not allowed. A value of zero in a cell means that you looked for that species in that site and did not observe it.

The second argument (`iweigh=0`) is a logical argument (0/1 meaning NO/YES) allowing you to decrease or "down-weigh" the contribution of rare species when obtaining the site and species scores. This is actually a bit of fossil code that was in Hill's original DECORANA program and that was left in the **R** version of `decorana()` in order to allow for comparison with the original version but you should leave this argument at its default value (i.e. no down-weighting). If you do use this down-weighting argument (`iweigh=1`) then those species whose total abundance in the data set is less than 20% of the total abundance of the most abundance species will have their abundances further decreased in proportion to their rarity.

Why is this almost always a poor choice? Remember that the site scores are already weighted means of the species scores. The weights are the relative abundances of each species in a given site. A species that is rare (i.e. with a low relative abundance) will already contribute very little to the site scores because its species score will be multiplied by a small number and so reducing an already small contribution makes little difference. It is true that the species scores of rare species will usually be based on the site scores of only a few sites (those few sites in which these rare species actually occur) and this means that the species scores of rare species will be more affected by chance occurrences of these rare species in particular sites.

The third argument (`iresc=4`) controls the number or rescaling cycles to use and the default value was chosen based on empirical experience. You should not change this value.

The fourth argument (`ira=0`) is a logical argument stating if you want a correspondence analysis (`ira=1`) or the detrended version (`ira=0`). Choosing `ira=1` is equivalent to using the `corresp()` function of the MASS library, except that the scaling of the axes differs by a constant.

The other arguments (`mk=26, short=0, before=NULL, after=NULL`) are also fossil code from the original DECORANA function and control the number of segments to use when rescaling the axes (`mk`), whether the rescaling is also applied to the shortest axis (`short`) and so on. These should be left at their default values.

3.6) An empirical example of DCA

The Wisconsin Plant Ecology Laboratory (PEL) database includes plant community survey and resurvey data of understory vegetation from forest stands throughout Wisconsin and its adjoining states. John T. Curtis and colleagues conducted extensive field surveys across the state of Wisconsin during the late 1940s and 1950s while visiting over 1000 sites to quantitatively assess plant community composition (Curtis 1959). Beginning in the early 2000s, Don Waller and colleagues resurveyed many of the original PEL sites (Waller et al. 2012). The core PEL vegetation dataset includes vegetation data from two periods for 293 sites (Figure 3.10). In addition to baseline Curtis-era surveys and contemporary resurveys, the database was expanded to include data on local edaphic conditions, environmental data, and landscape context as well as plant functional traits and a high-resolution molecular phylogeny for all vascular plants in Wisconsin. We will only use the vegetation survey data done since the year 2000 and then compare the resulting ordination with the local environmental data. We will only use a subset of 254 sites for which there was complete data on a series of environmental variables. A total of 575 plant species occurred in at least one sampling unit (a site). All of these species are either herbaceous plants, small shrubs or tree seedlings. The first four columns of the data set contain some basic information (the sampling year, the original community classification of Curtis for the site, a site code and the quadrat code). The remaining 575 columns contain the abundance of each species. Abundance was quantified as the number of 1m^2 quadrats per site in which a species was observed. This speciesXsite data set therefore has 254 rows and 579 columns. Each entry in this table is the estimated abundance (counts) of a given species in a given site. The abundance of a given species in a single site varied from 0 (the species was never observed in the site) to 156. The most common species had a total abundance of 5888 sightings over all sites and the rarest species had only a single sighting in the entire data set.

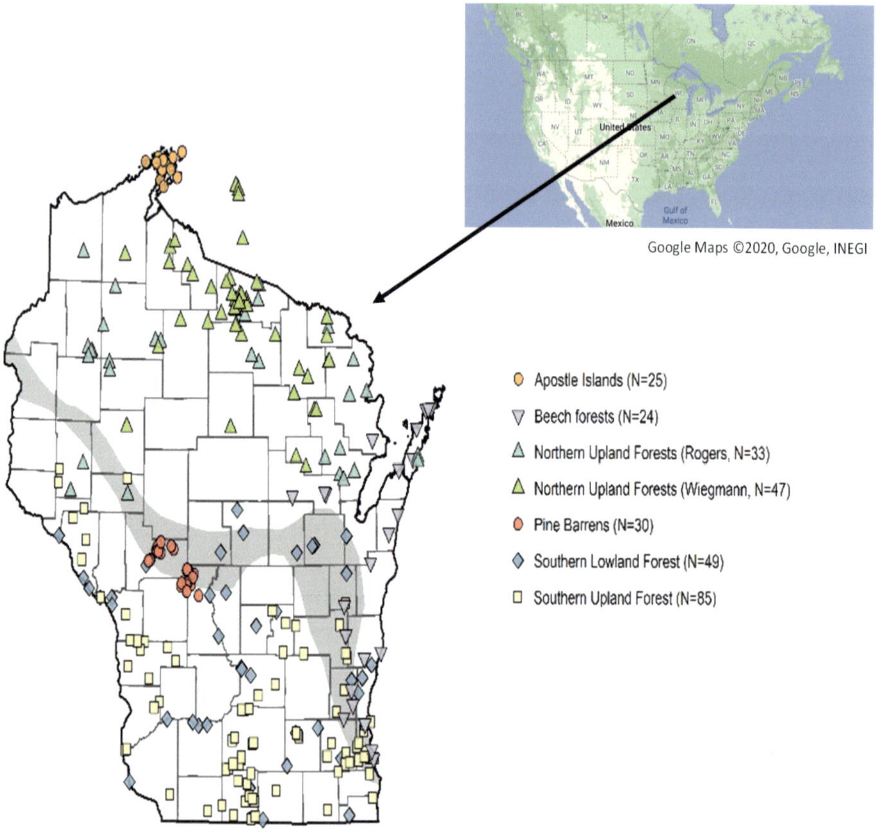

Figure 3.10. The location of the sampling units in the Wisconsin Plant Ecology Laboratory database. The names for the vegetation types follow the original classification of John Curtis. Figure curtesy of Don Waller.

We begin with the basic calls for a correspondence analysis and for a detrended correspondence analysis. Since the first four columns of the speciesXsite matrix contained information other than species' abundances, these columns must be excluded:

```
library(vegan)
#Correspondence analysis
wisconsin.ca<-decorana(veg=wisconsin[,-
c(1,2,3,4)],ira=1)
#Detrended correspondence analysis
wisconsin.dca<-decorana(veg=wisconsin[,-
c(1,2,3,4)],ira=0)
```

The result are two lists (`wisconsin.ca` and `Wisconsin.dca`) containing the results. The most important elements in these lists are `$evals`, which is a vector giving the total variance[24] accounted by each axis, `$rproj`, which is a matrix of the site scores on each axis, and `$cproj`, which is a matrix of the species scores on each axis. These three elements are also returned via the `summary()` function. The `summary()` function for the detrended version (i.e. ira=0) also returns the estimated "axis lengths". Remembering that DCA scales each axis such that the tolerance (or standard deviation of a Gaussian response curve) of each species is unity, the axis length is the number of standard deviation units comprised within it.

There is a basic plotting function[25] associated with the object output by the `decorana()` function. The distribution of the site scores (black) and species scores (red) on the first two axes, for both the CA and DCA ordinations, are shown in Figure 3.11. The arch effect is clearly visible in the CA plot (top, left) and so we can't interpret the second axis. The arch

---

[24] Actually, they are the eigenvalues of the eigenanalysis version of these methods and eigenvalues quantify the variance.

[25] You simply use the code `plot(obs)`, where obj is the name of the DCA object that you saved.

effect is removed in the detrended version. The length of the axes in the DCA can be obtained via the `summary()` function, giving a length of 6.20 and 4.20 SD for the first and second axes. We can expect 100% turnover of species between sites along a DCA axis after about four SD. However, one site, whose black point is circled in blue and indicated by an arrow, is clearly very different from all of the others. It also has an extreme environment since the soil pH (8.3) is the most alkaline of any of the sites. It also has unusually sandy soil and has the least soil organic matter of any of the sites. Since our goal is to reveal the general patterns of these understory plant communities, and since this one extreme site will dominate the resulting analysis too greatly, we will remove it before continuing. The plot in the lower left of Figure 3.11 shows the DCA ordination after removing this one site. The length of the first DCA axis (6.28 SD) has increased slightly while the length of the second DCA axis (3.26) has decreased more substantially after removing the extreme site.

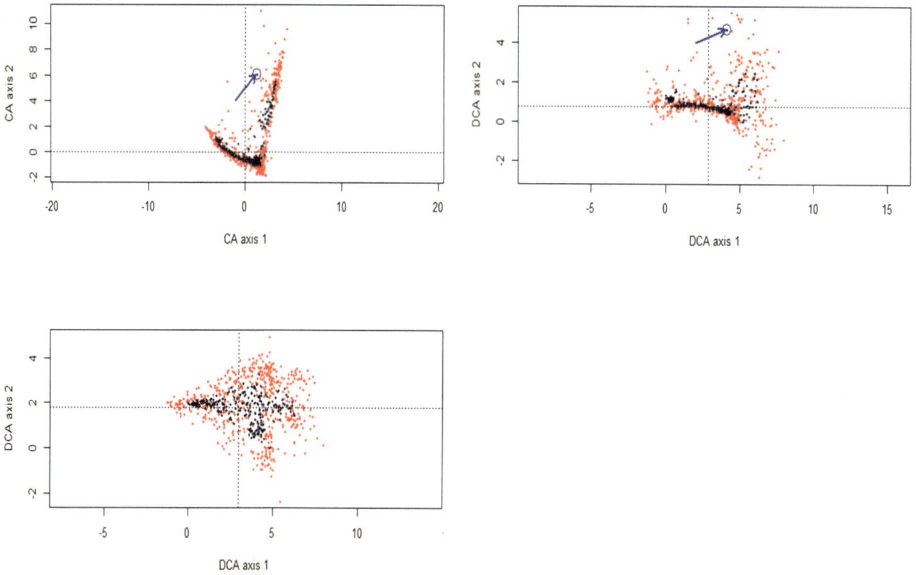

Figure 3.11. Site (black) and species (red) scores on the first two axes of a correspondence analysis (top, left) and a detrended correspondence analysis (top, right) of the Wisconsin PEL vegetation data. One extreme site is circled in blue and indicated by an arrow. Removing this one extreme site results in the DCA ordination on the lower left.

3.7) Interpreting site scores

Are the site scores on these two DCA axes associated with actual changes in the environments of these sites? The Wisconsin PEL data also included measurements of eleven environmental variables. There were eight variables related to soils: pH, percent of sand, silt and clay, percent organic matter content and concentrations in ppm of three base cations (Ca,

K, Mg). There were also two variables related to climate. Potential evapotranspiration[26] (PET, mm per month) is a measure of the ability of the atmosphere to remove water through a combination of evaporation and transpiration based on a standard crop (Allen et al. 1998) and is determined by a nonlinear function of average daily temperature above 0°C, day length and atmospheric relative humidity. PET therefore increases as the temperature during the growing season (i.e. temperature above 0°C) increases. In a temperate area like Wisconsin, PET increases from north to south and is strongly correlated with mean annual temperature. PET is also strongly correlated with potential net primary productivity of vegetation, or net primary productivity when plant growth is not limited by water availability (Rosenzweig 1968). The limitation of potential net primary productivity by water availability is the second climatic variable. This is second climatic variable is called the aridity index, which is the ratio of actual precipitation to PET. The aridity index therefore measures by how much a site is precipitation-limited *relative* to temperature. This index is difficult to interpret because it is essentially a ratio of supply (precipitation) to *potential* demand (the amount of water lost by evapotranspiration if plant production was not water-limited). Since this is a ratio, and since PET varies much more than does precipitation in Wisconsin, the aridity index is strongly negatively correlated to PET and can be approximately associated with the ratio of mean annual precipitation to mean annual temperature. The final variable is the basal area of the canopy trees, which is a proxy for

[26] https://cgiarcsi.community/data/global-aridity-and-pet-database/

the incident light reaching these understory plant species. I have transformed the abundances of the three base cations to their natural logarithms and I have transformed the sand, silt and clay values using Aitchinson's centered log ratio transformation, as explained in Chapter 2.

The first step is to look at the bivariate correlations between each environmental variable and the site scores on the first two DCA axes. I will calculate Spearman rank order correlations (`cor(x,y,method="spearman")`) so that we don't have to assume linearity between the environmental variables and the vegetation trends, although the Pearson correlation coefficients are not much different in this data set. Table 3.1 gives these Spearman correlation coefficients. Every one of these correlations in Table 3.1 are statistically significant at the 5% significance level except for the correlation between the soil clay content and the site scores on the second DCA axis.

Environmental variable	DCA axis 1	DCA axis 2
pH	0.70	-0.21
sand	-0.55	0.18
silt	0.52	-0.19
clay	-0.19	0.07
Organic matter (OM)	0.42	-0.12
Ca	0.81	-0.19
K	0.63	-0.15
Mg	0.83	-0.25
Tree basal area	0.51	-0.33
Potential evapotranspiration (pet)	0.66	-0.33

Aridity index (ai)	-0.70	0.17

Table 3.1. Values are the Spearman rank correlation coefficients calculated between each of the eleven environmental variables and the site scores on the first and second DCA axes.

At this stage, the statistical problem becomes one of interpreting the patterns of correlation between the environmental variables and how the vegetation is responding to these environmental variables, as revealed by the site scores of the DCA. There are two general approaches. We focus on correlations between each environmental variable and all chosen ordination axes or we can focus on correlations between each ordination axis and all chosen environmental variables.

Let's first focus on correlations between each environmental variable and all chosen ordination axes. This is like conducting separate multiple regressions (one for each environmental variable) of the environmental variable (e) on the chosen set of ordination axes (i.e. $\text{lm}(e \sim axis_i + axis_j + ...)$) and then looking at the partial regression coefficients of this multiple regression. In doing so, we are ignoring the pattern of correlations between the environmental variables since each multiple regression is done separately. These partial regression coefficients then shown graphically as arrows using the `envfit()` function of the vegan package. The directions of the arrows are determined by the signs of the regression coefficients. The numerical values of the regression coefficients are proportional their lengths. The actual lengths are stretched or reduced in order to fit onto the plots. The output[27] of this function can be used to add arrows to plots like those in Figure 3.11. There are two equivalent ways of calling this function using the eleven environmental variables and the first two DCA axes of the Wisconsin data:

```
wis.envfit1<-
envfit(ord=wisconsin.dca2,env=wis.env2[,-
1], choices=c(1,2))
```

envfit()

```
wis.envfit1<envfit(wisconsin.dca2~pH+sand+silt+clay+O
M+Ca+K+Mg+
basal.area +pet+ai, choices=c(1,2),data=wis.env2[,-1])
```

The first line of code separately specifies (i) the ordination object that holds the output of an ordination done using an ordination method in the vegan package (here, `ord=wisconsin.dca2`) and (ii) a matrix containing the environmental variables as columns and the sites as rows (here, `env=wis.env2[,-1]`). Note that this environmental matrix must not contain anything except the environmental variables, although these environmental variables can be factors; I excluded the first column (`wis.env2[,-1]`) because the first column contained the site codes and these are not environmental variables. The third argument specifies the ordination axes that you want to use and we are using the first two DCA axes (i.e. `choices=c(1,2)`). The second line of code does exactly the same thing but uses the "formula" version.

[27] The output also provides a permutation test of significance for each multiple regression. We will postpone discussion of this until Chapter 5.

Once you have saved the result and named it (here, `wis.envfit1`), you can add the resulting arrows to a pre-existing plot. Since the basic plotting function, used in Figure 3.11 is not very pretty, here is some code that produces a produce a nicer graphic that includes the arrows. The result is shown in Figure 3.12 (left).

```
#First, set up the plot
plot(wisconsin.dca2,choices=c(1,2),type="n")
#Next, add site numbers as text
text(wisconsin.dca2,display=c("sites"),col="black",ce
x=0.5)
#Then, add the species weighted means as red points
points(wisconsin.dca2,display=c("species"),pch=16,cex
=0.5,col="red")
#Now add the environmental correlations as arrows
plot(wis.envfit1)
```

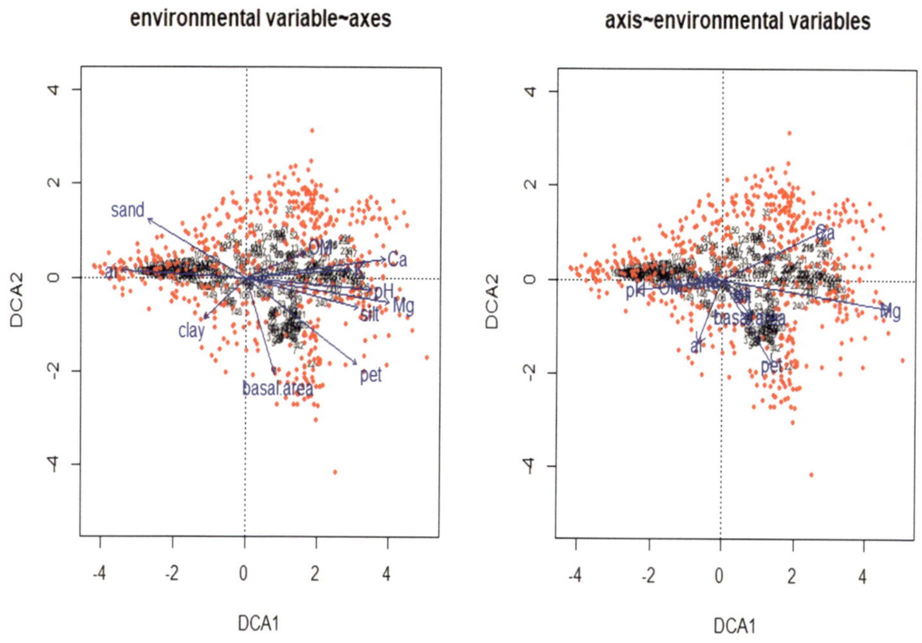

Figure 3.12. DCA plot of the Wisconsin data. The site scores are plotted as black numbers. Species weighted means are plotted as red circles. The arrows indicate the partial regression coefficients between each of the 11 environmental variables and the two DCA axes (left) or between each of the DCA axes and the 11 environmental variables (right).

Figure 3.12 looks like the biplots that we studied in Chapter 3, and the graphs are quite similar to biplots, but they are not exactly the same thing. The length of each arrow in Figure 3.12 (left), projected 90° to each axis, is proportional to the strength of the partial regression coefficients (i.e. the coefficients from $lm(e \sim axis_i + axis_j + ...)$) between each environmental variable (e) and the chosen DCA axes (here, axes 1 and 2). The direction of each arrow relative to each axis shows the sign of the correlation with each axis. For instance, Ca is strongly and positively correlated with the first DCA axis and weakly but positively correlated with the second DCA axis while tree basal area is weakly and positively correlated with the first DCA axis and more strongly, but negatively, correlated with the second DCA axis. Comparing these axes with Table 3.1, and remembering that the correlations in Table 3.1 are Spearman correlation coefficients rather than Pearson correlation coefficients, you will see that the arrows are approximately[28] a graphical representation of the correlation coefficients.

I said that there are two different approaches to producing such plots. The first approach was to focus on correlations between each environmental variable and all chosen ordination axes using the `envfit()` function. The result was Figure 3.12 (left). The first approach ignores the patterns of correlation between the environmental variables since the underlying multiple regressions, $lm(e \sim axis_i + axis_j + ...)$, are done separately for each environmental variable. The second approach consists of conducting multiple regressions of the type $lm(axis \sim e_i + e_j + ...)$ separately for each ordination axis. Note that the predictor and response variables in these two regressions have been reversed. The first approach

[28] Approximately because the arrows are proportional to the partial regression coefficients of each multiple regression of type `lm(ei~DCA1+DCA2)`, where e_i is the environmental variable.

asks how the combination of ordination axes predicts each environmental variable. The second approach asks how the combination of environmental variables predicts each ordination axis. The two approaches are complimentary but present different information. This second approach makes sense because, from a biological perspective, the environmental variables are the causes, not the consequences, of the patterns of association between species across the sites. In this second approach, the partial regression coefficients reflect the amount by which a unit change in each environmental variable would change the position of a site along that axis *if all of the other variables didn't change*. Although this second approach ignores the correlations between the axis scores, these scores are linearly independent by design and so we don't loose anything important[29]. There is no function in **R** that will do plot these second type of arrows and so I present a simple function, `ord.on.env.arrows()`, to do this at the end of this chapter. We will use the same lines of code to produce the plot of site and species scores as for Figure 3.12 (left) but then use this new function to add the arrows instead of `envfit()`. Here is the code:

ord.on.env.arrows()

```
ord.on.env.arrows(ordination.site.scores=wisconsin.dc
a2$rproj[,1:2],env.matrix=wis.env2[,-1],
arrow.col="blue",arrow.scale=5)
```

[29] This linear independence between axes will not be true when we use this method for ordinations produced by non-metric multidimensional scaling in Chapter 4.

The first argument is a matrix containing the site scores which, in this case are the site scores on the first two DCA axes (`wisconsin.dca2$rproj[,1:2]`). The second argument is a matrix containing the environmental variables (`wis.env2[,-1]`). The third argument specifies the colours of the arrows. The forth argument is a scaling factor for the arrow lengths to make them visually clear; the default for this fourth argument is to let the function try and find the best scaling factor. The result is shown in Figure 3.12 (right).

The differences between the left and right graphs in Figure 3.12 reflect the differences in the patterns of correlation between the environmental variables (left) and the patterns of partial correlation between the environmental variables (right). Such complicated patterns of correlation between environmental variables are the rule, not the exception, in real ecological data sets. From your knowledge of multiple linear regression you will know that predictor variables that are highly correlated, or "collinear", cause a number of problems. Besides problems related to hypothesis testing, which don't concern us here, highly correlated predictor variables can also produce unstable regression coefficients. In other words, small changes in the values of a few points can produce large changes in the regression coefficients associated with such highly correlated predictor variables, even to the extent of changing the signs of the regression coefficients, while hardly affecting at all the actual predictive ability of the regression model. Since our objective is to interpret the underlying environmental gradient to which the vegetation is responding, we need to avoid such problems of collinearity.

The first step is to look at the patterns of correlation between the environmental variables. You can do this by calculating the bivariate correlations between the environmental variables via the `cor()` function as in Table 3.1, or by looking at the set of bivariate scatterplots between the environmental variables via the `pairs()` function. The `pairs()` function has the advantage that you can identify any obvious nonlinear trends between the environmental variables. Figure 3.11 shows the scatterplot matrix of the environmental variables with the Pearson correlation coefficients on the subdiagonal. For ease of interpretation, correlation coefficients that are greater than 0.7 in absolute value are shown in red.

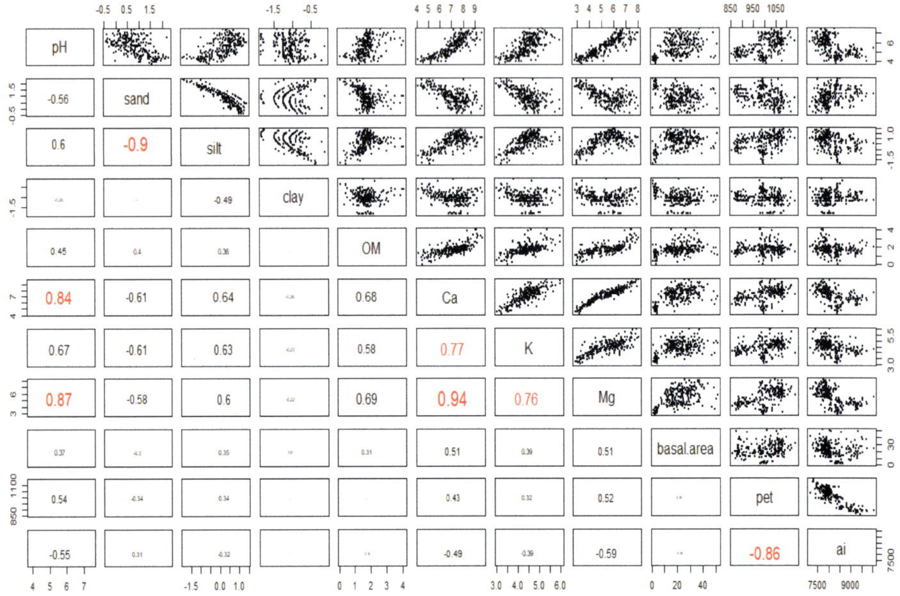

Figure 3.13. A scatterplot matrix of the 11 environmental variables measured on the Wisconsin PEL data set. The subdiagonal gives the Pearson correlation coefficients. Correlations greater than |0.7| are shown in red.

There are three groups of environmental variables for which the correlations within them are especially strong. The first one is the pair of climatic variables (`pet` and `ai`) with a correlation of -0.86. This strong correlation is not surprising since the aridity index (`ai`) is the ratio of precipitation to potential evapotranspiration (`pet`). Since `pet` is an index of potential net primary productivity while `ai` is an index of the decrease of actual net primary productivity from its potential, I will call this group "net primary productivity", or `NPP`. The second group is the triplet of sand, silt and clay. The correlation between sand and silt is -0.90 and such a strong correlation in not surprising either since the sum of the three soil texture variables must sum to unity. I will call this second group `soil.texture`. The third group involves the three base cation concentrations plus soil pH. The strong correlations in this third group are not surprising because, together, they determine soil base saturation. In acidic soils, the exchange sites on soil particles are mostly occupied by hydrogen ions rather than by base cations. As soil acidity decreases, and as soil pH increases, the exchange sites become occupied more by the soil cations (Ca, Mg, K, Na, NH_4) and less by hydrogen ions. Along with soil texture, these determine the cation exchange capacity of forest soils. I will call this third group `cation.exchange`.

Since the relationships between these highly collinear variables with the three groups are approximately linear, we can use our knowledge of principal component analysis to replace them with three new variables. Each new variable will be the site scores on the first principal component of each PCA. The PCA for `NPP` and for `cation.exchange` will be done on standardized variables since these involve variables having different measurement units. The PCA for `soil.texture` will be done on unstandardized variables since the three texture variables have the same measurement units (% soil dry mass). In the lines of code below, `wis.env2` is a data frame holding the environmental variables and the column numbers are those holding the relevant environmental variables for each PCA. The site scores on the first principal component axis of each are found in the **R** objects `cation.exchange$scores[,1]`, `soil.texture$scores[,1]` and `NPP$scores[,1]`. The Figure 3.14 shows the biplots of these three principal component analyses[30]. Since the first principal components of each of the three PCAs account for a large proportion of the variance, I have also written my interpretation of the first PCA axes in Figure 3.14.

```
cation.exchange<-
princomp(wis.env2[,c(2,7,8,9)],cor=T)
soil.texture<-princomp(wis.env2[,c(3:5)],cor=F)
NPP<-princomp(wis.env2[,11:12],cor=T)
```

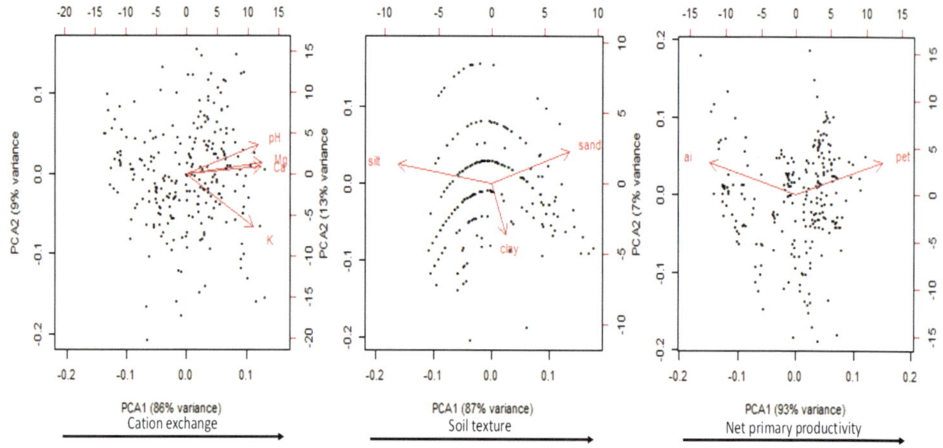

Figure 3.14. Biplots of three principal component analyses. The site scores on

[30] The biplot for soil texture has obvious nonlinear patterns that are caused by the fact that the three proportions must sum to unity.

the first principal component of each will serve as composite variables representing soil cation exchange capacity, soil texture and plant-available water.

Now, rather than using the actual environmental variables included in the three PCAs, we can use the scores of the sites on the first PCA axis of each. This allows us to replace these three sets of highly correlated variables with new composite variables that we can still interpret biologically and then redo Figure 3.12. The result is Figure 3.15.

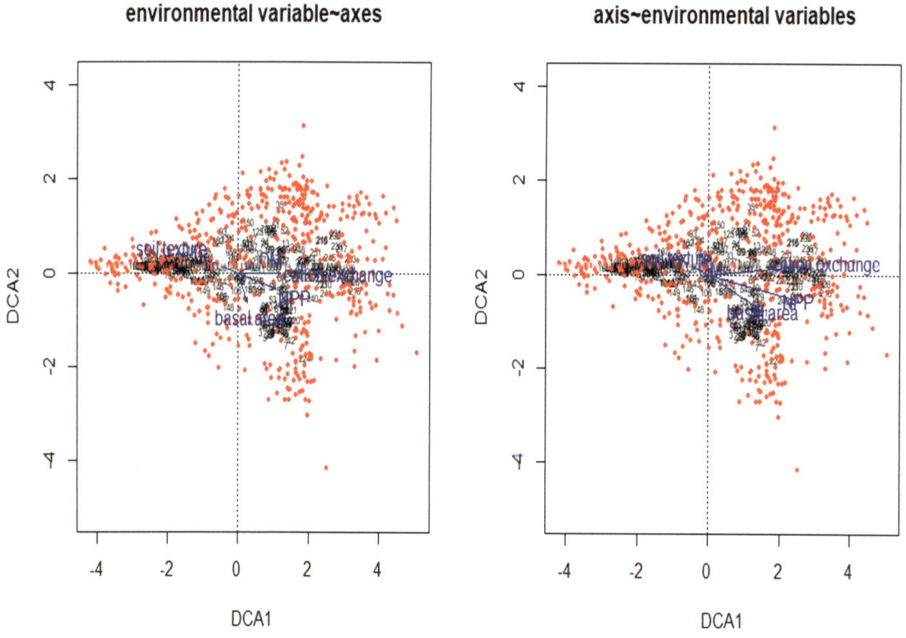

Figure 3.15. DCA plot of the Wisconsin data. The site scores are plotted as black numbers. Species weighted means are plotted as red circles. The arrows

indicate the partial regression coefficients between each of two measured environmental variables (basal.area and OM) and three latent environmental variables from PCA (soil.texture, cation.exchange and NPP) and the two DCA axes (left) or between each of the DCA axes and the environmental variables (right).

We are finally ready to interpret the two major trends in the structure and composition of the Wisconsin understory vegetation in terms of the underlying environmental gradients. The major underlying driver in organizing these understory plant communities, i.e. the first DCA axis, appears to be variation in soil conditions. Sites on the left of DCA axis 1 have sandy soil (`soil.texture`) that has low base cation exchange capacity (acidic pH, low concentrations of Mg, Ca and K). For sites that have the same placement on the first DCA axis, sites lower on the second DCA axis have a higher net primary productivity and forests with a higher basal area, i.e. forests with more and/or larger canopy trees.

A complimentary way of interpreting the variation in the site scores of this understory vegetation is to differentiate the sites based on the canopy forest types. Curtis (1959) had classified the forests of Wisconsin, as seen in Figure 3.10. Each site in the Wisconsin data was assigned to one of five forest types: Southern upland forests, Beech forests, Northern upland forests, Pine Barrens and Southern lowland forests. We can therefore plot the site scores on the first two DCA axes using different colours to represent these forest types. The result is Figure 3.16. Since the distance between sites in the DCA reflects the degree of difference between them in terms of the composition of the species in them, we can clearly see that the understory vegetation of sites within a single forest type is much more similar than is the understory vegetation of sites found in different forest types. The variation of the understory communities along the first DCA axis largely reflects the differences between the major forest types while the second DCA axis largely reflects the differences of the understory vegetation within each major forest type. It is understandable that the environmental variables could quite accurately predict the position of sites along the first DCA axis but could not predict very accurately the position of sites along the second DCA axis. In addition, we can see that variation in site scores on the second DCA axis was especially driven by variation between sites within two forest types: Southern upland forests and Beech forests.

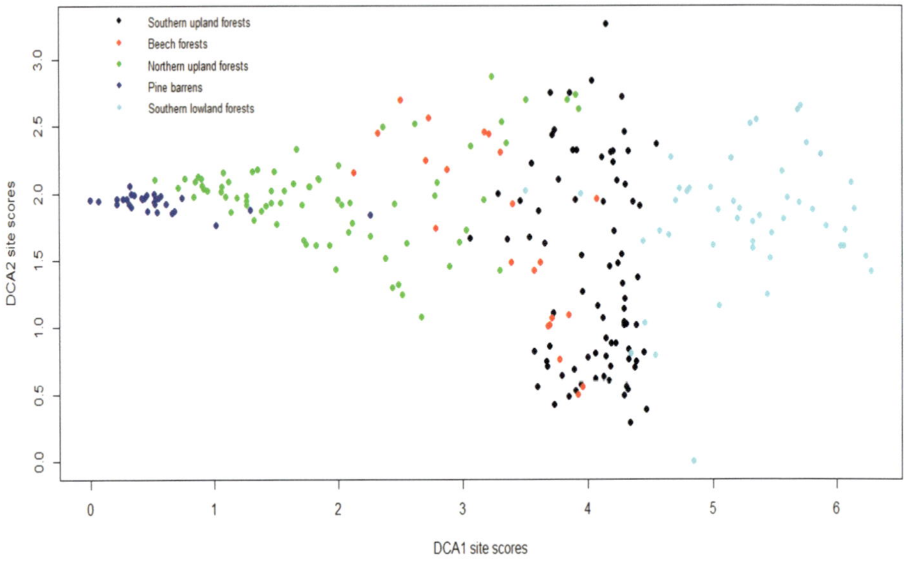

Figure 3.16. The site scores of the understory vegetation on the first two DCA axes are differentiated according to the original Curtis classification of the major forest types of Wisconsin.

The last step in our detrended correspondence analysis is to look at the species scores associated with each DCA axis. The species scores are the weighted means of each species along these axes and, now that we can relate these axes to underlying environmental gradients, these species scores can be interpreted as estimates of the habitat conditions at which each species reaches its maximal abundance within the range of habitats represented in the data. These species scores can be obtained either via the `summary()` function associated with the `decorana()` list object, or directly by specifying the `$cproj` element of the `decorana()` list. Typing `wisconsin.dca2$cproj[,1]` will give the species scores on the first DCA axis that are in the `wisconsin.dca2` list that was created via a `Wisconsin.dca2←decorana()` call. The `plot()` function associated with the `decorana()` object can project the species names onto a plot of the species scores, but there are far too many species in the Wisconsin PEL data set. Instead, I have identified a few common species in Figure 3.17; the point labelled *Abies balsamea* (Balsam Fir) only refers to the seedlings of this species since the Wisconsin PEL data only include the understory vegetation, but including seedlings of the tree species.

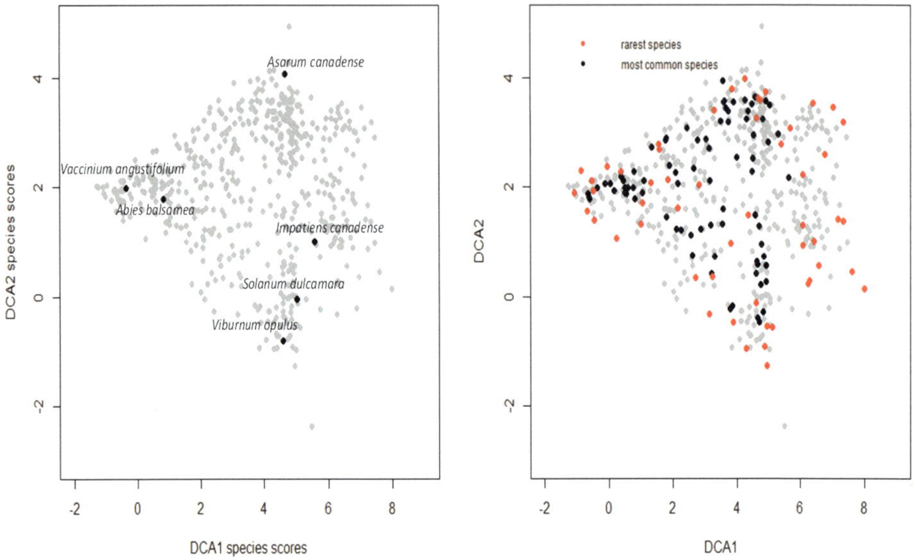

Figure 3.17. The species' scores of the Wisconsin PEL data are plotted on the first two DCA axes as grey dots. Left: the position of six common species. Right: the positions of the rarest species (red) that only occur in a single quadrat and the common species (black) that occur in at least 400 quadrats.

There is a clear pattern in the location of the rare species in Figure 3.17 that is commonly seen in detrended correspondence analysis: the rare species (here, defined as species found in only one quadrat) are located at the outer edge of the DCA axes. The most common species (occurring in at least 400 quadrats) are found throughout the cloud of points. Although such a pattern is sometimes interpreted as an artifact of detrended correspondence analysis, this is not necessarily the case. In terms of niche theory, which forms the basis of the reciprocal averaging algorithm of correspondence analysis, a rare species will be one whose realized niche is almost entirely outside of the range of habitats that occur in an area. If we were to plot the Gaussian response function of such rare species along the underlying environmental gradient of the DCA axes, then these response functions would have their maxima outside of the extremes of the axes with only the tails extending inside the observed values of the underlying environmental gradient. We should therefore expect the rare species to have their species' scores located at the outer edges of the DCA axes. Rare species whose species scores occur well within the cloud of points are perhaps more interesting. Such species either have habitat requirements that are not included in the measured environmental variables or else their rarity is determined by factors other than environmental conditions.

```
ord.on.env.arrows<-
function(ordination.site.scores,env.matrix,choices=c(1,2),arrow.sca
le=NA,arrow.col="red"){
 #adds arrows to 2D plots of ordination scores representing the
regression weights
 #of multiple regressions of additional variables on each ordination
axis.
 #
 #ordination.site.scores: matrix or data frame of numerical
ordination scores
 #env.matrix: matrix or data frame of environmental variables
 #choices: the axes for which the regressions are sought
 #arrow.scale: the scaling factor to fit arrows onto the ordination
plot

 #centre the ordination site scores, in case they have not already
been centered
 ordination.site.scores1<-ordination.site.scores[,choices[1]]-
   mean(ordination.site.scores[,choices[1]])
 ordination.site.scores2<-ordination.site.scores[,choices[2]]-
 mean(ordination.site.scores[,choices[2]])
 #Now, do multiple regressions and extract regression coefficients
 lm.axis1<-lm(ordination.site.scores1~scale(env.matrix))
 lm.axis2<-lm(ordination.site.scores2~scale(env.matrix))
 axis1<-coefficients(lm.axis1)[-1]
 axis2<-coefficients(lm.axis2)[-1]
 var.names<-colnames(env.matrix)
 n.vars<-dim(env.matrix)[2]
 #get range of axis scores and range of regression coefficients
 range1<-range(ordination.site.scores1)
 range2<-range(ordination.site.scores2)
 min.range<-
min(range1[2],abs(range1[1]),range2[2],abs(range2[1]))
```

```
a1<-max(abs(axis1))
 a2<-max(abs(axis2))
 arrow.range<-max(a1,a2)
 #Try and get a good scaling factor if not specified
 if(is.na(arrow.scale))arrow.scale<-min.range/arrow.range
 #add arrows to the pre-existing ordination plot
 for(i in 1:n.vars){
  x<-axis1[i]*arrow.scale
  y<-axis2[i]*arrow.scale
  arrows(x0=0,y0=0,x1=x,y1=y,col=arrow.col,length=0.1)
  text(x=x,y=y,labels=var.names[i],col=arrow.col)
 }
 #return the partial regression coefficients and the scaling factor
list(axis1=summary(lm.axis1),axis2=summary(lm.axis2),arrow.scale=ar
row.scale)
}
```

Chapter 4 Principal coordinates and multidimensional scaling

We will study types of ordination called "principal coordinates analysis" (PCoA) and "multidimensional scaling" (MDS) in this chapter. One important type of MDS is called "non-metric multidimensional scaling" (NMDS). The underlying mathematical methods of these two types of ordination are very different but the goals of both methods are quite similar. This similarity in goals is made more explicit using an alternative name for PCoA: "classical multidimensional scaling". Both methods are also presented together because both use the same type of input data matrix. This data matrix is an RxR matrix of "distances" (or "dissimilarities") between each of the R observational units or rows (R). An RxR distance matrix is different from the type of data matrix that we using in the previous chapters, which was an RxC matrix consisting of C variables (columns) measured on each of the R observational units (rows). Since both PCoA and MDS both use the same type of input data matrix, we will begin this chapter by looking at the many different types of distance and (dis)similarity measures that exist. After we have obtained an overview of these different measures, I will then provide an intuitive explanation of how PCoA and MDS work and how to carry out these analyses in **R**.

4.1) Distance and (dis)similarity measures

In Chapters two and three we used RxC data matrix with R rows and C columns. Each row represents a sampling unit and each column represents a variable attribute measured on each sampling unit. For instance, the worldwide leaf economic spectrum that we analysed in Chapter 2 had R=492 rows, in which each row (sampling unit) was a different plant species, and C=5 columns, in which each column was a leaf trait. In Chapter 3, we analysed the understory plant communities in Wisconsin forests. This data frame had R=253 rows, in which each row was a sampling unit (site), and C=575 columns, in which each column was the abundance of a different understory species. In this chapter, one also usually begins with the same general type of RxC data matrix but we then convert it to an RxR matrix of dissimilarities or distances between each pair of observational units. However, in some cases, the degree of difference between observational units can be measured directly without first obtaining an RxC data matrix.

The first step in PCoA and NMDS is to quantify the degree of difference, or "distance" between each pair of observational units, i.e. between each pair of rows in the RxC data matrix. Later on, I will make a distinction between a "distance" and a "dissimilarity" but, for now, we will ignore this distinction. The result is an RxR distance matrix. Each element (i,j) in this matrix quantifies by how much the ith and j^{th} rows (sampling units) of the RxC matrix differ from each other in terms of their variable attributes. Of course, the less two things are different, the more that they are similar. A "similarity" coefficient is simply the compliment of a "dissimilarity" coefficient.

Why would we want to convert our RxC data matrix into an RxR distance matrix? After all, such a conversion means that we lose information about the values of the variables in the columns of the RxC matrix. The reason is that this gives us more flexibility with respect to the types of variable attributes that we can use and the types of "difference" that we can measure. Both PCA and CA require that the variable attributes be measured in an interval or ratio scale; they have to be real values or integer counts. This limits the types of variable attributes that we can use. As you will see, by converting or RxC data matrix into an RxR distance matrix, we can incorporate other types of variables. Furthermore, the only type of difference between sampling units that PCA allows is the Euclidean distance. The only difference that a CA can use is the Chi-Squared distance. Depending on the biological question, the Euclidean or Chi-Squared distance might not properly represent what we mean by "difference".

Many different types of similarity and distance measures have been developed and it is easy to become overwhelmed by this literature. One of the most complete descriptions of these measures is given in chapter 7 of Legendre and Legendre (2012) and I will base this next section largely on that reference. The main criteria for choosing between the different types are (i) the mathematical nature of the variable attributes that are to be included and (ii) the biological interpretation of what constitutes "similarity" or "difference". The first criterion is straightforward. The second criterion requires a good biological understanding of the phenomena under study as well as clear and explicit hypotheses linking the chosen measure of similarity to the stated goal of the analysis. Many different similarity and distance measures have been proposed and they differ depending on the measurement scale of the variables (i.e. the columns of our original data set). The different measurement scales are classified as (i) *categorical nominal*, (ii) *categorical ordinal*, (iii) *measured on an interval scale* and (iv) *measured on a ratio scale*. Each type of variable has its own distance/similarity measures. After introducing some of the more common distance measures for each variable type, we will see how to combine these different types of variables together.

A *categorical nominal* variable is one that can exist in different discrete states but without these different states being ordered in any way. For instance, seed dormancy can be classified into three types based on its causes. The measured variable is called "dormancy type" and the three possible values (states) of this variable are "physical", "physiological" and "morphological" dormancy (Fenner and Thompson 2005). These three states are discrete. We know that they are different but not by how *much* they are different. Each type of seed dormancy is different from the others, but one type is not more or less dormant from the other two. An unordered binary variable is a special case of a categorical nominal variable since it can only exist in one of two states and these two states are not ordered. For instance, if you categorize each individual as either a male or a female then you are using a binary categorical nominal variable.

A *categorical ordinal* variable is one that can exist in different discrete states and we can also qualitatively, but not quantitatively, order these different states in some natural way. For instance, we might rank different animal species on a five-point scale in terms of how commonly they are associated with humans. The variable is "human association" and the five possible values of this variable are never (1), rarely (2), occasionally (3), often (4) or always (5). Not only can we say that "never" is different from the other four values but we can also rank them: never < rarely < occasionally < often < always. Although we can order the values, they don't have associated measurement units and so we can't quantify by how *much* each ordered level differs from the next. " In other words, the values of categorical ordinal variables are different and the differences are ordered, but the sizes of the differences between the values are unknown. In other words, we can't say that the amount of difference between "rarely" and "occasionally" is equal to the amount of difference between "occasionally" and "often".

A variable measured on an *interval* scale can be either discrete or continuous. An interval scale variable shows quantitative measurement, either as integers or as real numbers. Unlike ordinal categorical variables, the difference between values of an interval scale variable is also meaningful but the ratio between the values is not meaningful. In fact, the word "interval" means "distance between two entities". One can meaningfully add or subtract the values of an interval scale variable but one cannot meaningfully multiply or divide them. An interval scale is relative, not absolute. It measures the distance between entities but does not measure the distance between an entity and some natural lower bound. The value of "zero" in an interval scale is arbitrary, rather than natural, and so values can be either negative or positive. An example of an interval scale variable is temperature measured in degrees Celsius. Imagine that you have measured the subsurface temperature of the water in three different lakes and get readings of 0, 10 and 20 °C. These values have a natural order (0 < 10 < 20) and represent different amounts of heat. Furthermore, the difference between the first and second lakes is 10°C, the difference between the second and third lakes is 10°C, and the difference between the first and third lakes is equal to 10+10°C=20°C. We can meaningfully add and subtract between the values. Now imagine that your American friend took measurements at the same time but used a thermometer graduated in degrees Fahrenheit, which is another interval scale measurement. Her measurements would be 32, 50 and 68 °F. Her values are ordered in the same way as yours. The difference between the first and second lakes is 18°F, the difference between the second and third lakes is 18°F, and the difference between the first and third lakes is equal to 18+18=36°F. We can meaningfully add and subtract between her values as well. The result is the same as when using a Celsius scale. However, it makes no sense to multiply or divide the values on an interval scale. Why? The third lake is twice as hot as the second according to your Celsius scale (20/10=2) but it is only 1.36 times as hot according to your friend's Fahrenheit scale (68/50=1.36) even though the amount of heat hasn't changed in the two lakes. The relative physical meaning of a value is the same in both interval-scale variables (the amount of heat) but the absolute physical meaning of a value is not the same. This is because a value of zero does not mean the absence of heat but is rather an arbitrary choice used to define the scale.

As with an interval scale variable, a variable measured on a *ratio* scale can be either discrete or continuous; it can be either an integer or a real number. Furthermore, both the difference and the ratio between values are meaningful. One can both meaningfully add or subtract the values and meaningfully multiply or divide them. A ratio scale is absolute, not relative. It both measures the distance between entities and also the distance between an entity and some natural lower bound. Another way of saying this is that a value of "zero" on an interval scale means the absence of the measured property (heat, mass, length etc.) and so values can only be positive. The Kelvin temperature scale is a ratio scale of temperature because it is physically impossible to have a negative value; 0°K occurs when there is no kinetic or potential energy in an atom or molecule. Other examples of ratio scale variables are units of mass, (physical) distance, area or speed.

4.2) Measures for categorical nominal variables

In Chapter 2, we used PCA to describe the pattern of covariation between a series of leaf traits that define a "leaf economic spectrum". Escobar et al. (2020) wanted to do something similar with respect to a series of regeneration traits of 82 Brazilian savanna plants. Regeneration traits are traits that determine the ability of a seed to disperse, to survive and persist in the soil, to germinate at the appropriate time and for the seedling to establish. Among these regeneration traits is the ability of the plant to disperse vegetatively via rhizomes, stolons and so on. It is incorrect to analyse such data using PCA because several of the variables are not continuous variables. Although we will look at these data in more detail later, let's consider one of the variables, "dispersal syndrome", that they measured on each of their 82 species. A dispersal syndrome describes the way in which the seeds of a species are typically dispersed. Escobar et al. (2020) used four types of dispersal syndromes. "Endozoochory" is a dispersal syndrome in which the seeds are dispersed through the digestive tract of an animal species (thus "endo") and plants using this dispersal syndrome typically have seeds in fleshy fruits. "Exozoochory" occurs when the seeds are dispersed by animals on their body surface (thus "exo") and plants using this dispersal syndrome typically have fruits or seeds containing appendages that allow them to become attached to the animal. "Anemochory" refers to seeds that are typically dispersed by the wind and plants using this mode of dispersal typically have seeds with wings or other adaptations, are flat, or have very small and light seeds. Finally, "autochory" is used by species whose seeds are not dispersed by either wind or animal vectors.

How can we calculate a degree of similarity or distance between pairs of species with respect to a variable like "dispersal syndrome"? These four categories represent four different values, or states, of the variable "dispersal syndrome". Is anemochory more different from endochory than is exochory? The answer to this question is biological, not statistical. If the biological answer is "no" then "dispersal syndrome" is an unordered categorical variable: the dispersal syndromes are different but the degree of difference is unknown or meaningless. However, perhaps the biologist wants to use these categories to captures the physical distance that a seed will disperse from the mother plant. If these different dispersal syndromes can disperse seeds by greater or lesser distances, then this variable would not be unordered when making comparisons. This shows that the biological meaning of a variable is important when choosing measures of similarity. However, lacking such a biological interpretation, we will consider this variable as unordered. If two species possess the same type of dispersal syndrome then we count this as "similar" (i.e. 1) and any other combination of dispersal syndromes is counted as equally "different" (i.e. 0). If the two sampling units have several such unordered categorical variables then the degree of similarity between the two sampling units is the proportion of such unordered categorical variables that are identical. This is called the "simple matching coefficient" and varies between 0 and

1. A zero value means that the two sampling units do not have any unordered categorical variables that have the same value and a value of 1 means that the two sampling units are identical in all of their unordered categorical variables.

Some complications can occur when using unordered categorical variables. The dispersal syndromes defined by Escobar et al. (2020) results in mutually exclusive states; a species can only have one type of dispersal syndrome and so a pair of species can only either be similar (they are both in the same state) or are different (they are in different states). However, some unordered categorical variables do not have mutually exclusive states. For instance, if we had classified plant species by the type of animal vector that disperses their seeds (mammals, birds, ants or none) then it might be possible for the same plant species to have more than one dispersal vector. Maybe it is possible for the seeds to be dispersed by mammals, birds, ants or any combination of the three? In this case, the similarity between to plant species could vary between 0 and 3 (i.e. both are dispersed by all three animal vectors). In such an unordered categorical variable, you would want to divide the number of states shared between a pair of plant species by the total number of possible agreements for such a variable.

4.3) Measures for binary variables

Binary variables can exist in only one of two states: dead/alive, present/absent, male/female, and so on. Although a binary variable is simply an unordered categorical variable for which there are only two categories (states), many different similarity coefficients have been proposed for different purposes.

Each of the physical locations on a DNA molecule is occupied by either an adenine (A), cytosine (C), guanine (G) or thymine (T) nucleotide base. The C nucleotide may appear in most individuals at a specific base position in the human genome but the position might be occupied by an A nucleotide in a minority of individuals. This means that there is a single nucleotide polymorphism (SNP) at this specific position. The two possible nucleotide variations (C or A) are alleles for this specific position and represent the two possible states of the SNP. An SNP is a binary variable. We can compare a total of N positions on equivalent strands of DNA in two individuals (i and j) and count the number of locations in which both individuals have the same nucleotide (n_1) or a different nucleotide (n_2). An obvious measure of similarity is the simple matching coefficient, which is the proportion of locations in which the two individuals have the same nucleotide. In this example, the simple matching coefficient is $s_{ij}=n_1/N$ and a measure of difference (or distance) between them is $d_{ij}=n_2/N= 1-s_{ij}$. Two individuals having identical nucleotides at each position in which a

SNP resides would have no distance separating them (s_{ij}=1 and d_{ij}=0) and two individuals having different nucleotides at each position in which a SNP resides would have a maximum distance between them (s_{ij}=0 and d_{ij}=1). If we then sample R different individuals, and calculate this distance measure between each pair of individuals, then we would have an RxR matrix of distances. I presented the PCA analysis of Novembre et al. (2008) in Chapter 2, based on SNPs of native Europeans. The 1387 rows of that data set consisted of sampling units (individual European individuals) and the half a million columns consisted of the nucleotide (A, C, G or T) at each of the physical locations on their DNA SNPs. Although Novembre et al. (2008) analysed these data using PCA[31] they could have used PCoA on such a distance matrix.

In Chapter 3 we looked at a data matrix in which each of the R rows represent sites and in which each of the S columns represent the abundance of a particular species. In some situations is it too difficult to measure the abundance of each species and so an ecologist will only record a list of species that were present (1) or absent (0) in each site. Such a data set would be a siteXspecies matrix of zeros and ones. In order to compare the similarity or difference between each pair of sites (i, j), we would construct a contingency table (Figure 4.1). A general measure of similarity in this context is given in Equation 4.1 and the resulting dissimilarity measure is 1-s(i,j). The numerator counts the cases in which both sites display

---

[31] Remember that PCA can use binary variables.

similarities (present/present and absent/absent) while the denominator counts the total number of cases over all four possible combinations. The values of w are "weights" that we can assign to each combination.

$$s(i,j)=\frac{w_1n_1+w_4n_4}{w_1n_1+w_2n_2+w_3n_3+w_4n_4} \tag{4.1}$$

		Site j species		
		present	absent	Row sum
Site i species	present	n_1	n_2	n_1+n_2
	absent	n_3	n_4	n_3+n_4
				N=n_1+n_2+n_3+n_4
Column sum		n_1+n_3	n_2+n_4	

Figure 4.1. A contingency table summarizing the number of species that are present in both sites i and j (n_1), present in one and absent in the other (n_2, n_3) and absent in both sites (n_4).

Legendre and Legendre (2012) list 18 different similarity measures that have been proposed by ecologists based on such a contingency table involving binary variables. The differences between them are due to the assumptions (sometimes implicit) linking each similarity measure to the underlying biological phenomenon. For instance, an important goal in community ecology is to identify the underlying environmental variables that control the assembly of ecological communities. The fundamental assumption is that sites that are very similar in terms of their species composition will also be very similar in terms of the underlying environmental variables that control the assembly of their communities. The differences in the similarity measures used in community ecology mostly reflect different assumptions concerning how presence or absence of a species in a site does, or does not, reflect similarity in the underlying environmental variables that control community assembly. For example, if two sites both have the same species present (contributing to the value of n_1 in the contingency table in Figure 4.1) then does this increase the likelihood that the two sites also have similar values of the underlying environmental variables that control the presence of these species? If two sites both lack the same species (contributing to the value of n_4 in the contingency table in Figure 4.1) then does this also increase the likelihood of the two sites having similar values of the underlying environmental variables that control the absence of these species? If yes, then do both cases (presence/presence and absence/absence) provide equal evidence, or is one case a better indication of community assembly than the other case? Notice that these are ecological questions, not statistical ones. These questions boil down to deciding how to assign weights, i.e. the w values in

Equation 4.1, to each combination in the contingency table. Here are some examples.

Let's assume that a species can only maintain a viable population in a site if the environmental conditions of the site are within its realized niche; i.e. the range of environmental variables that allow its individuals to survive and reproduce. If the same species occurs in a pair of sites then, given this assumption, the environmental conditions of both sites must be within the realized niche of this species. The environmental conditions are not necessarily the same, but they can't be *too* different. Now imagine that several different species also occur in the same pair of sites. Since the realized niche of each species in a given site will be somewhat different, a greater number of species that occur in both sites (i.e. n_1) should indicate a greater degree of similarity in the underlying environmental conditions of both sites. On the other hand, knowing that a species is absent from a site only tells us that one or more important environmental conditions of the site is outside of the realized niche of that species given our assumption. Each missing species could be absent because of a completely different combination of environmental conditions. If so, then the number of species that are absent from both sites (i.e. n_4) doesn't tell us much about the degree of similarity in the important environmental conditions of the two sites. Alternatively, a species could be missing from one site simply because it hasn't yet dispersed to it. If so, then the number of species that are absent from both sites (i.e. n_4) doesn't tell us anything about the degree of similarity in the important environmental conditions of the two sites. In this case, the number of species that are absent from both sites should not be considered indicative of "similarity".

One common measure of similarity that agrees with this argument is Jaccard's coefficient, which amounts to assigning a weight of zero to n_4 and weight of one to each of n_1, n_2 and n_3 in Equation 4.1. In other words, Jaccard's coefficient ignores double zeros. However, Jaccard's coefficient also assumes that each other combination (presence/presence, presence/absence and absence/presence) provides equal amounts of information concerning the similarity of the underlying environmental conditions. The Sorensen coefficient assigns the following weights: $w_1=2$, $w_2=w_3=1$, $w_4=0$. This weighting scheme assumes, as does the Jaccard coefficient, that the combination (absent/absent) is meaningless but it also assumed that the combination (presence/presence) is twice as informative as the two combinations in which one species is present and the other absent. Table 7.2 of Legendre and Legendre (2012) lists several other similarity/distance measures for binary variables. These different measures, when assembled into an RxR matrix of distances between each pair of sampling units, will produce different PCoA and MDS ordinations. In order to know which measure is best for your purpose, it is important that you clearly state this purpose. Once you have clearly stated the purpose then you must explicitly relate the different weights to the underlying biological phenomenon. Since ordination of siteXspecies data is such an important topic in community ecology, we will look at the ability of these different measures to capture the underlying environmental gradients controlling community assembly in Chapter 5.

4.4) Measures for categorical multistate ordinal variables

The values of both categorical nominal and ordinal variables represent different discrete states in which a sampling unit can exist. The two sampling units are maximally similar in the two types of categorical variables if both are observed in the same state. The difference between categorical nominal and ordinal variables lies in how to interpret situations in which the two sampling units are observed in different states. In the case of nominal variables, all such differences are equally "different". In categorical ordinal variables, we can also qualitatively, but not quantitatively, order them in some natural way. Because of this, we can also qualitatively, but not quantitatively, order the degree of difference between pairs of sampling units when these are observed in different states. Consider again the example of the categorical ordinal variable that I gave earlier: the degree to which an animal species is typically found close to humans. We might apply the following classification: never, rarely, occasionally, often or always. Certainly, two species classified as "never" would be counted as "similar". What if species A was classified as "never", species B was classified as "rarely" while species C was classified as "always"?. Although species A and B are not in the same state, their degree of difference (never vs. rarely) is less that between species A and C (never vs. always). Because we can qualitatively rank a categorical ordinal variable, we want to take this qualitative ranking into account when we calculate degrees of difference.

The most complete treatment of calculated the degree of difference in ordinal variables is given in Podani (1999), who differentiates between the ways in which ordinal variables might arise. One way is when we measure a variable on a continuous scale (i.e. an interval or ratio scale variable) but then replace the continuous variable by ranks. For instance, we might simply estimate heights of trees by some imprecise method but then record the heights as "small", "medium" or "tall". In such cases, each sampling unit will have a different rank except for occasional ties resulting from rounding to significant digits. This is not a common procedure. It is almost always better to maintain the interval or ratio scale since the ranking of such variables results in a loss of information. More commonly, a ranking arises from a classification based on somewhat subjective assessments. The scoring of animal species on a five-point scale relative to how often they are found in association with humans is such an example. We would do this when we are confident that we can distinguish between the five levels of our scale but we have not actually taken quantitative measurements.

The most basic measure of similarity for categorical ordinal variables is to assign integer ranks to the categories. We could assign a rank of 1 to the "never" class, a rank of 2 to the "rarely" class and so on such that the ranking of the descriptors (never < rarely < occasionally < often < always) follows the ranking of the integers ($1 < 2 < 3 < 4 < 5$). The integer rank of the kth categorical ordinal variable for sampling unit i is indicated by $r_k(i)$. Once the categories have been converted to numerical ranks then the degree of similarity (called a metric similarity) or difference between the ith and j[th] sampling units is given in Equation 4.2.

$$d_k(i,j)=\frac{|r_k(i)-r_k(j)|}{max(r_k)-min(r_k)} \tag{4.2}$$
$$s_k(i,j)=1-d_k(i,j)$$

Since $r_k(i)$ is simply a rank, and not a number with measurement units, it is important to assign sequential integers. The denominator serves to standardize the similarity between 0 and 1 so that the degree of similarity is not affected by the number of different classes that we have chosen to use. Some variation exists in Equation 4.2; some authors standardize using the largest and smallest ranks observed in the data set rather than the largest rank that is defined.

4.5) Measures for quantitative variables

This next section applies to interval or ratio scale variables. Such variables always have an associated measurement unit and can be either discrete (counts) or continuous. As with binary variables, many different ways of measuring distance have been proposed and these differences reflect different biological assumptions concerning degrees of "difference".

Imagine that we have counted the number of eggs laid (Y) and the number fledglings produced (X) by four females of a particular species of bird, as given in Table 4.1.

Female	Fledglings produced (X)	Eggs laid (Y)
A	2	2
B	3	3
C	2	3
D	1	1

Table 4.1. number of eggs laid (X) and the number fledglings produced (Y) by four females of a particular species of bird.

Figure 4.2 plots these values. How different are each of these four females based on both of these two reproductive attributes? That depends on how we choose to define and measure the concept of "different". Here is one way to define the concept "different". Females A and C each produced two fledglings that successfully left the nest, and so they don't differ in this respect, but female C produced one extra egg than female A. It seems intuitive to say that females A and C differ by one unit in total.

Similarly, females B and C both produced three eggs, and so they don't differ in this respect, but female C only produced two fledglings. It also seems intuitive to say that females B and C differ by one unit. Comparing females A and B, we see that female B produced one extra egg and also one extra fledgling, and so it seems intuitive to say that females A and B differ by two units. If we express our intuition about the meaning of the amount of "difference", d_{ij}, between any two females (i and j) in the form of an equation, we would write: $d_{ij} = |x_i - x_j| + |y_i - y_j|$. Using this equation we obtain d_{AC} =|2-2|+|2-3|=1, d_{BC} =|3-2|+|3-3|=1 and d_{AB} =|2-3|+|2-3|=2. This is an example of a "Manhattan" (or "city block") distance. It is shown by the green arrows in Figure 4.2 and its general formula is given in Equation 4.3, where x_{ik} is the value of the kth variable attribute measured on sampling unit i and the summation is over the V different variable attributes measured on the two sampling units (i,j).

$$d_{ij} = \sum_{k=1}^{V} |x_{ik} - x_{jk}| \tag{4.3}$$

We then collect these values together into a distance matrix as given in Table 4.2. Here is how to obtain these values using the `vegdist()` function of the vegan package of **R**:

```
vegdist(test.data,method="manhattan",diag=T)
```

	A	B	C	D

A	0	2	1	2
B	2	0	1	4
C	1	1	0	3
D	2	4	3	0

Table 4.2. Manhattan or city block distances between female birds A, B, C and D of Table 4.1.

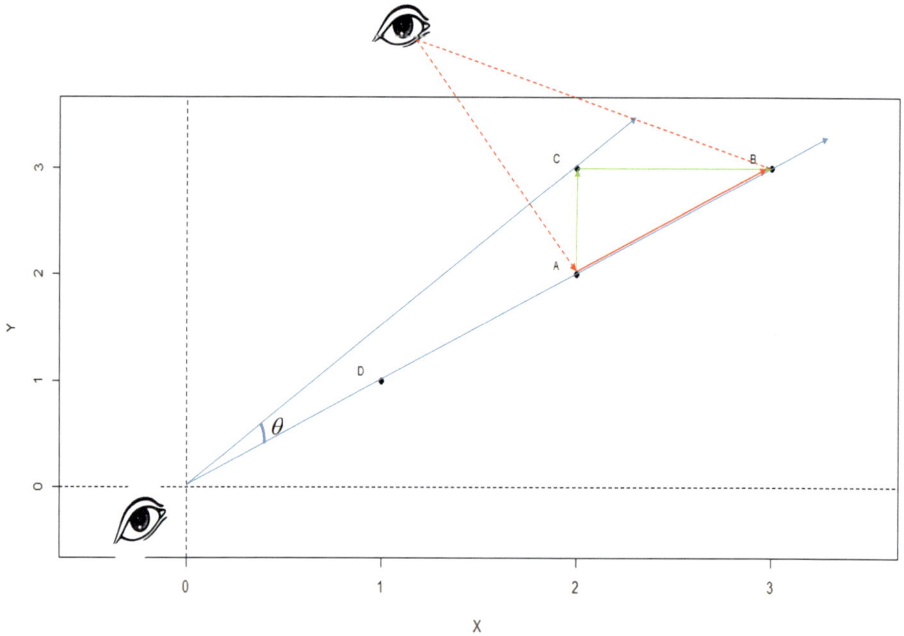

Figure 4.2. A metric space defined by two variables, fledglings produces (X) and eggs laid (Y), and three different distance measures defined on this space. The four points (A, B, C, D) are the values of four female birds. The Manhattan distance between points A and B are the sum of the green lines, the Euclidean distance between points A and B is the red line and the Chord distances between point C and points D, A and B are measured as angles.

Here is another way of defining the concept of "different". Imagine that we are the eyeball at the top of Figure 4.2 such that we are looking down on the two-dimensional Figure 4.2 from a third-dimensional vantage point, as shown by the dotted red lines. We could define the amount of difference between our female birds as the physical distance between each of the points on Figure 4.2. We already know, from Chapter 2, that this distance between points is given by the Euclidean formula (Equation 4.4). For instance, the physical distance between points A and B (the red arrow) is the hypotenuse of a triangle.

$$d_{ij} = \sum_{k=1}^{V} \sqrt{\left(x_{ik} - x_{jk}\right)^2} \tag{4.4}$$

We then collect these values together into a distance matrix as given in Table 4.3. Notice that several of these distances are different that the Manhattan distances. Here is how to obtain these values using the `vegdist()` function of the vegan package of **R**:

```
vegdist(test.data,method="euclidean",diag=T)
```

	A	B	C	D
A	0	1.41	1.00	1.41
B	1.41	0	1.00	2.83
C	1.00	1.00	0	2.24

D	1.41	2.83	2.24	0

Table 4.3. Euclidean distances between female birds A, B, C and D of Table 4.1 rounded to two digits.

Using either the Manhattan or Euclidean distances, every individual female is different from every other female and the largest difference is between individuals B and D. Here is a third way of defining the concept of "difference": Imagine that we are the eyeball that is sitting inside the two-dimensional Figure 4.2 at the origin. Because we are not looking from outside of the figure in the third dimension, the only way of measuring the distance between the points is by the number of degrees that we have to shift our angle of view in order to look at one point or another, as shown by the solid blue lines. This is analogous to us looking at the stars in the night sky; we cannot actually distinguish between the distances of these stars, only the differences in their angular position in the sky. This type of distance measure is called the "chord" distance in ecology; the angular distance in astronomy differs from the chord distance only by a constant. The chord distance is only concerned with proportional differences, not absolute differences. Consider the points D, A and B in Figure 4.1, whose (X, Y) coordinates are (1,1), (2,2) and (3,3). These coordinates can equivalently be written as (1,1), 2(1,1) and 3(1,1), showing that they differ in absolute value but not in relative value since the ratio of X to Y is always 1:1. The Euclidian and Manhattan distances between these three points are not zero but the chord distances are zero. The chord distances are zero because the angle (θ) between them is zero, meaning that you would not have to change your angle of sight from the origin in order to "see" each of these three points in Figure 4.2. On the other hand, you do have to change the angle of sight from the origin to look at point C, located at coordinates (2,3). The form of the equation for the chord distance that explicitly refers to the angle between a pair of point is given in the first version of Equation 4.5. The second version uses the trigonomic definition of a cosine. The third version makes clear its relationship to the Euclidean distance. In fact, it *is* a Euclidian distance but based on transformed values of the original data; specifically, each value (x_{ik}) of the original RxC data set is divided by the square root of the sum of squares ($\sum_{k=1}^{V} x_{ik}^2$) of that row. The effect of this particular transformation is to express each value in relative terms.

$$d_{ij} = \sqrt{2\left(1 - \cos\Theta_{ij}\right)}$$

$$d_{ij} = \sqrt{2\left(1 - \frac{\sum_{k=1}^{V} x_{ik} x_{jk}}{\sqrt{\sum_{k=i}^{V} x_{ik}^2 \sum_{k=1}^{V} x_{jk}^2}}\right)} = \sqrt{\sum_{k=1}^{V}\left(\frac{x_{ik}}{\sqrt{\sum_{k=1}^{V} x_{ik}^2}} - \frac{x_{jk}}{\sqrt{\sum_{k=1}^{V} x_{jk}^2}}\right)^2} \quad (4.5)$$

Here is how to obtain these values using the `vegdist()` function of the vegan package of **R**. First, you must transform the original RxC data matrix (`test.data`) using another function called `decostand()` that is also in the vegan package of **R**. This transformation is called "norm" or "normalize" in `decostand()`. You then calculate the Euclidean distance based on this transformed matrix. The resulting values are shown in Table 4.4.

```
vegdist(decostand(test.data, "norm"),
method="euclidean",diag=T)
```

	A	B	C	D
A	0	0	0.12	0
B	0	0	0.12	0
C	0.12	0.12	0	0.12
D	0	0	0.12	0

Table 4.4. Chord distances between female birds A, B, C and D of Table 4.1, rounded to two digits.

We will look at four other measures for interval or ratio variables that are often used in ecology. Three of these are Euclidian distances calculated on transformations of the original RxC data matrix while the fourth is not. The first one (Equation 4.6) is called the distance between "species profiles" since it is usually applied to siteXspecies data. As can be seen, this distance is simply the Euclidian distance calculated on the values in each row expressed as proportions (or relative frequencies) of the total value of that row. Since each row in such data refers to a site, this means that the abundances of each species present in a site is expressed as a proportion of the total abundance of all species (i.e. relative abundances) present in that site. The second one, (Equation 4.7) called a "Hellinger" distance, is actually the chord distance based on square-root transformed relative abundances and is important when analyzing siteXspecies data like those in Chapter 3. The third one is called the "Chi-Squared" distance (Equation 4.8). The last one (Equation 4.9) is often called the Bray-Curtis distance, although the correct name based on scientific priority is the Steinhaus dissimilarity coefficient (Legendre and Legendre 2012, p. 285). If the variables (x_{ij} and x_{ik}) are binary (presence/absence) then Equation 4.9 is the same as the Sorensen coefficient.

$$d_{ij} = \sqrt{\sum_{k=1}^{V}\left(\frac{x_{ik}}{\sum_{k=1}^{V} x_{ik}} - \frac{x_{jk}}{\sum_{k=1}^{V} x_{jk}}\right)^2} \tag{4.6}$$

$$d_{ij} = \sqrt{\sum_{k=1}^{V}\left(\sqrt{\frac{x_{ik}}{\sum_{k=1}^{V} x_{ik}}} - \sqrt{\frac{x_{jk}}{\sum_{k=1}^{V} x_{jk}}}\right)^2} \tag{4.7}$$

$$d_{ij} = \sqrt{\sum_{j=1}^{R}\sum_{i=1}^{R} x_{ij}}\sqrt{\sum_{k=1}^{V}\frac{1}{\sum_{i=1}^{R} x_{ik}}\left(\frac{x_{ik}}{\sum_{k=1}^{V} x_{ik}} - \frac{x_{jk}}{\sum_{k=1}^{V} x_{jk}}\right)^2} \tag{4.8}$$

$$d_{ij} = 1 - \frac{\sum_{k=1}^{V} min(x_{ik}, x_{jk})}{\sum_{k=1}^{V} x_{ik} + \sum_{k=1}^{V} x_{jk}} \quad (4.9)$$

Equation 4.9 can be obtained directly using `vegdist(…, method="bray")`. Here is how to obtain the values of Equations 4.6 to 4.8 using the `vegdist()` function of the vegan package of **R**. First, you must transform the original RxC data matrix (`test.data`) using another function called `decostand()` that is also in the vegan package of **R**. The transformation to choose differs for each case. You then calculate the Euclidean distance based on this transformed matrix.

```
#species profiles difference
vegdist(decostand(test.data, "total"),
method="euclidean",diag=T)
#Hellinger distance
vegdist(decostand(test.data, "hellinger"),
method="euclidean",diag=T)
#Chi-squared distance
vegdist(decostand(test.data, "chi.square"),
method="euclidean",diag=T)
```

4.6) Gower's distance

There are more distance/dissimilarity measures besides those that we have looked at so far. Each group of distance measure is specific to a particular type of variable (categorical nominal, categorical ordinal, interval and ratio) but we sometimes measure more than one type of variable for each sampling unit. How can we combine these different types of distance measures to give an overall measure of difference between sampling units? We do this using the Gower distance coefficient (Gower 1971). In what follows, I will present the version of this coefficient proposed by Legendre and Legendre (2012).

The general formula for the Gower similarity coefficient (s_{ij}) between sampling units i and j is given in Equation 4.10; the associated distance coefficient is 1-s_{ij} and varies between 0 and 1. The general idca is to first group together those variables that are of the same type. We then calculate the similarity (s_{ijk}) between the two sampling units separately for each of the k variables that are of the same type (binary, categorical nominal, categorical ordinal, interval or ratio variables) and using a similarity coefficient that is appropriate for that type of variable. Finally, we calculate the average similarity over the k variables. The w_{ijk} in Equation 4.10 are weights assigned to each variable (k) for that pair (i, j) of sampling units; we will look at these weights later. If each variable is assigned the same weight, w_{ijk}=1, then the Gower similarity coefficient reduces the Equation 4.11 where V is the total number of variables.

$$s_{ij} = \frac{\sum_{k=1}^{V} w_{ijk} s_{ijk}}{\sum_{k=1}^{V} w_{ijk}} \tag{4.10}$$

$$s_{ij} = \frac{1}{V} \sum_{k=1}^{V} s_{ijk} \tag{4.11}$$

If the variable, k, is binary then $s_{ijk}=1$ if it is the same for both sampling units (0/0 or 1/1) and $s_{ijk}=0$ otherwise. Alternatively, if you do not want to count double zeroes, then only the case of 1/1 is counted as "similar" with $s_{ijk}=1$. If the variable, k, is categorical nominal, then $s_{ijk}=1$ if the value is the same for both sampling units and $s_{ijk}=0$ otherwise. If the variable, k, is quantitative (interval or ratio variables) then $s_{ijk}=1-|x_{ik}-x_{jk}|/D_k$ where D_k is the largest difference between any two sampling units with respect to variable k. If the variable, k, is categorical ordinal, then different options can be used. The original suggestion of Gower (1971) was to treat the ranked values as if they were quantitative, as described above. Alternatively, you can use Equation 4.2 as proposed by Podani (1999).

An example will make these points easier to understand. Imagine that we have six variables measured over different species. Each species is a sampling unit. Table 4.5 gives the values of these six variables for one particular pair of species (A and B). Variable 1 (Annual) is a binary variable indicating if (yes/no) the species has an annual lifestyle. Variable 2 (vegetative reproduction) is also a binary variable indicating if (yes/no) the species can reproduce vegetatively. Variable 3 (rarity) is a categorical ordinal variable that classifies the species into five ordered levels (very rare, rare, average, common or very common). Variable 4 (height, cm) is a quantitative (ratio) variable that varies between 5 and 100 cm in the data set. Variables 5 and 6 are categorical nominal variables; variable 5 classifies the flower colour (red, blue, yellow) while variable 6 classifies the seed dispersal vector (mammals, birds, wind, water, none). The similarity and distance coefficients between these two species is shown in the last two rows of Table 4.5.

	Annual (1)	Vegetative reproduction (2)	Rarity (3)	Height (cm, 4)	Flower colour (5)	Seed dispersal vector (6)
Species A	Yes	No	Rare	22.3	Blue	Mammals
Species B	Yes	yes	Average	18.6	Blue	Birds
S_{ABk}	1	0	|2-3|/4=0.25	|22.3-18.6|/(100-5)=0.039	1	0
d_{ABk}	1-1=0	1-0=1	1-0.25=0.75	1-0.039=0.961	1-1=0	1-0=1

Table 4.5. A subset of a larger data matrix consisting of six variable attributes measured on two species (A and B). Note that these variables are a mixture of binary, categorical nominal, categorical ordinal and quantitative variables. The last two lines give the similarity and distance coefficients for each variable.

If we have assigned the same weight to each of these variables (i.e. Equation 4.10) then the Gower distance between species A and B is given in Equation 4.12. Doing the same thing for each unique pair of species in the data set would produce an RxR matrix of distances.

$$s_{ij} = \frac{1}{V}\sum_{k=1}^{V} s_{ijk} = \frac{1}{6}\left(0+1+0.75+0.961+0+1\right) = 0.6185 \qquad (4.12)$$

Gower (1971) originally used weights to take into account missing values. If one or both of the two sampling units (i, j) is missing a value for variable k then w_{ijk}=0. If both sampling units had a record for the value then w_{ijk}=1. This is a very useful property of the Gower coefficient. However, it is actually possible to assign any weight between zero and unity to a variable. This is useful when there are a series of variables that are all measuring something similar. For instance, imagine that we had replaced the binary variable "vegetative reproduction" with a series of more detailed binary variables related to different types of vegetative reproduction: stolons (yes/no), rhizomes (yes/no), bulbs (yes/no) corms (yes/no), tubers (yes/no). These are all types of vegetative reproduction in plants and it is possible for the same species to exhibit more than one type. Doing so would increase the relative importance of "vegetative reproduction" relative to the other variables. By assigning a weight of 1/6 to each of these variables, the total weight of "vegetative reproduction" would still be 1.0.

gowdis()

The most complete **R** function for calculating the Gower distance is given in the `gowdist()` function of the FD package (Laliberté and Shipley 2009). To use this function, you must have defined binary and categorical nominal variables as unordered factors, categorical ordinal variables as ordered factors and interval or ratio variables as numerical variables in **R**. The `gowdis()` function has the following arguments: `x` is the RxC data frame containing the original data, `w` is a vector giving the weights assigned to each variable (column in `x`), `asym.bin` is a vector giving the names of the variables that are to be treated as asymmetric binary variables (i.e. binary variables in which 0/0 is not counted as a similarity), and `ord` is a logical argument that chooses how to deal with ordinal categorical variables. By default, all variables have a weight of 1.

```
gowdis(x, w, asym.bin = NULL, ord = c("podani",
"metric", "classic"))
```

A recent example of the use of Gower's distance is the study by Escobar et al. (2020). Those authors studied traits related to the ability of 82 different Brazilian plants to disperse their seeds and for those seeds to germinate. Although several of these traits are the type of continuous numerical traits that could be used in a PCA[32] (seed mass, the percentage of germination, the mean germination time and its standard deviation), there were also three traits that were unordered categorical variables. Each species was classified by its seed dispersal syndrome: (i) dispores dispersed through the guts of animals (endozoochory)and having seeds in fleshy fruits, (ii) dispores dispersed by being attached to animals (exozoochory) and having seeds or fruits containing hooks or sticky substances, (iii) disapores dispersed by the wind (anemochory) and being dust-like, winged

[32] After transformation to insure linearity between them

or flat, and (iv) species without specialized diaspores for dispersal by animals or wind (autochory). Each species was also classified by the season in which its diaspores were dispersed (dry season, rainy season, dry-to-rainy transition or rainy-to-dry transition). Finally, each species was classified according to one of five dormancy types. Each species was also classified in terms of two other properties not strictly related to regeneration: its growth form and its dominant habitat type (open or closed savannas). Another important aspect of this data set was that no seeds germinated in four species. As a result, no values of germination percentage, mean germination time or its standard deviation could be recorded for these four species. All of these complications can be accommodated using the Gower coefficient.

The first step is to verify that each of the variables are properly defined in **R** as either numeric, unordered factors or ordered factors because `gowdis()` uses these variable definitions to determine how to calculate similarity, as explained above. Unordered factors are treated an categorical nominal variables, ordered factors are treated as ordinal variables, numerical variables coded as 0 or 1 are treated as binary variables and numerical variables having more than 0/1 coding are treated as interval or ratio quantitative variables. Although **R** assigns these types by default it is always a good idea to explicitly define them before calling the `gowdis()` function. Categorical nominal variables are declared by using the `factor(…,ordered=F)` or `as.factor(,ordered=F)` functions. The `ordered=F` argument is the default value and so factors are treated as categorical nominal variables by default. If you want a variable to be treated as ordinal then you must explicitly define the order using the `ordered=T` argument possibly along with the `levels=` argument. I said "possibly" because the increasing order of the ordinal rank is determined by the levels= argument and, by default, **R** simply ranks these levels in increasing alphanumeric order. As an example, if you think that the types of dispersal syndromes used by Escobar et al. (2020) represent increasing ability for long-distance dispersal (this was not the case in that paper), with the order being autochory < anemochory < exozoochory < endozoochory, then you would define this variable by typing:

```
factor(escobar$Dispersal_syndrome,levels=c("autochory
","anemochory", "exozoochory","endozoochory"),ordered
=T)
```

4.7) Multidimensional scaling

The goal of both PCoA (classical multidimensional scaling) and other forms of multidimensional scaling, is to find an ordination of observational units whose Euclidean distances in only a few dimensions best reproduces the degree of dissimilarity between these same observational units when using all of the variable descriptors and based on any chosen measure of dis similarity. If this sounds suspiciously similar to the goals of PCA (Chapter 2) and CA (Chapter 3) then it should! In fact,

PCA and CA ordinations of observational units are special cases of the ordination resulting from PCoA. If you conduct a PCoA after calculating the Euclidean distances between each unique pair of observational units involving only interval or ratio scale variables then the scores will be identical to those obtained from a PCA. Similarly, if you conduct a PCoA after calculating the Chi-Squared distances between each unique pair of observational units involving only interval or ratio scale variables then the scores will be identical to those obtained from a CA. Then why not simply forget about PCA and CA and always use PCoA while choosing distance measures that are appropriate for each? There are several reasons. First, the effect of replacing the original data by pairwise measures of distance, as done in PCoA and MDS, erases the information on the variable attributes of the sample units, although you can reclaim this information with a bit of extra work. This contrasts with PCA, for which the explicit link between observational units and variable attributes is maintained. Second, the reciprocal averaging algorithm of CA makes the link between the ordination diagram and our underlying understanding of niche relationships much clearer than viewing CA as simply a mathematical exercise in extracting "eigenvectors" from a matrix of Chi-Squared distances between sampling units.

However, the advantage of PCoA and MDS is that you are not restricted to Euclidean or Chi-Squared distances. You are free to choose between any of a large number of different distance/dissimilarity measures that best correspond to the meaning of "difference" or "similarity" in your specific biological problem. The underlying mathematics of PCoA and multidimensional scaling are quite different but the general goal is the same. The name "multidimensional scaling" is a general term that includes several different forms, which we will study latter in this chapter. Both methods start with an RxR distance/dissimilarity matrix containing the distance between each pair of sampling units in the data set. The elements (d_{ij}) of this RxR matrix measure the degree of difference between rows i and j of the original RxC data matrix. The diagonal elements (d_{ii}) of this matrix measure the difference between a row and itself and so these values will always be zero. The actual values of the off-diagonal elements will depend on the distance measure that we have chosen. Both methods produce a new RxA matrix in which each of the R rows refers to a sampling unit and each of A columns refers to an ordination axis. The values in this new matrix are the scores of each sampling unit on each of the new axes. In PCoA there are always R columns and so the resulting matrix is still RxR, although you would normally only consider the first few axes. You decide how many axes (A) will be in MDS and so A is typically much less than R. When we place the scores of the axes on a graph, two sampling units that have identical values of their variable attributes, i.e. that have zero differences, are superimposed on each other and so they have zero distance between them on the graph. The greater the degree of difference between the points on the graph, the greater the distance between them the corresponding sampling units.

	V1	V2	V3	V4
R1	1.01	0.70	0.40	0.87
R2	1.91	2.15	1.82	2.55
R3	2.31	3.19	2.19	3.38

Table 4.6. The values of four different variable attributes (V1 to V4) measured on each of four sampling units (R1 to R4).

For now, we will use the Euclidean distance while remembering that the Euclidean distance is not always the most appropriate one to use. You will recall from Chapter 2 that one goal of principal component analysis was to display a multivariate pattern of correlations between variables in such a way as to best represent the original Euclidean distances between the points (i.e. the sampling units) in a small number of dimensions. PCoA and MDS have the same goal. Table 4.6 shows the values of four variables (V1 to V4) measured on each of three sampling units (R1 to R3). Since Table 4.6 has four variables, we cannot perfectly plot these three sampling units (i.e. rows of Table 4.6) in two dimensions. Since these four variables are all quantitative variables, we could to do this via PCA and then plot the PCA scores on the first two PCA axes. However, we can also do this by first converting our RxC data set (Table 4.6) of sampling units and variables into an RxR matrix of pairwise Euclidian differences (Table 4.7).

	R1	R2	R3
R1	0	2.78	4.17
R2	2.78	0	1.44
R3	4.17	1.44	0

Table 4.7. A table of Euclidean distances between each pair of sampling units given in Table 4.6, based on all four variables.

If we do this for the three unique pairs of sampling units then we obtain the matrix of Euclidean distances in Table 4.7, rounded to two digits. The two sampling units of Table 4.7 that are most similar to each other in terms of the four variable attributes, or equivalently are the least distant from one another, are those in rows 2 and 3 ($d_{23}=d_{32}=1.44$). The two rows that are most different are rows 1 and 3 since $d_{13}=d_{31}=4.17$.

4.8) Principal coordinates analysis

Here is an intuitive and graphical explanation of a principal coordinates analysis. We start by placing one of the sampling units (i.e. rows) at the origin of a line; actually we can place it anywhere along the line but this makes things simpler. Figure 4.3 places R1 at the origin. We can always place a second sampling unit on a line at a point equal to the Euclidean distance between them; we will choose R3 and place it at 4.17 units along this line since $d_{13}=4.17$. If the three sampling units in Table 4.6

existed along a single dimension then we could place R2 on this same line. If this was true then $d_{12}=d_{13}-d_{23}$. However, this isn't the case since d_{12} is 2.78 which is slightly larger than $d_{13}-d_{23}=4.17-1.44=2.73$. Therefore, in order to plot R2 in such a way as to agree with the measured distances, we need to add a second dimension (i.e. a second axis). How must we place R2 on this second axis such that it agrees perfectly with the observed distances? In our simple example, we can obtain this position using basic trigonometric formulas since these three points define a triangle. We know the lengths of the three sides of this triangle (i.e. the three Euclidean distances) and this allows us to obtain the angle between the first axis (the line between R1 and R3) and the line between R1 and R2, giving us the coordinate of R3 on this second axis. Solving this trigonometric problem tells us that R2 must be placed at a value of ~2.76 on the first axis and ~0.30 on the second axis. The coordinates (or scores) of the points R1, R2 and R3 along the first axis are (0, 2.76, 4.17) and are (0, 0, 0.30) along the second axis.

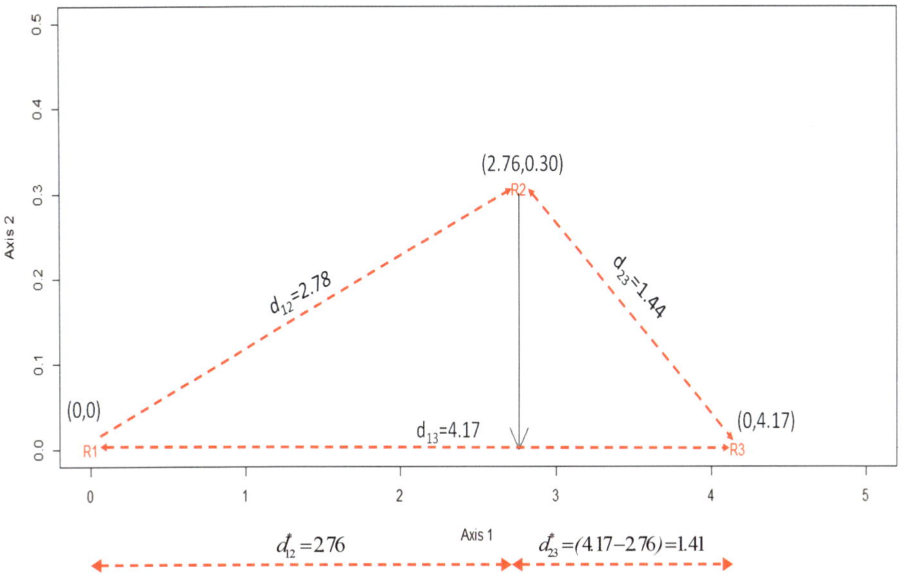

Figure 4.3. Placing the three sampling units, given in the rows of Table 4.X, in a two dimensional space so that the Euclidean distances between the points in this space agree with the measured distances in Table 4.7.

Notice that we had three rows in our data and that we could perfectly represent the Euclidean distances between these three rows using a triangle, which is a two-dimensional object. We can always produce a graph like Figure 4.3 that perfectly preserves the original distances between the sampling units using a total of R-1 axes (i.e. dimensions). Three rows produces a 2-dimensional object (a planar triangle). Four rows produces a 3-dimensional object. R rows produces an R-1 dimensional object. However, if there exist correlations between the original variable attributes from which the distance matrix was calculated then we can produce a graph that closely, but not perfectly, accounts for the measured distances using fewer than R-1 dimensions. The stronger the multivariate pattern of correlations between the original variable attributes, the fewer dimensions that we need to *approximately* recreate the original distances. For instance, the three sampling units in Figure 4.3can *almost* be placed in one dimension. If we ignore the second dimension, and so project point R2 onto the first axis, then it would be at the coordinates of (2.76, 0) instead of at (2.76, 0.30). By ignoring the second dimension, the distance between R1 and R2 on this axis would be 2.76 even though the true distance is 2.78 (an error of 0.02 units) while the distance between R2 and R3 would be 4.17-2.76=1.41 even though the true distance is 1.44, giving an error of 0.03. The total distance between these three points, when using both dimensions, is 4.17+2.78+1.44=8.39 units. The total distance between these three points, if we ignore the second dimension, is 4.17+2.76+1.41=8.34, giving an error of 100(8.39-8.34)/8.39=0.6%.

This intuitive explanation gives you an idea of how PCoA produces principal coordinates given a matrix of pairwise distances. I was able to solve this problem using simple trigonometric formulas because I only had three points that defined a 2-dimensional triangle. This strategy doesn't work with more than three points (sampling units). PCoA actually uses linear algebra and an eigen analysis just as like PCA but the logic is the same as I used in this simple example. We could approximately describe PCoA as a PCA conducted on a transformation of the distance matrix obtained from the original data set. In fact, the scores ("principal coordinates") obtained from a PCoA using Euclidean distances are identical to the scores obtained from a PCA on the covariance matrix of the original data (Chapter 2) and the scores from a PCoA using Chi-Squared distances are identical to the scores from a Correspondence Analysis (Chapter 3). If so, then why use PCoA? The answer is that PCA and CA require data in which all the variables are quantitative and based on a specific distance measure (Euclidean or Chi-Squared). PCoA, being based on a distance matrix, can be done using any distance measure and we are free to choose the distance measure that is appropriate for our variables.

The actual steps in a PCoA are as follows.

1. Choose an appropriate distance measure given the nature of the variables in the data set and the nature of the biological phenomenon.
2. Calculate the RxR distance matrix (**D**) between each pair of rows in the data set.
3. Transform matrix **D** to matrix **A**, where $a_{ij} = -0.5d_{1j}^2$.
4. Transform matrix **A** into matrix **B**, where $b_{ij} = a_{ij} - \overline{a}_{i.} - \overline{a}_{.j} + \overline{a}_{..}$. The second, third and fourth elements of this equation are the mean values of rows i ($\overline{a}_{i.}$), row j ($\overline{a}_{.j}$) and the mean of all of the values in the **A** matrix ($\overline{a}_{..}$).

5. Perform a PCA (i.e. obtain the eigenvalues and eigenvectors) of matrix **B**.
6. Rescale the eigenvectors by the square root of the associated eigenvalues to obtain the principal coordinates.

We have already looked at steps 1 and 2. Steps 3 and 4 are simply manipulations of the distance matrix to centre it in terms of its row and column sums. Step 5 involves details of matrix algebra that we will not discuss in this book but the result is the same as what we studied in Chapter 2: i.e. rigid rotations of the original axes in order to best capture the total "variation" in decreasing importance with each additional axis. Since we are now using a distance matrix rather than a variance – covariance matrix, we will use the term "inertia" rather than "variance", but the interpretation is the same. The eigenvectors are the scores of each observational unit on the new axes. The eigenvalues are the inertia (analogous to the variance) accounted by each axis. There is one additional step between steps 5 and 6 that is sometimes required with certain distance or dissimilarity methods. This step is required when some of the eigenvalues are negative and occurs with distance coefficients that are not "metric". To understand why this is important, and why "eigenvalues" can be negative in PCoA when using measures that are not "metric", we need to take a short diversion.

In PCA, we used scree plots to decide how many principal component axes to use, based on the proportion of the total variance that is captured by each successive axis. These proportions are calculated from an eigen analysis of the covariance or correlation matrix that is obtained from the original RxC data matrix. The eigen analysis produces eigenvalues for each axis. The proportion of the total variance that is captured by each successive axis is the ratio of the eigenvalue associated with that axis as a percentage of the sum of the eigenvalues associated with all of the axes. Since PCoA is simply an eigen analysis of an RxR distance matrix, we can do the same thing. However, if some of the eigenvalues are negative, which can occur if our distance measure is not metric, then this will artificially inflate the total inertia in the data that is captured by the first few axes. This, in turn, will produce incorrect scree plots and so we have to correct for this.

In mathematics, a "space" is a collection, or set, of points. A "metric" is a function that defines a concept of "distance" between points in this space. This metric function must obey certain basic properties. Given two points (A and B), these properties are:

1. The distance between A and B is zero if, and only if, A and B are the same point.
2. If points A and B are not the same point, then the distance between them must be positive.
3. The distance from A to B is the same as the distance from B to A.
4. The direct distance from A to B must be less than, or equal to, the indirect distance from A to B via some other point C.

If these four properties seem self-evident to you then that is probably because you are used to thinking about "Euclidean" distances. If you look again at Figure 4.3, which is based on Euclidean distances, you will see that each of the four properties is respected. In particular, the direct distance between points 1 and 3 (4.17) is less than the distance between points 1 and 3 when passing through point 2; i.e. 4.17<2.78+1.44. Because of this, we can draw these three points as a triangle and we say that these points respect the "triangle inequality": given any three points (i,j,k), $d_{ij} \leq d_{ik} + d_{kj}$. The relationship is a strict equality ($d_{ij} = d_{ik} + d_{kj}$) only when the three points lie along a single line; that is, when the three points exist in only one dimension. A Euclidean distance is a metric function and there are other types of distances that are also metrics, but there are also "distance" functions that do not agree with all four of the above properties, especially points 3 and 4. Strictly speaking, a function that does not obey the above four properties should not be called a *distance* at all, but rather a *dissimilarity* function, although this distinction is often ignored in the ecological literature. In this book I often use the words "dissimilarity" and "distance" interchangeably in order to avoid having to make the distinction each time. A dissimilarity function is one that violates the third property asymmetric while one that violates the fourth property is called a semimetric.

	Species A	Species B	Species C	Species D	Species E

Site 1	1	1	1	0	0
Site 2	0	0	0	1	1
Site 3	1	1	1	1	1

Table 4.8. The presence (1) or absence (0) of five species in three sites. Example taken from Legendre and Legendre (2012).

An example of a semimetric, taken from Legendre and Legendre (2012) is Sorensen's coefficient for binary data (Equation 4.13). The so-called Bray-Curtis, or Steinhaus, dissimilarity measure is the equivalent to Sorensen's index when used with ratio or interval data and so the Bray-Curtis dissimilarity is also semimetric. Sorensen's coefficient is a version of Equation 4.1 in which agreements (presence/presence) are given a weight of 2 and cases of absence/absence is given a weight of zero.

$$d_{ij} = 1 - \frac{2n_1}{2n_1 + n_2 + n_3} \qquad (4.13)$$

Table 4.8 shows the presence (1) or absence (0) of five species in three sites. Applying Equation 4.13, the values of Sorensen's coefficient are $d_{12}=1.0$, $d_{13}=0.25$ and $d_{23}=0.43$. The triangle inequality requires that $d_{12} \le d_{13} + d_{23}$ but this is not true since 1.0>0.25+0.43=0.68. Stated equivalently, the direct distance between 1 and 2 is larger than the indirect distance when going from 1 to 3 and then from 3 to 2. It is impossible for these points to form a triangle on a graph like in Figure 4.3. Because the Sorensen dissimilarity coefficient violates the triangle inequality, it is possible that some of the smallest eigenvalues would be negative if we were to perform a PCoA on such a distance matrix. In fact, even metric distance matrices can (rarely) produce negative eigenvalues, a condition that Gower (1982) called "non-Euclideanarity". The presence of missing values can also cause the smallest eigenvalues to be negative. When this happens then it is necessary to correct for such negative eigenvalues before determining the proportion of inertia accounted by each PCoA axis.

There are three different methods to deal with negative eigenvalues in PCoA. First, several distance measures can be made metric simply by taking their square root; Table 7.2 of Legendre and Legendre (2012) identifies these measures. Two other methods to remove negative eigenvalues (Cailliez 1983, Gower and Legendre 1986) consist of adding a constant to each of the non-diagonal elements of the RxR distance matrix. However, these two methods do not produce the same additive constant and so the resulting PCoA will differ slightly. The `cmdscale()` function in the base package of **R** only implements the Cailliez (1983) method while the `wcmdscale()` function in the vegan package includes both methods.

cmdscale()
wcmdscale()

There are different **R** functions that can perform a principal coordinates analysis. The `cmdscale()` function in the base stats package of **R** provides a basic analysis. In the presence of negative eigenvalues, this function only implements the method of Cailliez (1983), using the `add="cailliez"` argument. The vegan package provides the `wcmdscale()` function. This function allows you to choose between the Cailliez (1983) and the Gower and Legendre (1986) methods of dealing with negative eigenvalues via the `add=` argument and also allows for different weights to the observational units. There is also the `pcoa()` function in the ape package that can perform a PCoA and, in the presence of negative eigenvalues, it includes an argument that allows the user to choose between the two methods for dealing with negative eigenvalues.

The `vegdist()` function in the vegan package calculates many different dissimilarity measures. The general form of this function is:

```
vegdist(x,   method="bray",   binary=FALSE,
diag=FALSE, upper=FALSE,na.rm = FALSE, ...)
```

vegdist()

The first argument (x) is an RxC matrix of C variables (columns) measured on each of R observational units (rows). In the case of siteXspecies data, each site is a row and the abundance of a given species across these sites is a column. The second argument (`method=`) specifies one of twelve different distance/dissimilarity matrices, including most of the ones discussed in this chapter. The `decostand()` function in the vegan package is often used in conjunction with `vegdist()`. This function standardizes an RxC matrix before applying `vegdist()`. For instance, a Chi-Square distance is obtained by first conducting a Chi-Square standardization and then calculating the Euclidean distance on this standardized matrix. We will look at these measures in more detail in Chapter 6. Here is the general form:

```
decostand(x, method, MARGIN, range.global,
logbase = 2, na.rm=FALSE, ...)
```

decostand()

The first argument (x) is the RxC data matrix that is to be transformed. The second argument (`method`) chooses the type of standardization. The third arguments (`MARGIN`) specifies if the standardization is to be applied to each row or each column although the various methods of standardization usually specify this automatically. Here are the combinations of `vegdist()` and `decostand()` that produce some of the most common distance/dissimilarity matrices:

Euclidian distances: `vegdist(X,method="Euclidean")`

Hellinger distances: `vegdist(decostand(X,method= "hellinger),method= "Euclidean")`

Bray-Curtis dissimilarities: `vegdist(X,method="bray")`

Chi-Square distances: `vegdist(decostand(X,method= "chi.square"),method= "Euclidean")`

4.9) (Non-metric) multidimensional scaling

In PCoA, PCA and CA we reduce the dimensionality of our full data by focusing our attention on the first few axes that account for the majority of the multivariate patterns in our data. However, we do not *force* all of these multivariate patterns into (say) two dimensions. Rather, we simply choose to ignore the remaining axes. Multidimensional scaling, in both its metric and non-metric versions, is different in that we *force* the multivariate patterns in our data to be compressed into a chosen number of dimensions (axes) as best as possible.

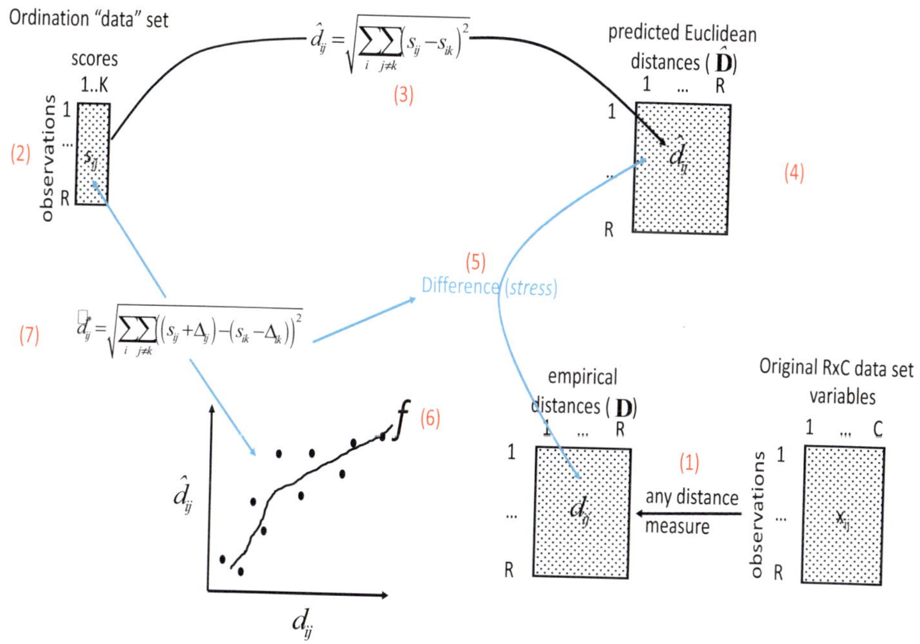

Figure 4.4. The general structure of a multivariate scaling problem: choose values of the scores (s_{ij}) on K new axes such that the resulting Euclidean distance matrix ($\hat{\mathbf{D}}$) is as close as possible (i.e. with the least possible stress) to the empirical distance matrix (**D**) derived from the original RxC data set.

The general structure of a multidimensional scaling analysis is illustrated in Figure 4.4. We start with our x_{ij} empirically measured values, collected in our original RxC data set of C variable attributes (columns) measured on R sampling units (rows). This RxC data set exists in C dimensions. From this, we (step 1) calculate our RxR distance matrix using whatever distance measure we think is most appropriate for our purpose. I will call this the "empirical" distance matrix (**D**) in which each pairwise distance is d_{ij}. As always, there are R(R-1)/2 unique elements in a distance matrix since the diagonal elements are zero (the distance of an observation to itself) and the distance between observations i and j is the same as the distance between observations j and i. This empirical distance matrix contains all of the multivariate patterns in our original data. Now, we create a new "data set" (step 2) having R rows but only K columns. We choose the value of K it is almost always less than C (the number of variables in the original RxC data set) and is usually only a small number (2 or 3). Each column in this new "data" set will contain a score (s_{ij}) on a new axis in the same way as a PCoA produces scores on new axes. We calculate the Euclidean distances (step 3) between each of the R sampling units based on these k new axes, which produces a second RxR Euclidean distance matrix (step 4). I will call this the "target" distance matrix ($\hat{\mathbf{D}}$) and I have written these distances as $\hat{d}_{ij}$ simply to remind you that these distances will not generally be identical to the empirically measured d_{ij} values; in fact, the values of $\hat{d}_{ij}$ can be very far from the empirically measured values at the start of the process. Next, we measure the difference between the empirically measured distances (d_{ij}) and the target Euclidean distances, $\hat{d}_{ij}$, obtained from our scores (step 5). This difference is measured by a "stress" function, which I will explain later. The smaller the stress (i.e. difference) between the two distance matrices, the more closely our target scores produce the same Euclidean distances as the original distances, measured by whatever distance measure you choose. The goal is to choose the scores (s_{ij}) on the new axes in such a way as to make the empirical RxR distance matrix (**D**), which exists in R dimensions, as close as possible to the target distance matrix ($\hat{\mathbf{D}}$), which exists in only K dimensions. Once we have succeeded then we will have produced the best possible representation of the empirical distance matrix in a fewer number (K) of dimensions. We do this (step 6) by choosing a function (*f*) that best relates our predicted distances ($\hat{d}_{ij}$) with our observed distances (d_{ij}). Different choices of this function result in different types of multivariate scaling.

How do we choose the scores (s_{ij}) on the new axes in such a way as to make the empirical RxR distance matrix (**D**) as close as possible to the target distance matrix ($\hat{\mathbf{D}}$)? We start the process with a "guess" about these scores, although some guesses will be better than others. When we reach step 7, we look for small changes in our MDS axis scores that make the stress value a bit smaller. Stated equivalently, we make small changes in the axis scores so that the resulting Euclidean distances between the sampling units, $\hat{d}_{ij}$, and therefore $\hat{\mathbf{D}}$, will be a bit closer to **D**. We then feed these new values of our scores (s_{ij}^*) back into step 1 to start the process over. We keep cycling through this iterative process, each time reducing the stress a bit more, until we reach a point at which any additional change to our scores will increase the stress. When we reach this point then we stop. This, in essence, is what multidimensional scaling does. The difference between "metric" and "non-metric" multidimensional scaling depends on our choice of the fitting function (*f*) at step 6. Metric multidimensional scaling requires a quantitative metric relationship between the original distance measures in **D** and the new Euclidean distances in $\hat{\mathbf{D}}$. Non-metric multidimensional scaling (NMDS) only requires a non-metric ordinal (rank order) relationship between the original distance measures in **D** and the new Euclidean distances in $\hat{\mathbf{D}}$. NMDS is more appropriate if the true distances in the empirical data are not metrics, as I defined above. A common example of this is when the triangle inequality does not hold in the original data; this is a common occurrence in speciesXsite data when sites span a wide range of underlying environmental variables, as explained in more detail in Chapter 6.

You already know how to do step 1. You also know how to go from steps 2 to 3 to 4 since this simply involves calculating the Euclidean distance. Step 5, calculating the difference between the empirical and predicted distance matrices (the "stress") is also quite straightforward. Several different stress measures have been proposed. The original measure, which we will call "stress 1", was proposed by Kruskal (1964) and is given in Equation 4.14. Another common stress formula is given in Equation 4.15 (stress 2), which I include since it is allowed in some of the **R** functions that I will describe later. Both of these stress measures vary between 0 and 1. A value of 0 occurs when the two distance matrices are exactly equal and increases as the two distance matrices become increasingly different. The only advantage of Equation 4.14 is in terms of interpretation since it is analogous to the equation for the proportion of unexplained variance in a regression context.

$$Stress1 = \sqrt{\frac{\sum_{i=2}^{R}\sum_{j=1}^{i}\left(d_{ij} - \hat{d}_{ij}\right)^2}{\sum_{i=2}^{R}\sum_{j=1}^{i} d_{ij}^2}} \tag{4.14}$$

$$Stress2 = \sqrt{\frac{\sum_{i=2}^{R}\sum_{j=1}^{i}\left(d_{ij} - \hat{d}_{ij}\right)^2}{\sum_{i=2}^{R}\sum_{j=1}^{i}\left(d_{ij} - \bar{d}\right)^2}} \tag{4.15}$$

The main difference between the different versions of multidimensional scaling arises in step 6 (de Leeuw and Mair 2009). This step consists of regressing the predicted Euclidean distances from step 4 on the empirical distances from step 1. This regression is used to improve our previous guess of the scores (s_{ij}) by making small changes (Δ_{ij}) to them. The different forms of regression reflect how we want to transform the measurement scales of the two distance measures (**D** and $\hat{\mathbf{D}}$). Remember that **D** is your chosen distance measure and is based on all C rows of the RxC data matrix while $\hat{\mathbf{D}}$ is always a Euclidean distance and based on only the K dimensions that you specified. Different types of regressions result in two general classes of multidimensional scaling, called "metric" and "non-metric" multidimensional scaling. Two of these different regression forms are illustrated in Figure 4.5.

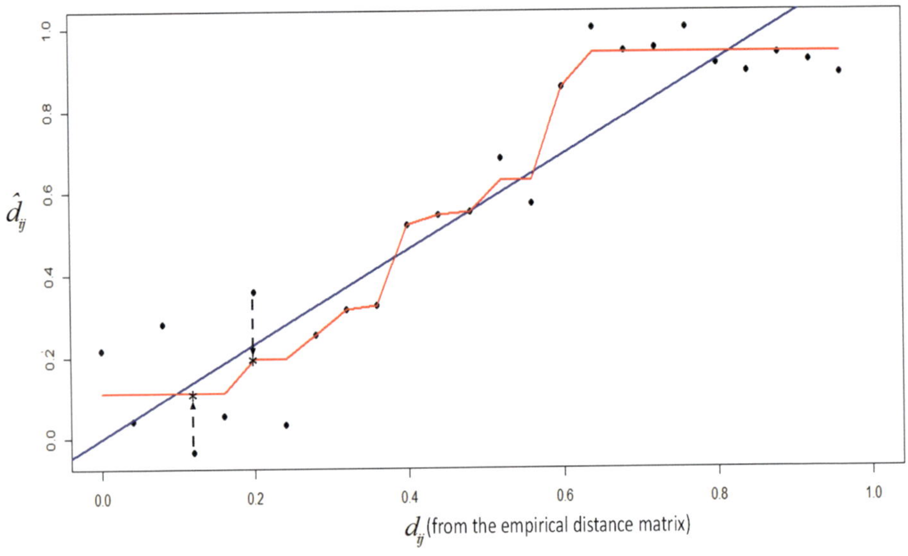

Figure 4.5. Linear regression through the origin (blue line) is used in metric multidimensional scaling with ratio scale transformation. Monotonic regression, based on ranks (red line) is used with non-metric multidimensional scaling with a rank scale transformation.

Metric multidimensional scaling (MDS) results when we assume a metric transformation between the empirical and predicted distance measures. The strictest assumption is that the two distance measures are ratio measures: a zero distance in one is the same as a zero distance in the other and increases in one are directly proportional to increases in the other. This assumption results in a linear regression of $\hat{d}_{ij}$ on d_{ij} with a zero intercept: $\hat{d}_{ij} = f(d_{ij}) = (0 + \beta d_{ij}) + \delta_{ij} = d^{*}_{ij} + \delta_{ij}$. Multidimensional scaling using this assumption is called Ratio Metric Multidimensional Scaling (Ratio MDS) and is shown by the blue line in Figure 4.5. The next strictest assumption is that the two distance measures are interval measures: increases in one are directly proportional in the other but a zero distance in one is not necessarily a zero distance in the other. This assumption results in a linear regression with a non-zero intercept: $\hat{d}_{ij} = f(d_{ij}) = (\alpha + \beta d_{ij}) + \delta_{ij} = d^{*}_{ij} + \delta_{ij}$. Multidimensional scaling using this assumption is called Interval Metric Multidimensional Scaling (Interval MDS). It is also possible to specify non-linear functions resulting in non-linear regression functions or even spline functions, but these will not be discussed in this book.

Non-metric multidimensional scaling results when we do not assume a metric transformation between the empirical and predicted distance measures. In other words, we will assume that if two empirical distance values can be ordered in increasing values (say, $d_{ij} \le d_{ik}$) then the best-fitting predicted distances must have the same ordinal rank (thus, $d^*_{ij} \le d^*_{ik}$) but we don't assume a particular numerical scaling between the two. For instance, if $d_{ij} = 0.5d_{ik}$ then d^*_{ij} can take on any value so long as it is less than, or equal to, d^*_{ik}. Stated equivalently, the rank order of the best-fitting values assigned to d^*_{ik} must agree with the rank order of the d_{ij} and so $\hat{d}_{ij} = f(d_{ij}) = (z_{ij}) + \delta_{ij} = d^*_{ij} + \delta_{ij}$ where z_{ij} is the rank order (an integer) assigned to the predicted Euclidean distance, $\hat{d}_{ij}$. The actual form of the function (*f*) in this regression is not fixed except that it has to either be increasing everywhere or decreasing everywhere, i.e. monotonic. Given this assumption, step 6 consists of a "monotonic" (sometimes called isotonic) regression (De Leeuw 2005) and is shown by the red line in Figure 4.5. A very basic **R** function to perform monotonic regression is the `isoreg()` function in the MASS package of **R**. Multidimensional scaling using a monotonic regression is called non-metric multidimensional scaling (NMDS). One additional detail to consider when using monotonic regression to produce a non-metric multidimensional scaling is how to deal with cases in which there are tied empirical distance values. One way is to require that tied values of d_{ij} require that the best-fitting values of $\hat{d}_{ij}$ also be tied. A less restrictive way is to relax this requirement.

The final step (7 in Figure 4.5) in multidimensional scaling is to use the fitted values of $\hat{d}_{ij}$ to improve our values of s_{ij} that will be fed back into step 1. If we look at the monotonic regression in Figure 4.5 then we can see by how much we must increase or decrease the predicted distances ($\hat{d}_{ij}$) in order to make them lie on the best-fitting red line; two of these points are shown as stars in Figure 4.5. Now that we have these best-fitting values, we must determine by how much we must change each of the scores (i.e. $s_{ij} + \Delta_{ij}$) so that the resulting Euclidean distances between the new scores agrees with the best-fitting values of $\hat{d}_{ij}$. This is done using a numerical method called "steepest decent" (Press et al. 1986). The mathematical details of this numerical method do not concern us but it is useful to have an intuitive understanding of how this method works. Imagine that you are blindfolded and find yourself in a landscape with several hills and valleys. Your goal is to walk to the lowest point on the landscape but you can't actually see the landscape around you. The best that you can do is to feel around with your foot and find a direction in which you are moving downwards most abruptly. You take one step in this direction and then again feel around for the next direction. Proceeding in this way, step by step, you will continue to move downhill until you reach a point on the landscape in which every direction that you try results in moving uphill. At this point you must stop because you have reached a valley. However, since you can't see anything, you have no way of knowing if you have reached the lowest possible point on the landscape or only a local valley. The best way of insuring that you have really reached the lowest point on the landscape is to (i) start your journey at a point on the landscape which you think is "close" to your final destination and (ii) repeat your voyage several times, each time starting at a somewhat different starting point. You would then choose the solution that results in the lowest point. The method of steepest decent, which is applied at step 7 in Figure 4.5, is the numerical equivalent of this blind person's exploration of the landscape. You begin (step 2) by providing initial values for the scores (s_{ij}). It is often a good idea to use the PCoA scores as starting values. You (or rather, your computer) then cycles through steps 2 to 6. The stress value (step 5) tells you how low you are on the stress landscape. The method of steepest decent then takes a small "step" on this landscape to see which directions of change (Δ_{ij}) in the scores (s_{ij}) results in the greatest decrease in stress. The scores are then changed by this amount and the process cycles again. When does the algorithm stop? Either when the decrease in stress becomes smaller than a chosen very small value, or when it has reached the maximum number of specified iterations. The first condition means that the algorithm has reached a valley bottom. The second condition means that the algorithm has given up trying.

There are several different **R** functions that can perform different types of multidimensional scaling. The most basic function is the `isoMDS()` function in the MASS package, which only does non-metric multidimensional scaling using the stress1 function and has no options for dealing with ties in

monoMDS()
metaMDS()

the input distance matrix. The `monoMDS()` function in the vegan package can perform both metric and non-metric multidimensional scaling using either of the two stress functions. The metric form is presumably an interval scale version but this is not entirely clear since the function description only calls it a "model with linear regression". The non-metric version of `monoMDS()` allows the user to choose how to deal with ties in the input distance matrix. The vegan package also has the `metaMDS()` function, which calls the `monoMDS()` function many times using different random starting scores, and chooses the one with the smallest stress. However, the most complete set of functions for multidimensional scaling is in the smacof package (de Leeuw and Mair 2009), and we will use this package.

mds()

The main function in the smacof package is `mds()`. You must always input the distance matrix via the `delta=` argument and the desired number of axes (dimensions) via the `ndim=` argument. The `type = c("ratio", "interval", "ordinal", "mspline")` argument specifies the type of multidimensional scaling that you want; the default is metric multidimensional scaling with a ratio scale and the non-metric version is chosen with the `type="ordinal"` argument. By default, the initial axis scores are obtained from the scores produced by a principal coordinates analysis on the distance matrix (this is what the `init = "torgerson"` argument means) but there are two alternatives to this default choice. You can choose random starting scores via `init = "random"` or else you can input your own starting values in the form of a matrix with R rows and the same number of columns as you specified in the `ndim=argument`. The other important argument in the `mds()` function specifies how to deal with tied distances in the input distance matrix. The `ties = "primary"` (the default) means that ties in the input distances do not require that the final scores to be equal while `ties = "secondary"` means that tied values of d_{ij} require that the best-fitting values of $\hat{d}_{ij}$ also be tied.

I will illustrate multidimensional scaling by using a simple toy example that has two advantages. First, it simulates observations from a physical object (a pyramid) so that you can easily visualize the process of projecting a three-dimensional object into only two dimensions. Second, I have chosen the "observations" so that the algorithm is particularly sensitive to changes in the starting values. Figure 4.6 (left) shows this pyramid, which has a square base whose sides are 1 unit long and whose apex is at the exact centre of the base and 1 unit high. The RxC data set (`pyramid`), consists of five observations (rows); these are the (x, y, z) coordinates of the five corners of the pyramid. Thus, the first observation (p1) is a vector (0, 0, 0), the second observation (p2) is the vector (1, 0, 0) and so on. This allows you to easily see the actual physical distances between each pair of points. I obtain the Euclidean distances between the five points (apexes) via the `vegdist()` function of the vegan package. The most basic call to `mds()` is as follows:

```
pyramid.ED<-vegdist(pyramid,method="euclidean",diag=T
)
def<-mds(pyramid.ED,ndim=2)
def

Call:
mds(delta = pyramid.ED, ndim = 2)

Model: Symmetric SMACOF
Number of objects: 5
Stress-1 value: 0.247
Number of iterations: 1
```

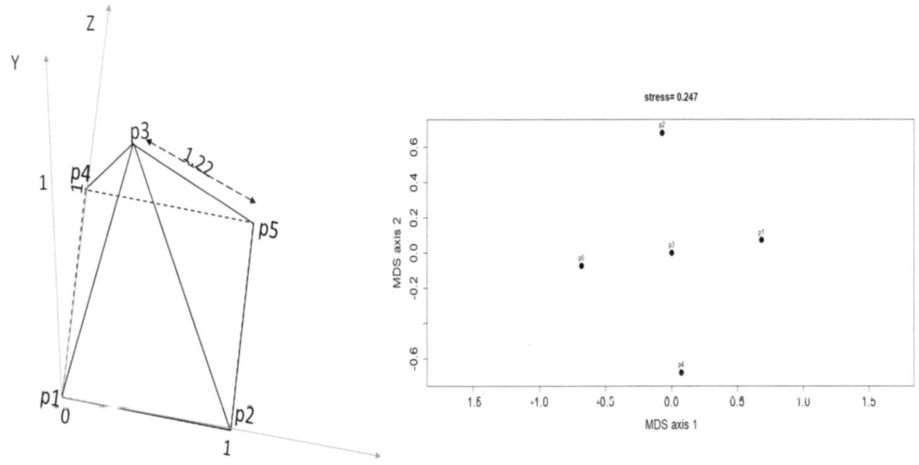

Figure 4.6. We "measure" the coordinates of five points (p1 to p5) on a pyramid (left) with a base having lengths of 1 and the apex dead centre and with a height of 1. The result of a metric multidimensional scaling of these five points using a linear (ratio) scaling is shown on the right.

Since this call to `mds()` uses all of the default options, it has performed a metric multidimensional scaling with ratio scaling. The output tells us that the stress 1 value of the final solution is 0.247 (remember that the stress varies between 0 and 1) and that the algorithm converged after only 1 iteration. We can see the scores on the two axes using the `summary()` function. The graph on the right of Figure 4.6 plots these scores. Since the distances between these points are always Euclidean distances, you could use a ruler to measure them but an easier way is to use the `vegdist(…, method=Euclidean")` function of the vegan package. The elements of Table 4.9 gives the Euclidean distances between the points in our 3D pyramid (below the diagonal) and in our 2D metric multidimensional solution (above the diagonal). The differences between the two, when input into the stress 1 formula results in the stress value of 0.247. As you can see from Figure 4.9, the result is that the apex of the triangle has been "squished" down to sit at dead centre of the base of the triangle. The largest errors in the 2D solution are with respect to the distances between point 3 (the pyramid apex) and the other points. If you apply the `summary()` function to the above mds object (`def`) then it prints out the percentage of the total stress that is caused by each observation. In this case, points 1, 2, 4 and 5 each contribute 15% to the total stress (i.e. 0.247) while point 3 contributes 40%. The distances between point 3 and the others in the two-dimensional MDS are therefore rather less well represented than are the distances between the other points.

	p1	p2	p3	p4	p5
p1		0.97	0.69	0.97	1.37
p2	1		0.69	1.37	0.97
p3	1.22	1.22		0.69	0.69
p4	1	1.41	1.22		0.97
p5	1.41	1	1.22	1	

Table 4.9. The Euclidean distances between each pair of points in the three dimensional pyramid are shown below the diagonal while the Euclidean distances in the two dimensional metric multidimensional scaling solution are shown above the diagonal.

The metric MDS solution in Figure 4.9 (right) is the one that results if you begin the iterative algorithm using the PCoA scores, which is the default option of the `init=` argument. In our toy example, this solution is neither stable nor the one that actually minimises stress. By "stable", I mean that if you use only slightly different starting values then you will not end up with the same final ordination. Using PCoA scores as starting values is often a reasonable approach (which is why it is the default option in the `mds()` function) but, as this example shows, it is certainly not guaranteed to provide the best solution. Remembering the analogy of the blindfolded person searching for the lowest valley in the landscape, the stress landscape defined by this pyramid is very rugged, with different hills leading to different valleys of varying depth. We need to more completely explore this landscape. Borg and Mair (2017) proposed such an exploration strategy which has been encoded as the `icExplore()` function of the smacof package.

icExplore()

We don't know where the lowest valley will be found in our rugged landscape. We also don't know the exact trajectory that the blindfolded algorithm will us to traverse the landscape, and both of these conditions can affect where the algorithm will stop. The first step of the strategy is to randomly sprinkle lots of blindfolded searchers onto the landscape. By default, `icExplore()` randomly generates 100 different starting configurations but this number can be changes with the `nrep=` argument. Increasing the value of this argument will better cover the entire landscape but `icExplore()` will take longer to complete the runs. The final stress value and the final configuration are saved for each of the runs and these are then available in the output if you include the argument `returnfit=T`. Here is a call demanding two axes (`ndim=2`) and a ratio metric multidimensional scaling (`type="ratio"`) using our toy example of the pyramid:

```
pyramid2D.explore<-icExplore(pyramid.ED,ndim=2,type="
ratio",ties="primary",returnfit = T)
```

The next step is to plot the results. To do this, we first eliminate meaningless differences between the different final configurations; i.e. differences that are not driven by the data. This is done using a "Procrustes analysis", which will be explained later. The distances between the remaining final configurations are then computed and the structure in these different final configurations are compared using a final metric multidimensional scaling. The resulting plot shows the position of each of the different final configurations in a two dimensional MDS space. Each final configuration is indexed by the number of the random run from which it began its trajectory. The stress value of each final configuration in this plot is indicated by the darkness and size of the index number with lower stress values having a darker font and a smaller size. The result (Figure 4.10) is a two-dimensional picture of the positions of the different valleys in the landscape. This figure allows you to quickly identify the final solutions that are identical in terms of their final configurations and that have the lowest stress. The call to produce Figure 4.10 is:

```
plot(pyramid2D.explore,main="Pyramid stress landscape"
,xlab="",ylab="")
```

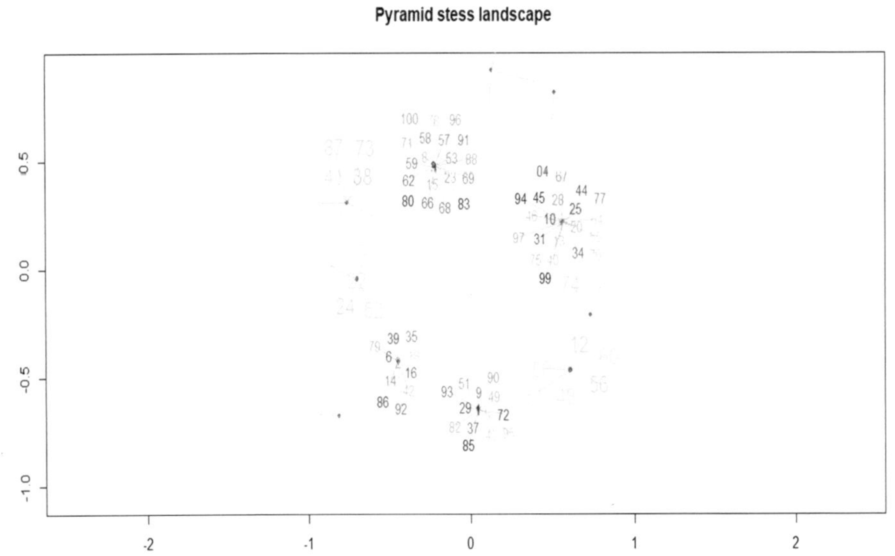

Figure 4.10. The different valleys in the stress landscape of the pyramid data based on a ratio metric multidimensional scaling.

Each of the 11 red dots in Figure 4.10 represents a different "valley" (i.e. a local minimum) in the stress landscape. Some valleys are lower than others, with the deepest valleys, representing the lowest stress values, being those with the smallest and darkest numbers. Each of the 100 numbers represents a different initial random configuration. If you change the default number of random initial configurations (`nrep=` argument) then the number of these numbers will also change. Those numbers pointing to the same red dot are the different initial random configurations that all resulted in the same final MDS solution. Since the smallest and darkest numbers are the solutions that produced the smallest stress, these are the runs whose final scores produced the closest two-dimensional metric representation of our three dimensional pyramid. Of the eleven different final solutions that were found, we can quickly see that there are four solutions that have lower stress than the others since these are the numbers having the smaller and darker fonts. Figure 4.11 shows the two metric MDS axes of four different initial runs (#3, 80, 83 and 96) that all ended up in the same valley. You will notice that they all are arranged in a circle of the same size and the ordering of the point around the circle, as well as the stress value, is identical. The fact that each circle can be perfectly mapped onto the others in Figure 4.11 by rotating it and/or by flipping it means that the solutions are said to be "topologically" identical.

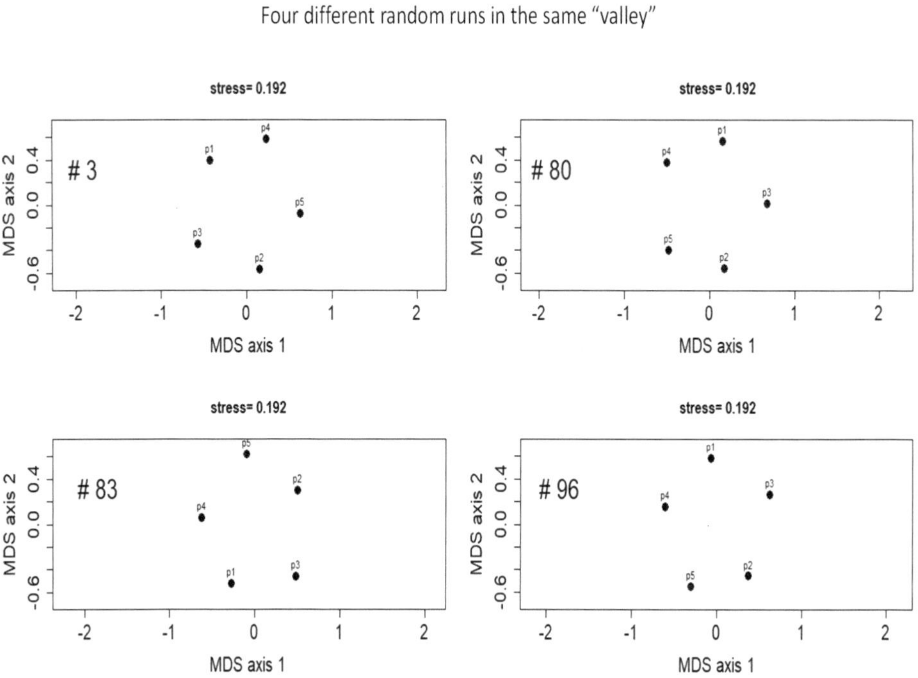

Figure 4.11. The scores of the five points on the pyramid in a two-dimensional ratio metric multidimensional scaling based on four different random starting configurations that all exist in the same "valley".

Let's now look at representative runs (#3, 10, 6, 1) that ended up in each of the four different "valleys" even though each has the same lowest stress value of 0.192 (Figure 4.12). The spatial ordering of the points in the final configurations are each topologically different although, since their stress values are equal, they are each equally good two-dimensional representations of our three-dimensional pyramid. It is perhaps surprising that the best 2D representation of a 3D pyramid is a circle but if you look at the spatial arrangement of the points in the first run (#3) you will see that we move around the circle in the order p1→p4→p5→p2→p3→p1. Looking at the positions of these points on the pyramid (Figure 4.6 left) you will see that we have traced a trajectory between adjacent points along the edges of the pyramid. The different solutions in Figure 4.11 are simply different ways of tracing such a trajectory when starting at a different corner while keeping the distances between the points as close as possible to the distances on the pyramid. It is as if our blindfolded searcher has randomly landed at one apex of the triangle and begun to search for a trajectory between the remaining apexes. Because the different possible directions that the blindfolded searcher can take are equally distant in each case, different trajectories are traced.

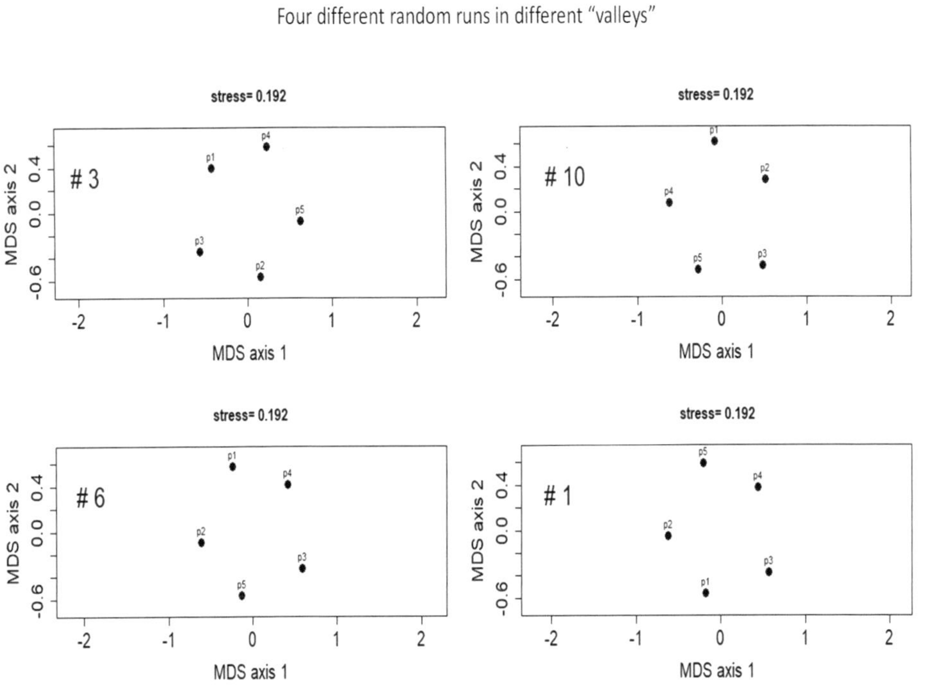

Figure 4.12 . The scores of the five points on the pyramid in a two-dimensional ratio metric multidimensional scaling based on four different random starting configurations that each end up in a different "valley".

The different solutions shown in my pyramid example are only equivalent because I purposefully chose it to illustrate such instabilities. Although real empirical data will not usually have several different solutions that give exactly equal minimum stress values, it is certainly possible to find cases in which the stress value of different final solutions are similar and in which the "best" solution can depend on just a few sampling units. It is therefore useful to know the sensitivity of the final solution to the actual data. We can explore this sensitivity by using the `jackmds()` function of the `smacof` library.

jackmds()

The `jackmds()` function implements a "jackknife" strategy based on the work by De Leeuw and Meulman (1986). The input object to the `jackmds()` function is the MDS object whose sensitivity is sought. This involves calculating "jackknifed" MDS axis scores. Jackknifed MDS axis scores are obtained by (i) removing one row (i.e. one sampling unit) from the original RxC data set, (ii) replacing the missing row by an estimate based on the remaining rows and then (iii) obtaining the row scores on the MDS axes via `mds()`. This is repeated separately for each row in the original data. Since there are R rows in our original RxC data set, we end up with R+2 different estimates of the scores of each row on each MDS axis: the R jackknifed estimates, the average axis scores estimated over all of the R jackknifed estimates plus the original estimates using the full data set. If the original MDS solution is quite stable – that is, if the effect of removing a single sampling unit doesn't change the final solution very much, – then each of the R+2 MDS solutions will end up close to each other. If not, then the different data sets will settle in quite different valleys, resulting in large differences in the axis scores. Why is this important? Your sampling units will almost always have been chosen randomly. This means that only dumb luck has determined if any particular sampling unit is (or is not) part of your data set. If removing one sampling unit results in a very different final MDS solution then you have to consider several of these different solutions.

Three different statistics are produced in the `jackmds()` function. The "stability" measures the variance between each of the jackknifed solutions relative to the total variance of the jackknifed solutions. A value of 1 (perfect stability) means that there was no variation between the jackknifed solutions: deleting one sampling unit makes no difference to the final solution. Smaller stability values indicate increasing instability. The "cross-validity" compares the predicted configuration of the axis scores for each row from the jackknifed solutions (i.e. the predicted configuration of that row when that row has been deleted from the data set) to the actual configuration of that row based on the full data set. A value of 1 (perfect cross-validity) means that the predicted solutions for every row are identical to the actual solution based on the full data set and smaller values indicate increasing instability in these predicted values. The total dispersion index combines the two previous ones and varies from 2 (perfect stability and cross-validity) to 0. Here is what happens if we use the `jackmds()` function on the metric multidimensional scaling object obtained from our pyramid data:

```
jack.pyramid<-jackmds(def)
jack.pyramid
```

```
Call: jackmds.smacofB(object = def)

SMACOF Jackknife
Number of objects: 5
Value loss function: 3.4532
Number of iterations: 3

Stability measure: 0.58
Cross validity: 0.0271
```

```
Dispersion: 1.3929
```

The output confirms what we already suspected: the effect of dropping a single row results in quite different MDS results (stability = 0.58). The axis scores of the sampling units when those sampling units (rows) have been deleted are very poorly predicted (cross-validity = 0.03). A plot of these jackknife solutions (Figure 4.13) shows the average scores of each row as labelled points and the lines radiating from each point show by how much the jackknifed MDS solutions differ from these average scores.

```
plot(jack.pyramid,col.p="black",col.l="black",main="J
ackknife plot of pyramid data")
```

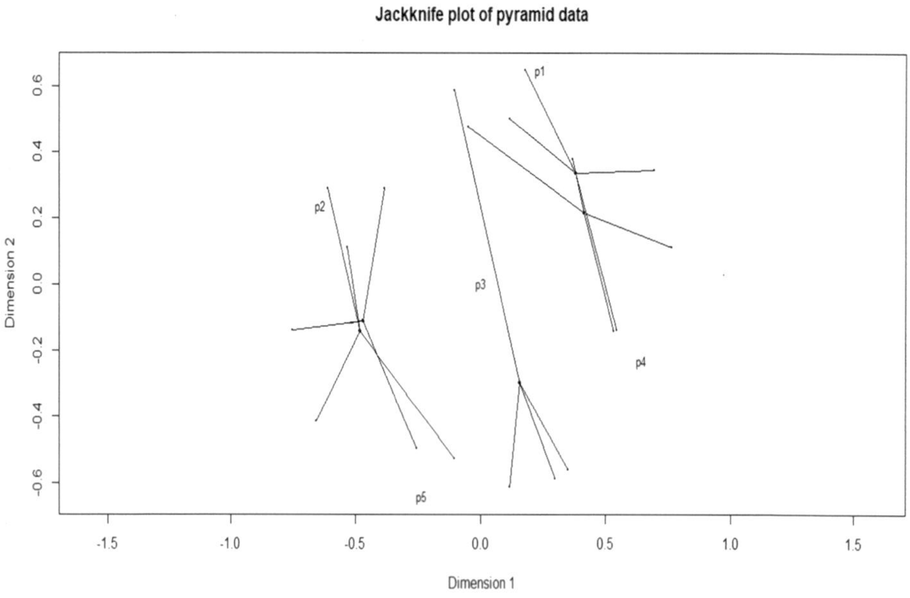

Figure 4.13. Each of the five dots show the average scores of the five sampling units (labelled p1 to p5) on the two ratio metric multidimensional scaling axes

over the five jackknife runs. The lines show the scores of each point in each of the four jackknife runs in which the point was included.

4.10) Procrustes analysis

Different starting values can result in different final configurations when using multidimensional scaling. Because of this, we need a method of comparing different configurations, of quantifying the degree of difference between these different final configurations, and of determining if these differences are significant. We do this using Procrustes analysis. To get an initial sense of how Procrustes analysis works, look again at Figure 4.11, which showed four different final MDS configurations of the five points on the pyramid that all reside in the same "valley". Although the scores of each point on the two axes are different in each plot, the distances between each of the points look the same. Because of this, we can perfectly superimpose these plots by some combination of rotating them, flipping them, or expanding/contracting them. We start with two different configurations (two matrices of scores for R rows and K axes). These two different configurations can be from two different MDS ordinations but they can also be from scores on PCA or CA axes or any other matrix having this structure. One configuration is the "target" and the other is the "comparison" configuration. Procrustes analysis consists of finding the combination of a rotation matrix, an expansion/dilation constant and a translation vector that makes the comparison configuration as close as possible to the target configuration.

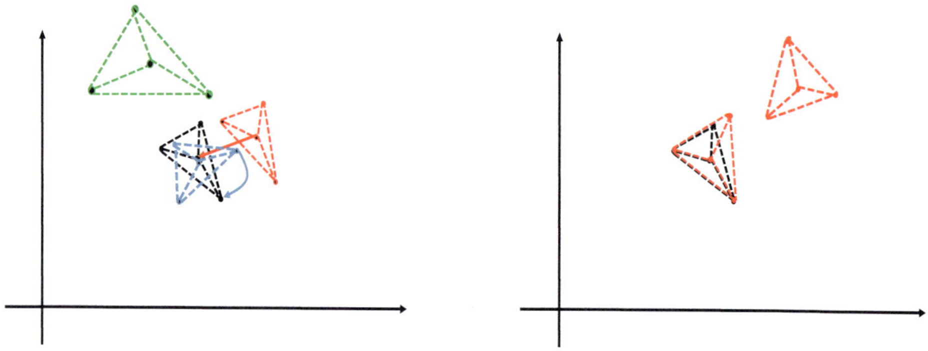

Figure 4.14. Examples of Procrustes analysis. The three configurations of four points on the left (red, green and blue) can be perfectly superimposed onto the target configuration (black) by some combination of a translation, a rotation or an expansion/dilution. The red configuration on the right cannot be perfectly superimposed onto the target configuration (black).

Figure 4.14 illustrates this idea. Imagine that we have four sampling units and have obtained the coordinates of these four sampling units on two axes. We therefore have a 4x2 matrix. Each row of this matrix gives the coordinates of a sampling unit in this two-dimensional space and so each row defines a point in this two-dimensional space. The series of four black points in Figure 4.14 is the target configuration to which we want to compare the three coloured configurations. I have drawn lines between each pair of points to represent the distances between each point.

These distances would be contained in a 4x4 distance matrix. The red, blue and green coloured points represent three different configurations of four points and the coloured lines show the distances between the points within each configuration. We want to know the degree of similarity between each of the coloured configurations and the black configuration; in other words, how closely do the interpoint distances in each of the red, blue and green configurations mirror the interpoint distances in the black configuration? By moving each of the four point in the red configuration in Figure 4.14 (left) in the same direction we can perfectly superimpose it onto the black target configuration; this is called a translation. Because the red points are perfectly superimposed onto the black points after this translation, the pairwise distances are equal. The blue configuration in Figure 4.14 (left) cannot be perfectly superimposed onto the black configuration by any translation but they can be perfectly superimposed by rotating each point by the same degree. The green configuration in Figure 4.14 (left) requires three changes in order to perfectly superimpose it onto the black configuration: it must be translated, rotated and also each of the points must be brought closer together (a dilation) by the same amount. Procrustes analyses consists of finding the combination of a translation, rotation and dilution/expansion of the comparison configuration that makes it as close as possible to the target configuration and then quantifies the remaining differences between the target and comparison matrices. For instance, the red configuration in Figure 4.14 (right) must be translated and rotated with respect to the black configuration but this cannot perfectly recreate the black target configuration.

After finding the combination of translation, rotation and dilution/expansion of the comparison RxC matrix that makes it as close as possible to the target RxC matrix, the Euclidean distances between the points (rows) in the comparison matrix are calculated. We now have two distance matrices: one for the target matrix and one for the modified comparison matrix that is now as close as possible to the target matrix. We now calculate the Pearson correlation coefficient between the distances in these two matrices. If the target and best-fitting comparison RxC matrices are identical then these two sets of distances will be exactly the same and the correlation between the two matrices will be 1. If they have absolutely no relationship to each other then the two sets of distances will be independent of each other. We now calculate the Pearson correlation coefficient (R) between the pairwise distances in the target and comparison configurations in order to quantify the degree of similarity between the two matrices. As an aside, the `protest()` function in the `vegan` package calculates the Pearson correlation coefficient between the target and comparison Euclidean distances and then performs a permutation test against the null hypothesis that the two distance matrices are independent, however such an inferential test is of little use in our context since we expect the two matrices to have some relationship to each other. The `Procrustes()` function in the `smacof` package calculates, with a bit of overkill, two indices of the

protest()
Procrustes()

strength of the fit between the two matrices: Tucker's congruence coefficient (i.e. Pearson correlation coefficient, R) and Guttman's alienation coefficient $\left(\sqrt{1-R^2}\right)$. Values of R that are close to 1 and values of $\sqrt{1-R^2}$ that are close to zero mean that the two distance matrices are almost identical. The `icExplore()` function uses the `Procrustes()` function to group together final MDS solutions that end up in the same "valley"; these are the ones whose Procrustes R^2 value are 1. Solutions that end up in different valleys have R values that are less than unity, even though they might have equal values of stress.

4.11) The Wisconsin understory plant communities

I introduced the Wisconsin understory plant communities in Chapter 3 and analysed these data using Correspondence Analysis and its detrended version. These data consisted of abundances (determined as counts) of each of 575 species of understory herbs and woody seedlings measured in 253 different sites throughout the American state of Wisconsin. Let's now reanalyse these data using the methods described in this chapter.

The first step is to choose an appropriate distance/dissimilarity coefficient. The consequences of different choices with respect to speciesXsite data will be explored in more detail in Chapter 6, when we look at simulation results. In this chapter I will use the Steinhaus (Bray-Curtis) dissimilarity coefficient, following the recommendation of Faith et al. (1987). Since this coefficient is not metric, and would result in negative eigenvalues in the PCoA, I will use the square root of these distances to render it metric (Legendre and Legendre 2012, Table 7.2). The following lines create the dissimilarity matrix using `vegdist()` and then fits the principal coordinates analysis using `wcmdscale()`. The `wisconsin[,-c(1:4)]` data frame holds the abundances of each of the 575 species (columns) in each of the 253 sites (rows); the first four columns must be excluded because these do not refer to abundances. I included the `eig=T` argument in order to include the eigenvalues in the output.

```
library(vegan)
wisconsin.bray<-sqrt(vegdist(wisconsin[,-c(1:4)],meth
od="bray"))
PCoA.wisconsin<-wcmdscale(wisconsin.bray,k=2,eig=T)
```

We must manually calculate the percent inertia (equivalent to the percent variance) captured by each axis. Remember that there will always be R-1 axes (and therefore eigenvalues) in a PCoA and the total inertia in the data will be distributed among all of these axes. Consequently, it is quite common for only a relatively small percentage of the total inertia to be captured by the first few axes when analysing the very large dissimilarity matrices produced by siteXspecies data since each site produces one column and row in the dissimilarity matrix. For instance, the `wisconsin.bray` dissimilarity matrix has 253 rows and columns. The output of the `wcmdscale()` function includes a list element called "GOF" that gives the percent of the total inertia that is captured in total by the number of axes requested (here, k=2 axes). We can also obtain the inertia (~variance) that is captured by each axis via the `$eig` object. Thus, `PCoA.wisconsin$eig[1:2]` outputs the inertia captured by the first and second axes and `sum(PCoA.wisconsin$eig)` calculates the total inertia captured by all of the axes. The second line of code therefore calculates the percentage of the inertia captured by the first and second axes. The first two axes of the PCoA of the Wisconsin data account for 11.6% and 6.4% of the total inertia, giving a total of 18.0%. Here is how to obtain this information:

```
PCoA.wisconsin$GOF
[1] 0.1804413 0.1804413

100*PCoA.wisconsin$eig[1:2]/sum(PCoA.wisconsin$eig)
[1] 11.620411  6.423721
```

We can find out how the measured environmental variables relate to these two PCoA axes in the same way as we did for DCA in Chapter 3. We already looked at these environmental variables in Chapter 3 and obtained transformations of these. There were also three sets environmental variables that were highly correlated within each set and that could be biologically interpreted and so we combined these to form composite variables using the site scores on the first axes of three principal component analyses (Chapter 2). These four environmental composite variables are "cation exchange" (the site loadings on the first PCA axis of soil pH, Mg, K and Ca, accounting for 84% of the variation), "soil texture" (site loadings on the first PCA axis of soil sand, silt and clay, accounting for 87% of the variation), and "net primary productivity" (the site loadings on the first PCA axis of potential evapotranspiration and aridity, accounting for 93% of the variation). Figure 4.15 shows the relationships between the environmental variables and the first two PCoA axes using the `envfit()` and `ord.on.env.arrows()` functions that were introduced in Chapter 3. I have also used different colours to show sites within each of the five vegetation groupings.

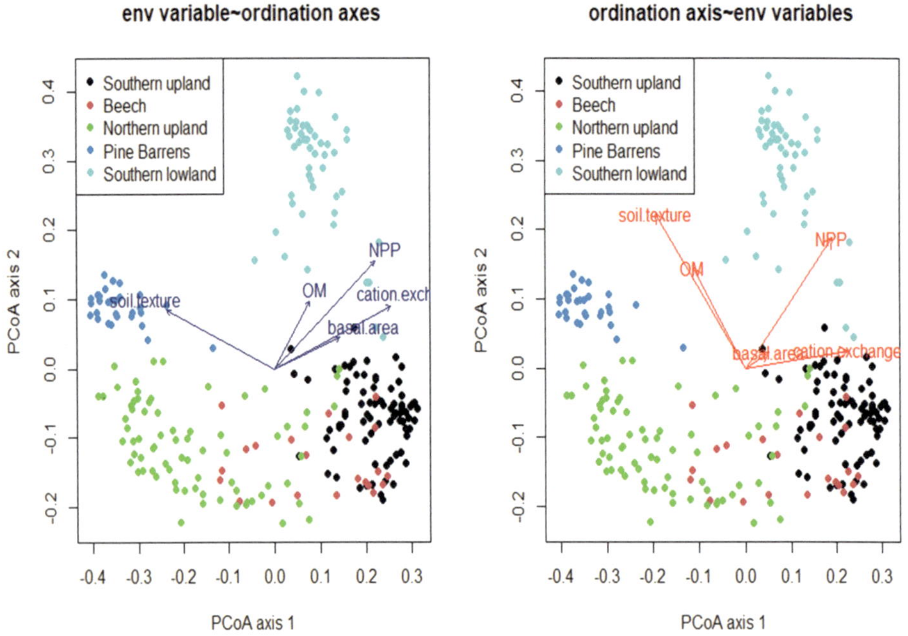

Figure 4.15. The site scores of the Wisconsin understory vegetation on the first two principal coordinate axes. Left: the regression coefficients predicting the value of each environmental variable given the site scores on both axes. Right:

the regression coefficients predicting the value of the site score on each axis given the values of all of the environmental variables.

Remember that `envfit()` tells us how to predict the value of each environmental variable given the site scores on both ordination axes while `ord.on.env.arrows()` tells us how to predict the site score on each ordination axes given all of the environmental variables. From Figure 4.15 (left), using `envfit()`, we see that as we increase the site scores on the first PCoA axis, we simultaneously increase the cation exchange capacity (less acidic and more base cations) and organic matter content of the soil, the net primary productivity (i.e. sites are both warmer and wetter) and the basal area of the canopy trees, while decreasing the relative amount of sand. The same tendencies occur as we increase the site scores on the second PCoA axis. Figure 4.15 (right), using `ord.on.env.arrows()`, generally shows the same information, but with some differences. For instance, soil organic matter content (OM) decreases along the first PCoA axis. Why? Because the right hand figure only looks at each ordination axis separately but considers all of the environmental variables together while the left hand figure does the opposite: it looks at both ordination axes together but considers each environmental variable separately while ignoring the other environmental variables.

What happens if we instead use nonmetric multidimensional scaling on this same Bray Curtis dissimilarity matrix? The first step is to choose the number of dimensions in which to force the resulting Euclidean distances to approximate the full 253X253 matrix of Bray-Curtis dissimilarities. To choose the number of dimensions we successively fitting a series of NMDS models, each time using the site scores on the first two PCoA axes as starting values, while successively increasing the number of dimensions. We then look at how the stress value decreases as the number of dimensions (axes) increases using a scree plot (Figure 4.16). Be sure to specify `type= "ordinal"` (i.e. using an ordinal regression) and `ties= "primary"` to allow tied Bray-Curtis dissimilarity values to result in untied Euclidean distances, as required by NMDS.

```
wisconsin.nmds.stress<-rep(NA,10)
for(i in 1:10){
  wisconsin.nmds.stress[i]<-mds(wisconsin.bray,ndim=i
,type="ordinal",ties="primary")$stress
}
```

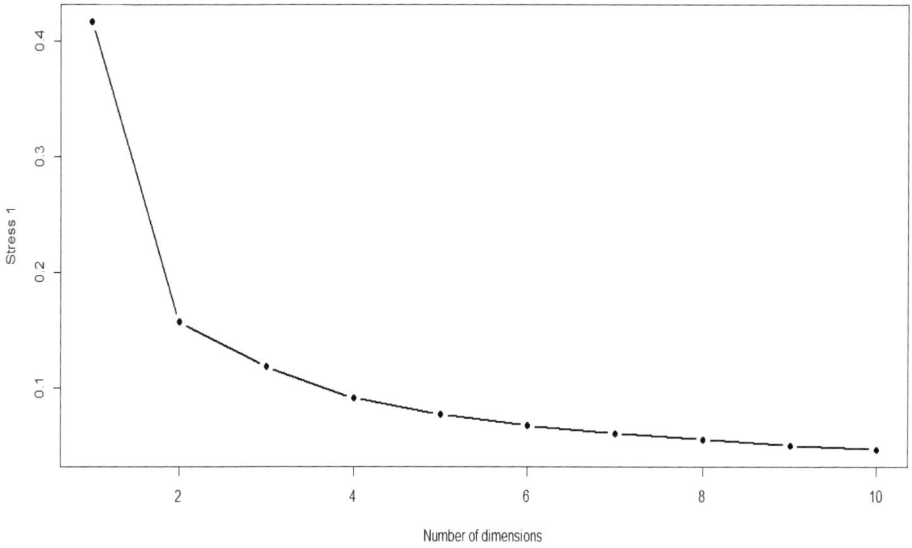

Figure 4.16. A scree plot showing how the stress on a NMDS analysis of the Wisconsin understory plant community data changes as we increase the number of dimensions (axes) allowed.

We see that there is a large decrease in stress as we increase the dimensions of the NMDS analysis from one to two dimensions, and then a more gentle and continuous decrease. This suggests that a two-dimensional NMDS, with a stress of 0.1565, would be sufficient. We will save this NMDS result as `wisconsin.nmds.default` and then use the `icExplore()` function to look for alternative solutions that might have even lower stress values, saving the output in an object called `wisconsin.explore`. By default, this function generates 100 random starting configurations and, given the size of our Bray-Curtis dissimilarity matrix, it will take a few minutes for this function to complete. Simply typing the name of the saved output from this function will provide a summary. Of the 100 random runs, values of stress varied from a low of 0.1564 to a high of 0.1766. The last line of the **R** code below identifies the solution, in this case run number 43, with the lowest stress value. Notice that this "best" solution (stress = 0.1564) has almost the same stress as our default NMDS solution (stress = 0.1565). Figure 4.17, produced by plotting the `wisconsin.explore` object, shows the result.

```
wisconsin.ndms.default<-mds(wisconsin.bray,ndim=2,typ
e="ordinal",ties="primary")
wisconsin.explore<-icExplore(wisconsin.bray,ndim=2,ty
pe="ordinal",ties="primary",returnfit = T,verbose=T)
wisconsin.explore
plot(wisconsin.explore,main="icExplore Wisconsin data"
)
#this best run is #43
c(1:100)[wisconsin.explore$stressvec==min(wisconsin.e
xplore$stressvec)]
[1] 43
```

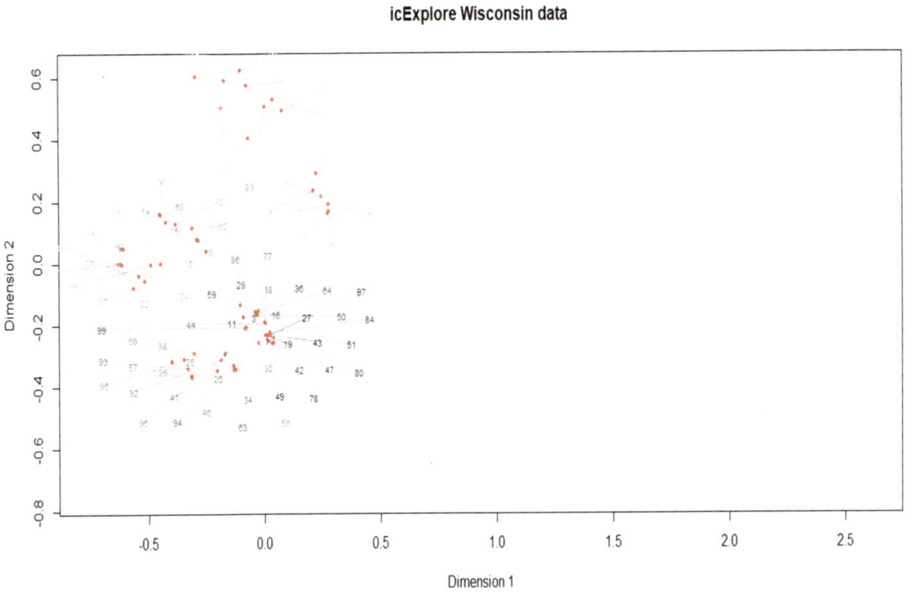

Figure 4.17. A graphical view of 100 NMDS runs (numbers) of the Wisconsin understory plant community data. Each red dot is a different final solution and darker numbers indicate lower stress values.

Since there are different NMDS solutions, a more complete analysis would require us to took at those solutions (each one is stored in the `wisconsin.explore` object) that have approximately equal and low stress values. These alternative solutions are those with the dark numbers that are associated with different red dots in Figure 4.17. However, since all of these solutions have stress values that vary only slightly, and since the default solution (`wisconsin.ndms.default`) had almost identical stress relative to the best solution (`wisconsin.explore$mdsfit[[43]]`), we will first compare these two solutions using a Procrustes analysis. If the solutions are very close then we have little to gain. In fact, the resulting congruence coefficient (i.e. the R value) is 1.00 and the alienation coefficient (i.e. 1-R^2) is 0.02. This means that the two solutions are virtually identical. Given that, I will use the solution contained in `wisconsin.mdms.default` that is obtained using the site scores on the first two axes of the PCoA as the starting configuration.

```
Procrustes(Y=wisconsin.explore$mdsfit[[43]]$conf,X=wi
sconsin.ndms.default$conf)
```

The next step is to see how sensitive this configuration is to each site value using the jackknife. In other words, how different is the final configuration if we leave out each site in turn? To find out, we call the `jackmds()` function and then simply type the name (`jack.wisconsin`) of the object which holds the results. Given the size of our Bray-Curtis dissimilarity matrix, this call takes several seconds to complete. The result is that the stability measure and the cross validity measure are both 1.0, indicating that the results does not differ to within rounding error between the jackknifed runs. It is useful to plot the result, as shown in Figure 4.18. The black numbers are the positions of each site in the two-dimensional NMDS ordination and the red dots, with red lines jointing each dot to a black number, shows by how much the site scores for that site changed when the NMDS analysis was conducted while leaving out that point. Those sites for which the red dots are directly over the black numbers are the sites whose site scores did not change at all and these are the vast majority of sites although a few sites (#93, 107) did change a bit.

```
jack.wisconsin<-jackmds(wisconsin.ndms.default)
plot(jack.wisconsin,main="Jackknife plot of Wisconsin
data",col.p="black",col.l="red")
```

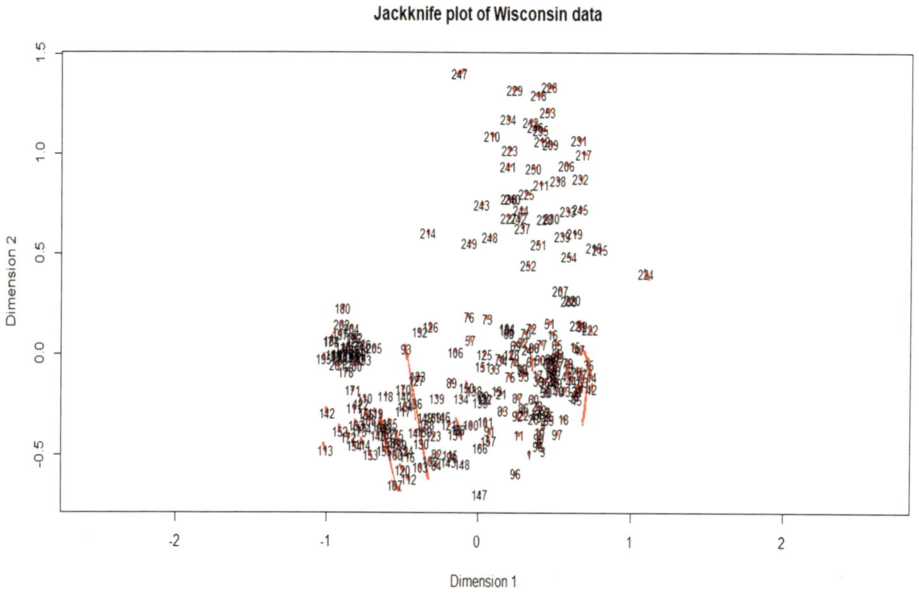

Figure 4.17. The black numbers are the positions of each site in the two-dimensional NMDS ordination and the red dots, with red lines jointing each dot to a black number, shows by how much the site scores for that site changed when the NMDS analysis was conducted while leaving out that point in the jackknife analysis.

Given all of these results, we can be confident that the NMDS solution that we obtained when using the site scores on the first two PCoA axes as starting values is at (or is extremely close to) the best NMDS solution and that this solution is very stable. Before looking at this solution, and the relationships between the environmental variables and the site scores on the two axes of this solution, it is insightful to look at a "Shepard" diagram, which shows how the original Bray-Curtis dissimilarities have been converted into Euclidean distances in our two-dimensional NMDS ordination (Figure 4.18).

Shepard diagrams

```
plot(wisconsin.ndms.default,plot.type="Shepard",xlab=
"Bray-Curtis dissimilarities",ylab="Euclidean distance
s (NMDS)",pch=16,main="Shepard diagram NMDS of Wiscons
in data")
```

The Euclidean distances in the NMDS solution increase approximately linearly with increasing Bray-Curtis dissimilarities until these Bray-Curtis dissimilarities reach around 0.7, although the Euclidean distances are all much closer to zero than are the Bray-Curtis dissimilarities. For pairs of sites whose Bray-Curtis dissimilarity were above about 0.7, the associated Euclidean distances in the NMDS solution begin to increase rapidly and almost exponentially such that Bray-Curtis dissimilarities of about greater than ~0.9 result in Euclidean distances of greater than 0.9. In other words, pairs of sites that are rather similar in terms of their species composition (i.e. whose Bray-Curtis dissimilarity is small) are placed even closer together in the NMDS solution while pairs of sites that are quite different in terms of their species composition (i.e. whose Bray-Curtis dissimilarity is large) are placed even further apart in the NMDS solution.

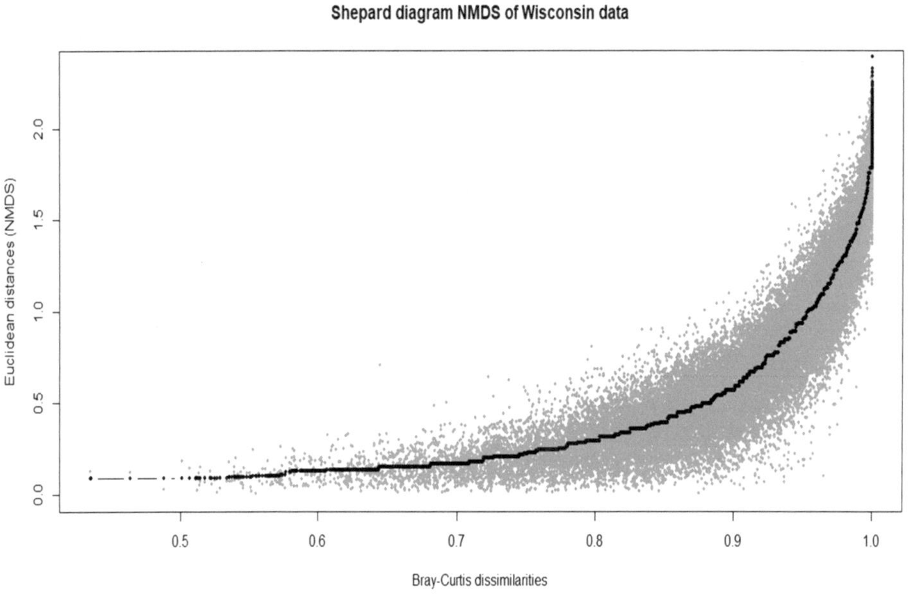

Figure 4.17. A Shepard diagram showing how the Bray-Cutis dissimilarities measured in the empirical data are converted into Euclidean distances in the non-metric multidimensional scaling solution.

Finally, we can use `envfit()` and `ord.on.env.arrows()` to see how the environmental variables relate to these NMDS axes, in the same way as we did for the PCoA ordination, remembering that the site scores on these NMDS axes are held in `wisconsin.ndms.default$conf`. Figure 4.18 shows the result.

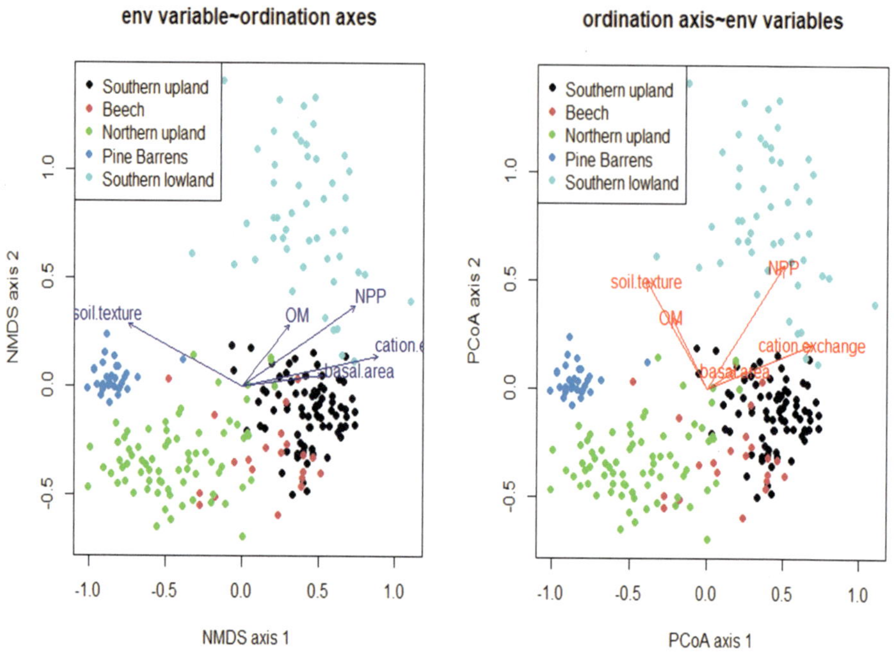

Figure 4.18. The site scores of the Wisconsin understory vegetation on two non-metric multidimensional scaling axes. Left: the regression coefficients predicting the value of each environmental variable given the site scores on both axes. Right: the regression coefficients predicting the value of the site score on each axis given the values of all of the environmental variables.

Comparing the NMDS ordination (Figure 4.18) with the PCoA ordination (Figure 4.15), we see that they are very similar, although the environmental variables accounted for slightly more of the site scores in the NMDS ordination that they did for the site scores in the PCoA ordination. We can make this comparison more numerically exact by comparing the two ordinations using Procrustes analysis:

```
Procrustes(X=PCoA.wisconsin$points,Y=wisconsin.ndms.d
efault$conf)
```

The two ordinations are indeed very similar, with an R^2 value (congruence coefficient) of 0.973.

Chapter 5 Constrained ordination techniques

5.1) Exploratory vs. inferential statistical methods

You may have noticed that the ordination methods described so far do not generally involve tests of statistical significance. Most of the statistical methods that you might have learned before reading this book involve concepts like "null hypotheses" and "significance levels" but these terms have been virtually absent so far in this book. Why? Statistical methods that involve tests of statistical significance are designed to test some *a priori* hypothesis, usually a null hypothesis, whose rejection allows you to consider the alternative hypothesis. For instance, here is what happens (slightly modified) when you do a simple linear regression using the `lm()` function in the base library of **R,** followed by the associated `summary()` function:

```
fit<-lm(y~x)
summary(fit)

Coefficients:
             Estimate Std. Error t value Pr(>|t|)
(Intercept)  -0.3635     0.9839  -0.369   0.7214
x             0.3877     0.1586   2.445   0.0403 *

Residual standard error: 1.44 on 8 degrees of freedom
Multiple R-squared:  0.4277,    Adjusted   R-squared:
0.3561
F-statistic: 5.978 on 1 and 8 DF,  p-value: 0.04025
```

You not only obtain an estimate of the slope (0.3877) and intercept (-0.3635) of the regression line, (called parameter estimates), you are also be given a Student's t-value associated with each parameter estimate as well as a null probability associated with each t-value. For instance, the slope estimate has a t-value of $t = |0.3877 - 0| / 0.1586 = 2.445$. The numerator compares the observed value of the slope (0.3877) to the value (zero) that the slope would have if the null hypothesis of independence was true. The probability (0.0403) associated with this t-value is the probability of observing at least this great a difference between the observed and hypothesized slope estimates by chance if the null hypothesis was true[33]. If the null probability is sufficiently small (below the "significance" level) then you would conclude that your estimate of the slope is very unlikely to arise purely by chance and so you would reject the hull hypothesis (i.e. that the slope is zero in the statistical population). You have not encountered these concepts so far in this book is because the ordination methods that you have learned so far are not concerned with testing inferences obtained from *a priori* hypotheses. Rather, their goal is to describe and summarize patterns, not to test hypotheses about how these patterns are generated. The methods that we will study in this chapter have a different goal because they combine the exploratory aspects of ordination with the testing of *a priori* hypotheses.

[33] And assuming that the residuals of the regression follow a normal probability distribution.

Each time we used PCA, CA, DCA, PCoA or NMDS, we subsequently looked for additional variables that might explain these patterns. For instance, after observing how the human SNP DNA data in chapter 2 was aligned along the first two PCA axes, we hypothesized that these first two axes reflected two different episodes of mass migration and population mixing with pre-existing populations in European prehistory. However, this hypothesis did not determine the PCA scores or loadings in any way. In fact, the direction of causality is exactly the contrary since the PCA scores suggested the hypothesis. The testing of this hypothesis would start where the PCA ended: with the resulting scores of each individual on the first two PCA axes. Similarly, each time we completed an ordination of the Wisconsin understory plant communities, we then attempted to interpret these ordination axes by regressing different environmental variables onto the site scores. However, the inclusion of these new variables only ever affected the subsequent *interpretation* of the ordination results; these new variables never affected the axis scores, and thus the ordinations, themselves. We could perform inferential tests of significance on such regressions to decide if these environmental variables are "significantly" associated with the ordination axis scores but such inferential tests would only apply to hypotheses concerning our interpretation of the ordination axes, not to the axes themselves. The types of ordinations that we have studied so far are examples of "exploratory" statistics. The purpose is to explore the data; that is, reveal and quantify patterns in the data and to subsequently propose new hypotheses.

Exploratory statistics are contrasted with inferential statistics. The goal of inferential statistics is to measure the strength of evidence for or against some *a priori* hypothesis so that we can decide to either accept or reject it.

What should we call the set of related methods that are described in this chapter? One term for the set of methods that combine ordination and regression is "canonical" analysis. In particular, "canonical correspondence analysis" is the version of correspondence analysis (Chapter 3) that includes this regression step. The term "canonical", originally referring to a law, regulation or dogma that is officially decreed by a Catholic Church council, more generally refers to some fundamental or formal set of rules governing actions or procedures. In mathematics, a "canonical" form is one that provides the simplest (for mathematicians) representation that allows a system of equations to be identified in a unique way. The regression – ordination versions of PCA and CA and PCoA, called redundancy analysis (RDA), canonical correspondence analysis (CCA) and distance-based redundancy analysis (db-RDA), are forms of "canonical" analysis because they can be formulated purely in terms of a type of linear algebra called eigen analysis. However, I won't use this terminology for three reasons. First, it emphasizes the mathematical properties of these methods rather than their biological goals. Second, the notion of a "canonical form" in mathematics is not generally understood outside of that discipline and this book is written for biologists, not mathematicians. Finally, the NMDS version of "canonical" analysis does not use eigen analysis at all but has the same goals as RDA, CCA db-RDA.

Instead, I will refer to this class of ordination methods as "constrained ordinations" because the act of including a regression step results in the ordination being forced, or "constrained", to agree with certain patterns in a second set of data, just like in regression.

5.2) Setting the scene

The testing of hypotheses in unconstrained ordinations, like those we have studied in Chapters 2, 3 and 4, start where the ordination ends. In each case, we first completed the ordination and then we compared the scores on the resulting ordination axes to other variables. We chose these other variables based on hypotheses concerning how to interpret the ordination axes. Such hypotheses do not determine the ordination results (the scores or loadings) in any way. This is not true with the constrained ordination methods that we will study in this chapter. To set the scene, and to better understand what "constraining" an ordination means, we will first look at an example of combining a regression and an ordination that will allow me to illustrate the main point before we look into the details.

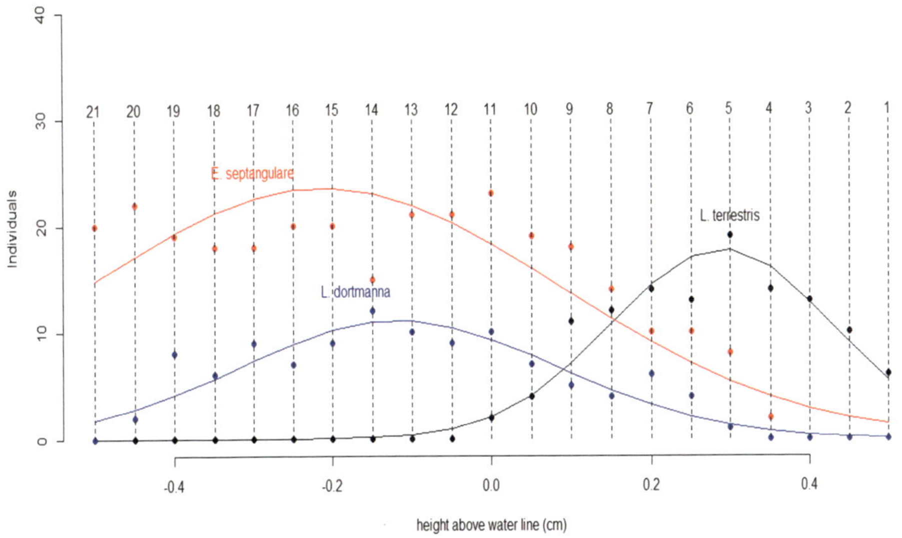

Figure 5.1. The abundances (points) of three species in a series of contiguous "quadrats" (numbered 1 to 21) that were explicitly chosen to follow a gradient of height relative to the water level.

In Chapter 3 we briefly looked at an example of direct gradient analysis involving the abundances of three shoreline species occurring in Axe Lake. Many more species also grew on the shoreline of this lake but I will pretend that the entire shoreline plant community consisted of just these three species to illustrate a point. The data were obtained by running narrow band transects perpendicular to the shoreline of the lake, starting at 50 cm above the water level and stopping at 50 cm below the water level. The number of plants of each species were counted in each 5 cm height increment in order to estimate their abundances. Although these contiguous height increments are not "quadrats" of fixed areas, we will pretend that they are so that we can obtain a siteXspecies data matrix with 21 rows indexing the sites (quadrats) and 3 columns indexing the abundances of each of the three species. Figure 5.1 recreates Figure 3.3 except that I have added the site numbers. The points are the observed abundances of each species. There are three points for each of the 21 sites since there are three species. Site 1 is at the highest part of the shoreline at the extreme right. Since the sites were actually contiguous "quadrats" that start at the highest point and descend down the shoreline and into the water (right to left), the site numbers are already ordered as a function of shoreline height. This 21X3 matrix of siteXspecies data does not include any environmental data (here, the site elevation on the shoreline). Figure 5.2 (left) shows the first axis of the ordination of this matrix using correspondence analysis. Notice that the order of sites along this axis (i.e. their site scores) agrees with the actual ordering of the heights of the sites along the shoreline. The spacing between successive sites along this axis does not agree with the actual 5cm spacing of these sites in terms of changes in shoreline elevation, which is why there is a non-linear pattern with sites 12 to 21 having almost the same site scores. This should not surprise you since you already know that such ordinations do not maintain the same measurement units as the actual physical gradient. All of the (unconstrained) ordination methods in Chapters 2, 3 and 4 answer the following question: *What is the arrangement of rows and columns that best represents the patterns of association in the data*? The correspondence analysis shown in Figure 5.2 (left), based on species abundances of shoreline plants, answers the more specific question: *What is the arrangement of quadrats and species that best represents the patterns of association in the observed abundances of these species*? Please remember this question in what follows.

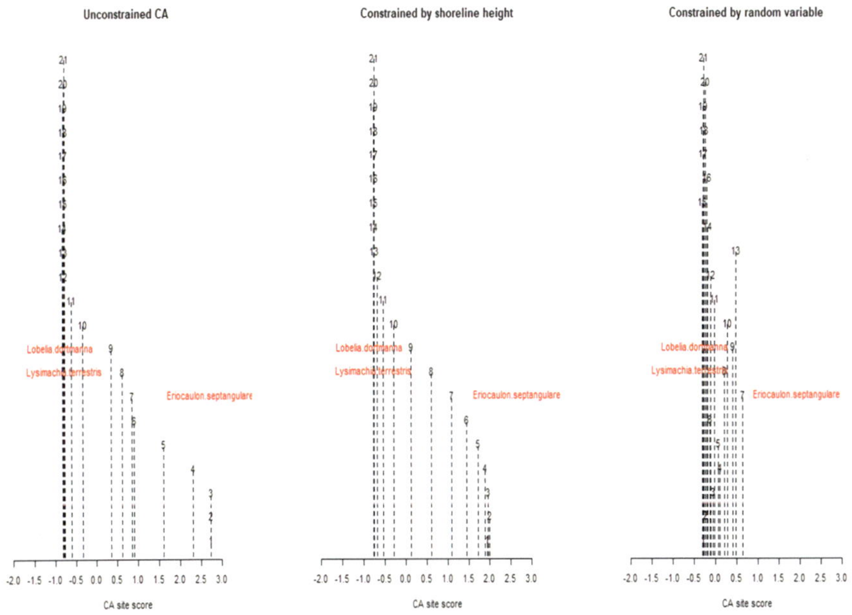

Figure 5.2. The site and species scores of a correspondence analysis of the data shown in Figure 5.1 based on the measured abundances (left: Unconstrained CA), based on the predicted abundances given the Gaussian response functions relative to shoreline height of Figure 5.1 (middle: Constrained CA), and based on the predicted abundances given the Gaussian response functions relative to a random variables (right: Constrained CA).

In Chapter 3 I also performed three quadratic linear Poisson regressions, which I called Gaussian response functions. In each regression (one for each of the three species of shoreline plants in Axe Lake) I regressed the logarithmically transformed abundance of that species against a quadratic function of the elevation of the site on the shoreline at which the abundance was estimated. These three regressions produce three sets of predicted abundances (one set for each species) at each elevation along the shoreline. These predicted abundances are shown as lines in Figure 5.1. The predicted abundances of each species are quite close to the observed values (the dots) since the Pearson correlations between the observed and predicted values are quite high (0.92, 0.98 and 0.89) Collecting these predicted abundances together gives us a second 21X3 matrix of siteXspecies. This second matrix does not contain the actual abundances of each species; instead, it contains the abundances that each species *would* have if they responded to the environmental variable (shoreline height) exactly as the Gaussian response functions require. These predicted abundances are therefore "constrained" to agree with this *a priori* hypothesis about how the abundance of each species will change as a function of shoreline height. Notice that this second siteXspecies matrix does not explicitly include any measurements of shoreline height but it does implicitly include this information because the predicted abundances are constrained to agree with the Gaussian response functions. If we then perform the same correspondence analysis on this second siteXspecies matrix, the result is as shown in Figure 5.2 (centre). This second ordination answers the following question: *What is the arrangement of quadrats and species that best represents the patterns of association in the abundances of these species* ***given the hypothesis that such abundances are a Gaussian response function of shoreline height***? The first part of this question uses the same wording as when we did the unconstrained correspondence analysis using the observed abundances but the part in bold font is new. The words in bold font make clear the difference between unconstrained and constrained ordinations. This second correspondence analysis is constrained by an *a priori* hypothesis that may, or may not, be correct. From Figure 5.1 we can see that the actual abundances of the three species do closely follow Gaussian response functions of shoreline height. Since this hypothesis is included in our second correspondence analysis and it is actually quite accurate, the constrained ordination that assumes this hypothesis (Figure 5.2 centre) looks very similar to the first ordination (Figure 5.2 left) that did not make such an assumption.

What would happen if we constrained the ordination by an *a priori* hypothesis that was wrong? If we use an environmental variable that has no relationship the abundances of the three species then we can find out. To do this, I generated 21 random values, representing 21 values of an environmental "variable" that has no relationship to the observed abundances, and performed the same three regressions of abundance on this environmental variable, again using a Gaussian response function. As we would expect, these three regressions provided much poorer predictions since the Pearson correlations between the observed and predicted abundances of each species were much lower (0.18, 0.36 and 0.26). Collecting these new predicted abundances together into a third 21X3 matrix, I perform a correspondence analysis on these new predicted abundances. The site and species scores of this new constrained correspondence analysis along the first axis are shown in the right-hand panel of Figure 5.2. This result looks quite different from the first one based on the correspondence analysis using the actual abundances. It also looks quite different from the second correspondence analysis using the predicted abundances given shoreline height. The reason is that the *a priori* hypothesis that constrains this third ordination is wrong. We forced the abundances to be a Gaussian function of a variable that had nothing to do with the actual process that controlled the distribution of these species.

This simple analysis illustrates a general property of constrained ordinations. A constrained ordination will never account for more variation (or inertia) in your data that the equivalent unconstrained version. In fact, it will always account for *less* variation than the equivalent unconstrained version if the constraining variables do not perfectly predict the observed values in your data. The difference in the variation that is captured between the constrained and unconstrained versions measures the degree to which your hypothesized constraints – that is, your *a priori* hypothesis – agrees with your data. If your goal is to capture the patterns of association in your data then you should use unconstrained ordination methods. If your goal is to test an *a priori* hypothesis about what is generating the patterns of association in your data, then you can use constrained ordination methods.

5.3) Redundancy Analysis (RDA)

Rao (1964), in section 8 of that paper, introduced an extension of PCA that he called "principal components of instrumental variables". This method was later called "redundancy analysis' by van den Wollenberg (1977). The word "redundancy" refers to the degree to which the information in one set of variables is also contained in a second set of variables. "Redundancy analysis" is not a very informative name because the quantification of "redundancy' is only one part of the method. However, this name has become standard and redundancy analysis (RDA) is the first type of constrained ordination that we will study.

Redundancy analysis is really a marriage between regression and PCA. The first step is a series of regressions. The second step consists of performing a PCA on the *predicted* values of these regressions. Because the resulting principal components are based on these predicted values, these principal components are *constrained* by the results of the regressions. This is different from the PCA that we studied in Chapter 2 because, in that chapter, the principal components were not constrained. I will first illustrate these steps using some simulated data and then use data from a published empirical study.

We start with a matrix **Y** of variables (each variable in a separate column of this matrix) measured on different observational units (each observational unit in a separate row of this matrix). **Y** is the matrix of response variables. In Chapter 2 we would have used PCA on this **Y** matrix in order to reveal the different principal component axes that describe the patterns of variation and covariation between the variables in **Y**. RDA uses a second matrix, **X**, consisting of a different set of variables that are measured on these same observational units. **X** the matrix of predictor variables. You choose these predictor variables because you believe that they determine the patterns of covariation in the **Y** matrix; that is, because you wish to test an *a priori* hypothesis linking the variables in **X** to the variables in **Y**. The first step of RDA (the regression step) is to separately regress each of the response variables in **Y** (Y_1, then Y_2, etc.) on the full set of the predictor variables via linear multiple regression based on normally distributed residuals. In other words, it is as if we conducted a series of multiple regressions of the form $Y_i \sim X_1+X_2+\ldots$. Actually, other statistical methods that are based on a linear model could be used, but these are not supported in the `rda()` function in the vegan library of **R** that we will use. These other methods include mixed model regression and generalized linear models. Since this book doesn't assume that you are familiar with these other methods, I won't say anything further about this extended version of RDA here. However, if you do have a good knowledge of these other methods, and if you can program in **R**, then you can use the code in Borcard et al. (2018 pp. 253-255) to write your own version of this extended version of RDA; you will simply have to replace the section of code identified as "`##2. Computation of the multivariate linear regression`" with code for the type of regression that you prefer.

This book does assume that you already know about linear multiple regression, which is what the `rda()` function uses. In the notation that I will use, the first index identifies the observation (i.e. a row of **Y** or **X**) and the second index identifies the variable (i.e. a column of **Y** or **X**). Therefore, Y_{ij} or X_{ij} means the ith observation (row) of the jth variable (column) in either the **Y** or the **X** matrix. If Y_{ij} is the ith observation of the jth response variable in the **Y** matrix, then linear multiple regression assumes that this observation is a normally distributed random variable that is a linear function of two things. The first thing is a set of p observed variables; in this case, the values of predictor variables X_{i1}, $X_{i2},\ldots,X_{jp}$ which are found in the ith row of the **X** matrix. The second thing is all of the other unobserved (and probably unknown) variables that also cause this variable to vary independently of the predictor variables. We group all of these unobserved causes together and call this group the residual variation, or e_{ij}. The general form of such a multiple linear regression is given in Equation 5.1. Both the response and the predictor variables are centred about their means. If these variables are measured using different units of measurement then they are usually standardized as well (i.e. we divide each variable by its standard deviation) so that the numerical values of the resulting analysis can be compared across variables. Each multiple regression with have p+1 parameters whose values must be estimated: p partial slopes (β_{ij}) and one residual variance (S_i). There is no intercept term since our variables are centred. The notation $N(\mu = 0, \sigma = S_i)$ in Equation 5.1 means "a normally distributed random variable (i.e. N) whose population mean (i.e. μ) is zero and whose population standard deviation (i.e. σ) is S_i". Of course, if there is only one variable in **X** (i.e. p=1) then this reduces to simple linear regression based on normally distributed residuals. The notation $\hat{Y}_{ij}$ in the second line of Equation 5.1 means "the predicted (not the observed) value of Y_{ij} given the values of the predictor variables in **X**", which is different from the observed value of Y_{ij} independent of the values of the predictor variables in **X**. Notice that we have three different types of Y variables in the second line of Equation 5.1: the observed Y values (Y_{ij}), the predicted Y values ($\hat{Y}_{ij}$) and the values of Y that are not predicted by the predictor variables (the residual values, e_{ij}). We can't actually know the exact values of these residuals because they are random, but we can estimate them as $e_{ij} = Y_{ij} - \hat{Y}_{ij}$. The third line of Equation 5.1 combines the observed response values (**Y**), the observed predictor variables (**X**), the regression coefficients (**B**) and the residuals (**E**) in matrix form.

$$Y_{ij} = \left(\alpha_i + \beta_{1j} X_{1j} + \cdots + \beta_{mj} X_{mj}\right) + e_{ij}; e_{ij} \sim N(\mu = 0, \sigma = S_j)$$

$$Y_{ij} = \hat{Y}_{ij} + e_{ij} \tag{5.1}$$

$$\mathbf{Y} = \mathbf{BX} + \mathbf{E} = \hat{\mathbf{Y}} + \mathbf{E}$$

In our simulated data (code below), the first four columns contain the four response variables (y1 to y4 which forms the **Y** matrix) and the next two columns contain the two predictor variables (X1 and X2 which forms the **X** matrix). In these simulated data, each of the four response variables is linearly caused by the two predictor variables (X1 and X2) plus random residual variation. The correlation between each response variable and each predictor variable in the statistical population is 0.25. The two predictor variables can either be mutually independent or correlated with a correlation coefficient of 0.25. Here is the **R** code to generate these simulated data:

```
sim.RDA<-function(n=100,correlated.predictors=F){
  set.seed(101)
  X1<-rnorm(n)
  if(!correlated.predictors)X2<-rnorm(n)
  if(correlated.predictors)(X2<-
0.5*X1+rnorm(n,0,sqrt(1-0.5^2)))
  y1<-0.5*X1+0.5*X2+rnorm(n,0,sqrt(1-2*0.5^2))
  y2<-0.5*X1+0.5*X2+rnorm(n,0,sqrt(1-2*0.5^2))
  y3<-0.5*X1+0.5*X2+rnorm(n,0,sqrt(1-2*0.5^2))
  y4<-0.5*X1+0.5*X2+rnorm(n,0,sqrt(1-2*0.5^2))
 data.frame(y1,y2,y3,y4,X1,X2)
}
```

I first generate a data set (`sim.dat1`) in which the two predictor variables are uncorrelated:

```
#Generate 100 observations with uncorrelated predictors
sim.dat1<-sim.RDA(n=100)
```

I then do an RDA on these data:

rda()

```
library(vegan)
sim.rda1<-
rda(sim.dat1[,1:4]~X1+X2,data=sim.dat1)
```

Notice that the general syntax of the `rda()` function is very similar to the `lm()` function that you already know. It states that you want to (i) regress each of the variables in the **Y** matrix (here, `sim.dat1[,1:4]`) on a set of predictor variables from the **X** matrix (here, X1 and X2) and then (ii) perform a PCA on the predicted values of these regressions. You can specify that all of the variables (columns) can be scaled to unit variance in `rda()` by using the `scale=T` argument and you can use only a subset of rows by using the `subset=` argument. The above code, using a formula object, shows the easiest way of specifying the regression step in `rda()`, which uses the same syntax as in the `lm()` function. However, the `rda()` function can also be used without a formula object by using arguments that specify the matrix or data frame holding the predictor variables (`X=`), the response variables (`Y=`) and any conditioning variables (`Z=`); this last matrix will be explained later. Thus, the call to `rda()` can equivalently be given (remembering that X1 and X2 are in columns 5 and 6 of `sim.dat1`) as:

```
sim.rda1<-
rda(Y=sim.dat1[,1:4],X=sim.dat1[,5:6],data=sim.dat1)
```

5.4) Obtaining output from rda()

Typing `summary(sim.rda1)` gives the main results. Since the resulting output contains many different types of information, I will break it down. The first result that is reported is:

```
Call:
rda(formula = sim.dat1[, 1:4] ~ X1 + X2, data = sim.dat1)

Partitioning of variance:
              Inertia Proportion
Total           4.091     1.0000
Constrained     2.340     0.5719
Unconstrained   1.752     0.4281
```

You could replace the word "inertia" with the word "variance" since, for the `rda()` function, they are the same thing. This table decomposes the total variance in the collection of response variables in the **Y** matrix (not each response variable separately) into two parts. The first part ("Constrained") is the variance in the collection of response variables that has been captured by the set of predictor variables in the **X** matrix (not each predictor variable separately). The second part ("Unconstrained") is the variance in the collection of response variables in the **Y** matrix that remains unexplained. This output is telling you three things. First, the total variance in the **Y** matrix of response variables is 4.091 units, which is why the proportion of the total variance, which is reported in the third column, is 1.0 (i.e. 100%). If I do an unconstrained PCA on the **Y** matrix, like in Chapter 2 (i.e. `summary(prcomp(sim.dat1[,1:4]))`), then it would tell me the same thing[34]: that the total variance in the **Y** matrix is 4.091 units. Second, the variance in the **Y** matrix that is accounted for by the two predictor variables ("constrained") is 2.340 units and this represents a proportion of 0.5719 (i.e. 57.19%) of this total variance; together, X1 and X2 (the predictor variables) explain just over 57% of the total variance in the collection of the four response variables (y1 to y4). Third, the residual variance in the **Y** matrix is 1.752 units and this represents a proportion of 0.4281 (i.e. 42.81%) of this total variance. Notice that when we generated our simulated data, we specified that the two predictor variables in our simulated data each account for 25% of the variance of each independent response variable in the statistical population; i.e. that X1 and X2 together would determine 50% of the total variance in each independent variable. The difference between 50% in the statistical population and 57.19% in our sample of 100 observations is due to sampling variation.

The next part of the summary output is the following:

```
Eigenvalues, and their contribution to the variance

Importance of components:
                        RDA1     RDA2    PC1    PC2    PC3
PC4
Eigenvalue            2.3066 0.032894 0.5279 0.4694 0.4338
0.32046
Proportion Explained  0.5638 0.008041 0.1290 0.1147 0.1060
0.07833
Cumulative Proportion 0.5638 0.571864 0.7009 0.8156 0.9217
1.00000
```

[34] The summary function of prcomp() gives the standard deviations of each of the four PCA axes and so you have to sum the squares of these standard deviations to get the total variance.

RDA1 and RDA2 are the two PCA axes based on the predicted values (not the observed values) of the four response variables. There are always as many of these axes as there are predictor variables. This is because the *predicted* values of these response variables ($\hat{Y}_{ij}$) are entirely determined by the predictor variables. In our case, we have two predictor variables (X1 and X2) and so we have two RDA axes. The row named "Eigenvalue" gives the variance associated with each axis. The first RDA axis accounts for 2.3066 units of the total variance. The total variance is 4.091 units, obtained by summing all values in this row or by looking at the previous output ("Partitioning of variance"). Therefore, the proportion of the total variance that is captured by RDA1 equals 2.3066/4.091 = 0.5638, which is the value given in the third row. This means that the first axis of the PCA based on the predicted response variables (i.e. RDA1) accounted for 56.38% of the total variance in the observed response variables. The second RDA axis (RDA2) accounts for only 0.0329 units of variation and this is 0.0329/4.091 = 0.0080, or 0.8% of the total variance in the observed response variables. The sum of the variances (eigenvalues) accounted for by the two RDA axes, and therefore the two predictor variables X1 and X2, is 2.3066+0.0329 = 2.340 (i.e. 57.19% of the total variance) and this is the same value that was reported before as the "constrained" variance in the "partitioning of variance" output. This means that the first PCA axis based on the predicted values of these response variables ($\hat{Y}_{ij}$), which is RDA1, accounts for virtually all of the variance that is explained by the predictor variables while the second PCA axis based on the predicted value of the response variables (RDA2) accounts for almost none of this "constrained" variance. This leaves 100%-57.19% = 42.81% of the total variance unexplained. This unexplained variance is the difference between the total variance of the observed response variables (the **Y** matrix) and the predicted response variables (the $\hat{\mathbf{Y}}$ matrix), which is variance of the residual $e_{ij} = Y_{ij} - \hat{Y}_{ij}$ matrix.

The next four rows of the `summary()` output (PC1 to PC4) report the variance associated with the PCA axes of this residual matrix; that is, the variance in the four response variables that is *not* accounted for by the two predictor variables (X1 and X2). There are always as many of these axes as there are response variables in the **Y** matrix. In our case, we have four response variables (Y1 to Y4) and so we have four such PCA axes. The output therefore gives the variance, and the proportion of the total variance, that is associated with these four PCA axes of residuals in decreasing order of importance. You could do a scree plot using the variance associated with each PCA axis. The values for the variance associated with these four PCA axes of residuals are about the same size in this particular example, and this tells us that there is no real structure remaining in these residuals. If you sum these four values (0.5279+0.4694+0.4338+0.32046= 1.752) then you get the same value as the "unconstrained" variance reported in the first part of the output, which is 42.19% of the total variance of the four observed response variables.

Putting this all together, we can say that the two predictor variables (X1 and X2) jointly account for 57.19% of the total variance in the four response variables (Y1 to Y4). This is the "constrained" variance. The joint effect of these two predictor variables is almost entirely captured by the first RDA axis. The remaining 42.19% of the total variance of the four response variables is not accounted for by the two predictor variables but no PCA axis based on these residuals captures particularly more of the patterns of correlation among these residuals than another. This is the "unconstrained" variance. Unless your chosen predictor variables are perfect predictors of each of your response variables – something that never occurs in real biological data – the constrained variation (i.e. the RDA axes) will always be less than the total variation that is captured by the axes of an unconstrained PCA that you studied in Chapter 2.

The next part of the output of the summary is the following:

```
Accumulated constrained eigenvalues
Importance of components:
                        RDA1    RDA2
Eigenvalue            2.3066 0.03289
Proportion Explained  0.9859 0.01406
Cumulative Proportion 0.9859 1.00000
```

This information is a source of confusion for many users of RDA. We read that RDA1 explains a proportion of 0.9859 (i.e. 98.59%) of the variance. This does *not* mean that the RDA1 axis, which is determined by the two predictor variables (X1 and X2), accounts for 98.59% of the total variance in the *observed* response variables! Rather, it means that the RDA1 axis accounts for 98.59% of the total variance in the *predicted* response variables (i.e. the constrained variance). Remember that the total variance accounted for by the two predictor variables was 57.19% and so RDA1 accounts for 98.59% of the 57.19% (i.e. 56.38%) of the total variance in the four observed predictor variables. It is entirely possible for an RDA axis to account for a large proportion of the *constrained* variance (i.e. the variance in the observed response variables that is accounted for by the predictor variables) even though this constrained variance is a small fraction of the total variance in the observed response variables. This distinction is essential when interpreting the resulting constrained ordination.

Is my explanation somewhat divorced from what you already learned about principal components analysis in Chapter 2? In fact, you already know how to recreate everything that I have presented so far given your knowledge of principal components analysis. To prove it, I will show you how you can obtain the same results using the `prcomp()` function from Chapter 2. The first step is to centre our response variables and our predictor variables:

```
#centre the response and predictor variables:
Y.obs<-scale(sim.dat1[,1:4],center=T,scale=F)
X.obs<-scale(sim.dat1[,5:6],center=T,scale=F)
```

The second step is to do the four multiple regressions and collect together (i) the predicted values of the four response variables ($\hat{Y}_{ij}$, y1 to y4) and (ii) the residuals:

```
#Predicted and residual matrices
Y.pred<-matrix(NA,nrow=100,ncol=4)
Y.res<-matrix(NA,nrow=100,ncol=4)
for(i in 1:4){
  fit<-lm(sim.dat1[,i]~sim.dat1[,5]+sim.dat1[,6])
  Y.pred[,i]<-predict(fit)
  Y.res[,i]<-residuals(fit)
}
#Residual matrix
Y.res<-Y.obs-Y.pred
```

The third step is to do three principal component analyses using the `prcomp()` function[35]: (i) the PCA of the observed variables (`Y.obs`), (ii) the PCA of the predicted values (`Y.pred`), and (iii) the PCA of the residuals (`Y.res`):

```
PCA1<-prcomp(Y.obs)
PCA2<-prcomp(Y.pred)
PCA3<-prcomp(Y.res)
```

We are now able to recreate that part of the output of the RDA that we have seen so far. The total variation of the response variables (the **Y** matrix) is given by the eigenvalues of the principal component analysis on the observed variables (PCA1):

```
total.var<-sum(PCA1$sdev^2).
```

The variation of the predicted values of the response variables (the $\hat{\mathbf{Y}}$ matrix), i.e. what the `rda()` function output calls the "constrained"

[35] You could also use the princomp() function.

variation, is given by the eigenvalues of the principal component analysis of the predicted response variables (PCA2):

```
constrained.var<-sum(PCA2$sdev^2).
```

The variation of the residual values of the response variables (the **E** matrix), i.e. what the `rda()` function output calls the "unconstrained" variation, is given by the eigenvalues of PCA3:

```
unconstrained.var<-sum(PCA3$sdev^2)
```

We can collect these results together to produce a table just like the output from `rda()`:

```
data.frame(source=c("Total","Constrained","Unconstrai
ned"),
         variance=c(total.var,constrained.var,unconst
rained.var),
         proportion=c(1,constrained.var/total.var,
         unconstrained.var/total.var ))

         source variance proportion
1         Total 4.091039  1.0000000
2   Constrained 2.339518  0.5718641
3 Unconstrained 1.751521  0.4281359
```

In general, the number of principal component axes obtained from a PCA of the original values in **Y** (here, the first four columns of `sim.dat1`) is always equal to either the total number of variables (i.e. columns) in **Y** or to the total number of observations in **Y** (i.e. rows), just as we learned in Chapter 2. This is not true for a principal component analysis done on the predicted values in $\hat{\mathbf{Y}}$. The number of principal component axes obtained from a PCA of the matrix of predicted values (

$\hat{\mathbf{Y}}$) is always equal to the number of predictor variables (i.e. the columns in **X**) that is used in the RDA. In the example above (`rda(sim.dat1[,1:4]~X1+X2,data=sim.dat1)`) we only used two explanatory variables (X1 and X2) and so there can only be two principal component axis. These two axes are called "RDA1" and "RDA2" in the summary output of the `rda()` function that was shown earlier. This makes sense since the matrix of predicted values ($\hat{\mathbf{Y}}$) are entirely determined by the values of the explanatory variables.

5.5) Hypothesis testing with rda()

An unconstrained ordination like principal components analysis simply repackages the existing patterns of variation and covariation in the response variables in a form that allows us to view it in only a few PCA axes. The purpose of conducting a constrained ordination like redundancy analysis, rather than an unconstrained ordination like principal component analysis, is because we want to test an *a priori* hypothesis about the generating process that creates these patterns of variation and covariation in the response variables. This *a priori* hypothesis is expressed in the form of the constrained axes of the RDA. Our analysis of the simulated data has so far only partitioned the variance into the part that is captured by the constrained axes (RDA1 and RDA2) and the variance that is not captured by the constrained axes (PC1 to PC4). Before proceeding any further, we need to know if our two constrained axes actually capture real effects of our two predictor variables (X1 and X2) on our response variables (Y1 to Y4). If this is not the case, then any further interpretation of our constrained axes in terms of how X1 and X2 relate to our response variables is meaningless. In other words, we now need to conduct an inferential test of significance.

This test of significance involves comparing the variance accounted for by the constrained axes to the variance not accounted for by the constrained axes (i.e. the residual variance). This should sound familiar to you because it is exactly what Fisher's F-test does in a multiple regression. In fact, the test of significance that we will use is exactly equal to Fisher's F-test if our response variables are normally distributed in a multivariate context. However, since biological data are often not normally distributed, we will use a non-parametric version of Fisher's F-test that is called a "permutation" test. The basic idea is quite simple if you look again at the general form of a multiple regression (Equation 5.1). As this equation states, the observed values of each response variable (Y_{ij}) is composed of two parts: a constrained part that is entirely determined by the predictor variables ($\hat{Y}_{ij}$) and an unconstrained part that is entirely determined by causes that are unidentified but that are independent of the predictor variables (e_{ij}); this is the residual of Y_{ij}. Each residual is, by definition, a random value that is unrelated to the constrained part. Therefore, we can randomly reassign these residuals to different constrained values. After all, since we have randomly chosen our j=1,2,…,n observations (i.e. our Y_{ij} values) then we *could have* chosen different observations since our choice was entirely random. A statistician would say that our j=1,2,…,n observations forms a random sample from a much larger statistical population. The relationship between each response variable (Y_{ij}) and the predictors, that is, the $\alpha_i + \beta_{1j}X_{1j} + \cdots \beta_{mj}X_{mj}$ part of the equation, is the same for every observation in this larger statistical population but the relationship between each response variable and the residual changes randomly between observations.

A permutation ANOVA consists of three steps. First, it calculates Fisher' F statistic[36] in the actual data. Next, it creates a new hypothetical "data set" by randomly reassigning the residuals to the predicted values and calculating the F-statistic for this new "data set". It does this a large number of times (the default number of times is 999) and this generates an empirical distribution of F-values. The larger the number of times this is done, the closer this empirical distribution approximates the theoretical distribution without having to assume any particular theoretical distribution like Fisher's F distribution. Finally, one compares the large number of simulated F-values to the observed F-value in the actual data. The proportion of times that the simulated F-values is at least as large as the observed F-value is an estimate of the null probability of observing at

[36] Actually, the `rda()` function calculates a pseudo-F statistic that is only proportional to a real F-statistic but, since this is done for each permuted data set, the proportionality constant is not needed.

least this large. This, or course, is the same thing that a parametric F-test does except that the permutation version is equally applicable to any probability distribution, not just a normal distribution.

To conduct this permutation version of an F-test, we use the `anova()` function that is included in the vegan package; I have increased the number of permuted data sets from the default of 999 to 9999 in order to show you how to change the default choice:

anova() for rda()

```
anova(sim.rda1,permutations=how(nper
m=9999))
```

```
Permutation test for rda under reduced model
Permutation: free
Number of permutations: 9999

Model: rda(formula = sim.dat1[, 1:4] ~ X1 + X2, data = sim.
dat1)
         Df Variance      F Pr(>F)
Model     2   2.3395 64.782  1e-04 ***
Residual 97   1.7515
```

This result tells us that there is less than one chance in 10000 (1e^{-4}) that the proportion of variance associated with the two constrained axes (RDA1 and RDA2) would be this large (i.e. 2.3395) purely by chance. We conclude that these two constrained axes do not represent purely random variation. We further conclude that these two constrained axes together, and therefore the two predictor variables (X1 and X2) which define this two constrained axes, really represent systemic (i.e. non-random) patterns of variation and covariation among the four response variables (Y1 to Y4). This conclusion allows us to proceed and interpret these constrained axes further. If we had concluded the contrary, i.e. that the constrained axes do not capture non-random patterns of variation and covariation, then we would not go any further.

Since there are two of these constrained axes in our simulated data, which is why there are two model degrees of freedom in the `anova()` output, we next want to know if both of these constrained axes are capturing real non-random patterns in the response variables. We can modify the `anova()` call to answer this question by inserting the `by="axis"` argument:

anova(,by="axis") for rda()

```
> anova(sim.rda1,permutations=how(nperm=9999),by="axis
")

Permutation test for rda under reduced model
Forward tests for axes
Permutation: free
Number of permutations: 9999

Model: rda(formula = sim.dat1[, 1:4] ~ X1 + X2, data = sim.
dat1)
         Df Variance        F Pr(>F)
RDA1      1  2.30662 127.7419 0.0001 ***
RDA2      1  0.03289   1.8217 0.1258
Residual 97  1.75152
```

In fact, this output is telling us that the first constrained axis (RDA1), but not the second axis (RDA2), is capturing real patterns amongst the response variables. Why? Because the null probability associated with the second constrained axis (RDA2) is quite large (0.1258) and above the traditional significance level of 0.05. What are the real patterns amongst the response variables that RDA1 is capturing? Remember that each constrained axis, including RDA1, is entirely determined by the two predictor variables (X1 and X2). If you go back to the **R** code that we used to generate our simulated data, you will see that each of the response variables was a linear function of both X1 and X2. For example, this was the line of code that defined the values of the first response variable: `y1<-0.5*X1+0.5*X2+rnorm(n,0,sqrt(1-2*0.5^2)`. The values of the two slopes (0.5, 0.5) were the same for each of the four response variables. The equation linking each of the four response variables to each constrained axis is held in the `rda()` object called `$CCA$v`:

```
sim.rda1$CCA$v
```

```
         RDA1        RDA2
y1 -0.5165904 -0.06811884
y2 -0.4959246 -0.72120264
y3 -0.4904166  0.11220498
y4 -0.4966737  0.68017396
```

The values in the first row are the coefficients linking each of the four response variables to the first constrained axis: RDA1=-0.52y1-0.50y2-0.49y3-0.50y4. The values in the second row give the same information for the second constrained axis (RDA2) although we know that this second axis does not capture any real patterns. If we go back to the line of code that defined the values of the first response variable, we could rewrite it as: `y1<-0.5*(X1+X2)+rnorm(n,0,sqrt(1-2*0.5^2)`. If we let RDA1=-1(X1+X2), then we would have: `y1<-`

`0.5*RDA1+rnorm(n,0,sqrt(1-2*0.5^2).` Written this way[37], we see that RDA1 is expressing the common effect of both X1 and X2 on each of the four response variables. Of course, if you change the code that we used to generate our data then you would change the resulting output of the `rda()` function. In the output from the `summary()` function associated with `rda()` you will see a table that is cryptically called "species scores". The values in this table actually hold the same information as does the `$CCA$v` object except that, being used for the graphical displays that will be discussed later, they are multiplied by a constant. The reason why the `rda()` function calls these values "species scores" is because the authors of the function assumed that users would be analysing community ecology data based on a siteXspecies data matrix. In fact, as I will show in Chapter 6, redundancy analysis is not always the best way of analysing such data. I urge you to play around with different ways of generating these simulated data in order to see how different ways in which the predictor variables link to the response variables will be reflected in the `rda()` output.

anova(,by="terms") for rda()

The `by="axis"` argument in the `anova()` function specified that we wish to test the significance of each axis. It is also possible to test for the significance of each predictor variable in your

[37] Actually, I have skipped a few steps. First, remember that X1 and X2 have been centred about their mean values. Second, the actual slopes must reflect the scaling between these predictor variables and the response variables. The actual slopes can be obtained via `coef(sim.rda1)`.

model by replacing the `by="axis"` with the `by="terms"` argument. If you specify `by="terms"` then the significance of each term (i.e. predictor variable) is tested in the order in which you entered these terms into your model. At this point, we must make a short detour to discuss the different types of ANOVA with respect to inferential tests of the terms when the data is "unbalanced". Good references for this topic are (Shaw and Mitchell-Olds 1993, Langsrud 2003).

A data set is unbalanced with respect to a model if the number of observations in each unique combination of values of the predictor variables is not equal. If all of the predictor variables have unique values then the data set is always balanced because each unique combination of values of the predictor variables has the same number of observations: one. When the data set is completely balanced then the different ways of conducting the test for terms are equivalent. However, if some of the predictor values are factors or are integers that repeat, then the data can be unbalanced. In such cases, different ways of conducting an ANOVA will result in different answers and so it is important that you understand how these different ways of conducting an ANOVA change the answer. In every case, the inferential test of terms in an ANOVA consists of comparing a model that excludes the term in question with a model that includes it. The different types of tests of terms occur because there are different ways of constructing the model that excludes the term in question. Consider a multiple regression like `lm(Y~X1+X2+X1:X2)`. A sequential

ANOVA test (also called Type I sums of squares) compares the following series of models; this is the default test in **R**.

```
m1<-lm(Y~1)
m2<-lm(Y~1+X1)
m3<-lm(Y~1+X1+X2)
m4<-lm(Y~1+X1+X2+X1:X2)
```

The test for the first term (X1) compares the residual sum of squares of model `m2` (which includes X1) with that of model `m1` (which doesn't). If variable X1 does not capture real variation in Y then the inclusion of the term X1 in `m2` will have about the same amount of residual variation as does the model `m1` that doesn't include X1. The test for the second term (X2) compares the residual sum of squares of model `m3` (which includes X2) with that of model `m2` (which includes X1 but not X2). The test for each term therefore compares a model without that term but including all of the variables *appearing before it as entered into the **R** command* with a model including that term plus all of the variables appearing before it. This is why it is called a sequential test. A sequential test therefore answers the following question: does the inclusion of this term significantly reduce the residual variation relative to a model that includes all terms in the model before it but excludes this one plus all terms after it? However, if the order in which you entered the variables into the model is arbitrary[38], then the result will depend on this arbitrary choice.

> *anova(,by="margin") for rda()*

To avoid drawing conclusions based on such a purely arbitrary choice, you can instead test the significance of each term by comparing two models: a model including all other terms except the one in question and a model including all terms. This is called a "marginal" ANOVA test or a Type III ANOVA test. A marginal test therefore answers the following question: does the inclusion of this term significantly reduce the residual variation relative to a model that includes all terms except this one? For instance, to test the significance of variable X2 alone (i.e. without considering its interaction with X1), a marginal test would compare the following two models:

```
m3<-lm(Y~1+X1+X1:X2)
m4<-lm(Y~1+X1+X2+X1:X2)
```

Before completing this short detour concerning different types of ANOVA tests for terms, I should mention that there are convincing reasons that marginal (Type III) tests are also not always the best choice when testing the significance of interaction terms (Langsrud 2003). In such

[38] the order in which you entered the variables into the model is usually arbitrary except in certain types of experimental designs.

situations, it is better to use "Type II" tests but the `anova()` function for constrained ordinations does not implement this method.

Before we begin looking at graphical methods of displaying a redundancy analysis, we need to look at two more statistics: the adjusted proportion of variance explained and variance inflation factors. Whenever you use the `summary()` function on an `lm()` object in **R** to do a linear regression you get two different measures of the proportion of the total variance in the response variable that is captured by the predictor variables. The first one (R^2) is equivalent to the square of the Pearson correlation coefficient (also called the "coefficient of determination") between the predicted and observed values of the response variable. The R^2 of a regression is the same thing as the constrained variance as a proportion of the total variance in RDA. However, the `summary()` function on an `lm()` object also reports an adjusted version of the coefficient of determination (R^2_{adj}) because the R^2 statistic is a biased estimate. To understand why, you have only to ask yourself the following question: If one regressed the response variable on a set of predictor variables that are all completely independent of the response variable, what value of R^2 would you expect to find? If you answered "zero" then you are thinking of what would happen in an infinitely large sample of data, or what a statistician would call the statistical population. In any random sample of finite size, there will always be some slight correlation between the predicted and observed values just by chance. Since a regression model is designed to maximize the correlation between the observed and predicted values, the resulting R^2 value will always be somewhat greater than zero. The smaller the sample size (i.e. the smaller the total degrees of freedom), the greater the resulting R^2 value that can be found by chance. The more predictor variables that are included in the regression model (i.e. the larger the model degrees of freedom), even if these predictor variables are completely independent of the response variables, the greater the resulting R^2 value that can be found by chance. The adjusted version of this statistic (R^2_{adj}) therefore corrects for the bias that is introduced based on the degrees of freedom associated with the model and the residuals. The same phenomenon occurs with RDA and so we should always concentrate on the R^2_{adj} value rather than on the R^2 value. This adjusted version is not output in the `summary()` function of an RDA object but we can still easily calculate it via the `RsquareAdj()` function of the vegan package. Applying this function to our RDA object, we see that the adjusted version is only slightly smaller than the non-adjusted version; this is because we had a fairly large sample size (100 observations) and only two predictor variables (X1 and X2). However, the difference between the two values can be much larger if you have included many predictors in a small sample of observational units.

RsquareAdj()

```
RsquareAdj(sim.rda1)

$r.squared
[1] 0.5718641
```

```
$adj.r.squared
[1] 0.5630366
```

The second statistic that we will need reports variance inflation factors. You are probably already familiar with variance inflation factors from your knowledge of multiple regression. The regression step of RDA involves calculating the partial slopes of each predictor variable. These partial slopes, being estimates, have associated standard deviations that quantify their precision. If two or more predictor variables are correlated then the resulting standard deviations of their slopes will increase and so they will each be less precise. This is the well-known problem in multiple regression of collinearity. If the correlations between the predictors get too strong then the estimates of their partial slopes become very sensitive to small changes in the correlated predictor variables and this makes the biological interpretation of these slopes, and therefore the RDA, more difficult. As with multiple regression, the degree to which this is a problem in RDA can be measured by variance inflation factors via the `vif.cca()` function in the vegan package. As a rule of thumb, variance inflation factors of 10 or greater indicate a high level of collinearity and you might want to remove such predictor variables before proceeding. The VIF of our two predictors (X1 and X2) in the simulated data are both close to 1 (indicating no collinearity) since we generated them that way. Here is how to do this using our simulated data:

vif.cca()

```
vif.cca(sim.rda1)

      X1       X2 
1.01176 1.01176 
```

You cannot use qualitative (i.e. factor) response variables (i.e. the **Y** matrix) in RDA although you can include binary response variables if they are numeric and coded as 0/1. This constraint does not apply to the predictor variables (i.e. the **X** matrix). To understand why, remember that the response variables are regressed on the predictor variables in RDA and multiple regression can accommodate qualitative (i.e. factor) predictor variables. A simple linear regression in which the lone predictor variable is a factor is an example of a 1-way analysis of variance. A multiple linear regression in which all of the predictor variables are factors is an n-way analysis of variance. A multiple linear regression in which there are both qualitative and quantitative predictor variables is an analysis of covariance. All of these different versions of the linear model can be incorporated into RDA by the simple expedient of defining each predictor variable as being a numeric or a factor variable.

5.6) Graphical output of rda() via triplots

We used biplots in Chapter 2 for graphically display of the results of a PCA. The graphical presentation of the results from an RDA is more complicated than the biplots of PCA because we have more information to present: the response and predictor variables, the row scores based on the observed response variables and the row scores based on the predicted response variables. Each can be presented based on the type 1 or type 2 scaling of biplots that you already understand from Chapter 2. Unfortunately, the explanation for interpreting the graphs is also complicated by the confusing terminology adopted by the `rda()` function because this terminology mixes together concepts from correspondence analysis and redundancy analysis.

PCA produces "loadings" and "scores". As you already know, the loadings are the coefficients linking the variables (i.e. columns of the data matrix) to the new axes that have been rotated and translated and the scores are the values of each observational unit on the new axes. Such rotated and translated principal component axes are those that best follow the major trends in covariation between the variables and the original axes. RDA also produces loadings and scores. The loadings of an RDA are the coefficients linking the columns of the matrix of *predicted* response variables to the rotated and translated original axes of the response variables; these are the constrained axes. However, the loadings in RDA are called "species scores" in the output of the `summary()` function. This unfortunate choice of words arises because the authors of the `rda()` function assumed that such redundancy analyses would be based on the types of siteXspecies matrices that are obtained in community ecology.

Since the columns (and thus variables) of such a matrix are abundances of each species, the coefficients are therefore called "species scores".

The "scores" of a PCA are the values of each observational unit (i.e. each row of the data matrix) on the new rotated and translated axes. RDA also has scores for each observational unit however, since RDA has both predicted ($\hat{Y}_{ij}$) and observed (Y_{ij}) values of the response variables, there are two types of scores. The scores of each observational unit based on the predicted values of the response variables are called "site constraints" or "linear combinations of constraining variables" in the `summary()` output of `rda()`. The term "site constraints" again comes from the assumption that the RDA is performed on a siteXspecies matrix from community ecology since each "site" in such a matrix is also one row of the matrix. The scores of each observational unit based on the observed values of the response variables are called "site scores" or "weighted sums of species scores"; the second name arises from the fact that the same `summary()` function is used for canonical correspondence analysis. These site scores are a compromise between the scores of an unconstrained ordination (i.e. a PCA) and the site constraints of the constrained axes.

Triplots via plot()

We used biplots in Chapter 2 because a PCA produces two types of information: row scores ("sites" in RDA parlance) and column, or variable, scores ("species" in RDA parlance). Triplots are produced via the `plot()` function associated with rda objects. For instance, typing `plot(sim.rda1)` will produce a triplot using the default settings. However, these default settings are not very esthetically pleasing and they do include all of the information that we often want to display and so we will follow the recommendations of Borcard et al. (2018) when producing such triplots. In fact, you can download[39] the `triplot.rda()` function from the website for that book but it is also possible to create equivalent plots by changing the default arguments of the `plot()` function in the vegan package. The row scores (the values of the observational units on the constrained axes) are shown as black points (either dots or numbers). Numeric predictor variables are shown as blue arrows while factor predictor variables are shown as points (centroids). Response variables are shown as broken red lines. Table 5.1 summarizes how to interpret these triplots.

Property	Type 1 scaling	Type 2 scaling
Angles between the constrained axes and the response variables (lines) or the predictor variables (arrows)	meaningless	Reflect the correlations between them
Angles between response variables (lines) and numeric predictor variables (arrows)	Meaningless	Reflect the correlations between them

[39] http://adn.biol.umontreal.ca/~numericalecology/numecolR/

Angles among response variables (lines)	Meaningless	Reflect the correlations between them
Angles among the predictor (variables (arrows)	Meaningless	Reflect the correlations between them
Projecting a point or centroid onto an arrow or line	Shows the position of the point or centroid on that numeric predictor (arrow) or response (line) variable	Shows the position of the point or centroid on that numeric predictor (arrow) or response (line) variable
Distances between point or centroids	Approximate the Euclidean distances between them	Approximate the Mahalanobis, not the Euclidean, distances between them

Table 5.1. A summary of how to interpret each property of an RDA triplot when using either type 1 or type 2 scaling.

You first have to decide which constrained axes you want to display by specifying the `choices=` argument within the `plot()` function. For instance, if you want to display the first and the third constrained axes then you would specify `choices=c(1,3)`. If you omit this argument then the default is to display the first two constrained axes.

You next have to decide which types of loadings and scores on the constrained axes that you want to include in the plot. You do this using the `display=` argument. The loadings (i.e. the coefficients by which the original response variables are rotated and translated to produce the constrained axes) are called "species" in the confusing nomenclature of the `plot()` function; an easy way to remember this is to note the structure of a typical siteXspecies matrix in community ecology. The columns of such a matrix index each of the variables which, in the context of a siteXspecies matrix, are the abundances of each species. Thus, including the argument `display="species"` specifies that only the loadings of the constrained axes will be plotted[40]. If a predictor variable is a categorical variable and defined as a factor in the `rda()` function then the mean value, or "centroid" of each of the different levels of this factor variable will be displayed as points by using the `display="cn"` argument. The scores are the values of each observational unit, are called "sites" in the confusing nomenclature of the `plot()` function because the authors of this function assumed that people would be using siteXspecies data and "sites" are represented by its rows. There are two different types of scores of the constrained axes in RDA. First, there are the scores of the *predicted* values of observational units on the constrained axes. These scores are specified using the `display="lc"` argument ("lc" is an acronym for "linear combinations" of predictor variables). Second, there are the scores of the *observed* values of the observational units on the constrained axes. These are specified using the `display="wa"` argument ("wa" is an acronym for "weighted averages" of predictor variables). You can combine these different options together; for instance `display=c("sp","wa","cn")` argument states

---

[40] The argument `display="reg"` allows you to display the actual partial slopes of the regressions of each response variable on the predictor variables, but this option is not often used because these partial slopes cannot be interpreted individually but only together.

that you want to plot the loadings of the response variables ("sp"), the scores of the observed values of the observational units ("wa") and the centroids of the predictor factor variables ("cn").

Next, you have to decide on which type of scaling you want to use as you already learnt in Chapter 2. This is done using the `scaling=` argument of the `plot()` function. Table 5.1 explains how to interpret triplots using either type 1 (`scaling=1` or `scaling="species"`) or type 2 (`scaling=2` or `scaling="sites"`). The final argument of the `plot()` function that we will use is `type=`. If we specify `type="points"` then loadings and scores will be shown as points but if we specify `type="text"` then the loadings and scores will be shown as text labels. If we specify `type="none"` then nothing will be plotted; this is useful since we will want to modify our triplots.

A good way of producing esthetically pleasing, and more easily interpreted, triplots is to first use the `type="none"` argument in the `plot()` function and then use some combination of the `points()`, `arrows()`, `text()` and `scores()` functions to control exactly how you want the loadings and scores to be plotted. When doing this, it is essential that you explicitly specify the same type of scaling as used in the `plot()` function or else the resulting triplots will be wrong. For instance, here is how to create a triplot of our simulated data using the first two constrained axes (`choices=c(1,2)`), type 1, or distance-preserving, scaling (`scaling=1`) and displaying the predicted row scores as black points, the two predictor variables that form the two constrained axes as blue arrows and the four response variables as red labels linked to red broken lines:

```
out<-
plot(sim.rda1,choices=c(1,2),scaling=1,type=c("none")
,
          display=c("sp","lc"))
title(main="Type 1",sub="distance-preserving")
text(sim.rda1,choices=c(1,2),display="bp",scaling=1,
col="blue")
points(sim.rda1,display="lc",scaling=1,col="black",
pch=1)
text(sim.rda1,display="species",scaling=1,col="red",
cex=0.75)
for(i in 1:dim(out$species)[1]){

lines(x=c(0,out$species[i,1]),y=c(0,out$species[i,2])
,
  col="red",lty=2)
}
```

Figure 5.3 (left) shows the resulting triplot of our simulated data using type 1 scaling. The first line sets up the triplot by specifying the constrained axes to use (`choices=c(1,2)`), the type of scaling (`scaling=1`), and what types of loadings ("sp") and scores ("lc"), i.e. `display=c("sp","lc")`. It also specifies that plot remain blank (`type=c("none")`). The third line adds the blue arrows and the labels for the two predictor variables. The forth line adds the predicted row values (the scores, thus `display="lc"`) as black points. The fifth line adds the loadings of the response variables (rows) as red text labels. Finally, the loop adds the broken red lines jointing the names of the response variables to the origin. Changes to the arguments will result in different information presented in the triplots so long as you remember to make the changes consistently in all of the lines. For instance, if you want type 2 scaling then you have to change from `scaling=1` to `scaling=2` everywhere. Similarly, if you want to display the observed row scores rather than the predicted row scores then you have to change from `display="lc"` to `display="wa"` everywhere. Figure 5.3 (right) shows the triplot using type 2 scaling. Using Table 5.1, we see the overwhelming importance of the first RDA axis using type 1 scaling, which preserves the Euclidean distances of the original data, since the scatter of points is almost entirely arranged along the first RDA axis. The patterns of correlation between the response variables (y1 to y4) and between the predictor variables (X1 and X2) is seen using type 2 scaling (Figure 5.3 right). We see that all four response variables are strongly and positively correlated with each other because the four red lines are all close to one another and therefore with small angles between them. These four response variables are also positively, but only moderately correlated with the two predictor variables because the angles between each red line (a response variable) and each blue arrow (the predictor variables) are larger. Finally the two predictor variables are not correlated at all with each other because the angles between the two blue arrows is almost 90°.From Chapter 2 you will remember that type 2 scaling approximates the Mahalanobis distance between the points, which means that the distances are scaled in standard deviations from the origin. This is why the distances between the points along the second RDA axis have been expanded while the distances between the points along the first RDA axis have been contracted.

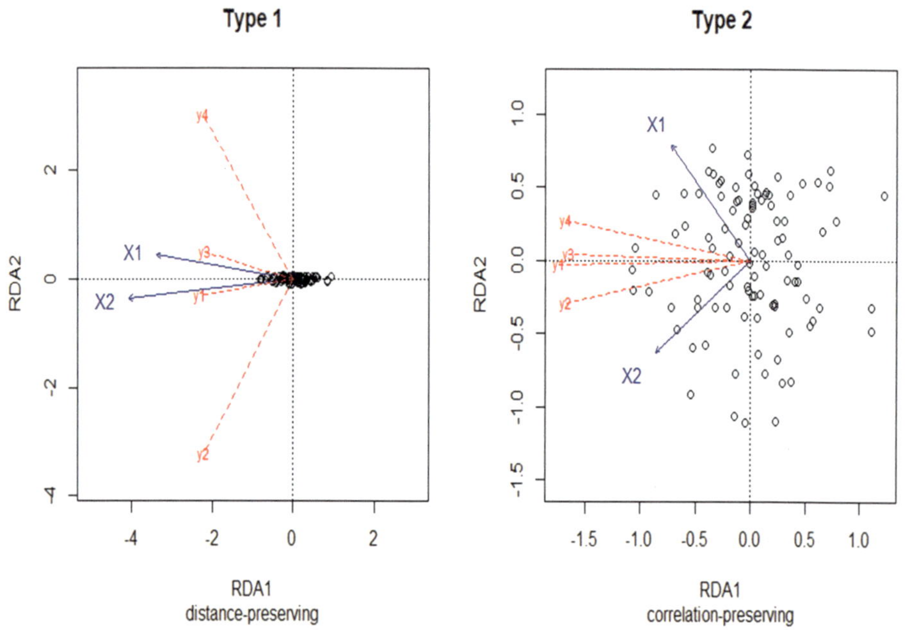

Figure 5.3. The same RDA triplot of the simulated data showing the different types of information that is provided by type 1 and type 2 scaling.

5.7) An empirical example of rda()

We will now look at an empirical example that will illustrate some additional steps in RDA. This example uses data from Daou and Shipley

(2019) in which four plant species (*Festuca rubra, Trifolium pretense, Triticum aestivum* and *Arabidopsis thaliana*) of contrasted ecology were grown under controlled environmental conditions as individual plants in 76 different natural soils taken from a wide selection of grassland areas in southern Quebec (Canada). Six soil properties were measured in each soil: nitrate and phosphate flux rates, pH, water holding capacity, log-transformed organic matter concentration and the log-transformed ratio of sand to clay content. The relative growth rate (RGR) of each species in each soil was also measured. RGR is the proportional increase in dry mass of an average individual of a species per day and so is a measure of growth efficiency in a specific environment which, here, means a specific soil.

An important assumption of RDA is that any systematic relationships between each predictor (here, the soil properties) and each response variable (here, the RGR values of each species) are linear or, at least, not strongly nonlinear. This is because RDA involves a linear regression of each response variable on the set of predictor variables. In this example, this translates to assuming that the variation in RGR of each species changes linearly with respect to each of the soil properties. If I thought that RGR had a more complicated nonlinear response with any of the soil properties then I would either have to find a transformation that would render the relationship approximately linear or else remove such a soil property from the analysis[41]. One way of evaluating this assumption is to plot each response variable as a function of each predictor variable if there is any good biological reason to suspect a nonlinear response. I won't show all of these plots but none of them displayed any obvious nonlinear patterns. The first step is to obtain the RDA object. The **Y** matrix, holding the RGR values of the four species measured in 76 different soils, is `Daou[,3:6]`. The `Daou` data frame also holds the **X** matrix that contains the six measured soil properties and I have chosen to enter these using the formula version of `rda()`:

```
RDA.Daou<-rda(Daou[,3:6]~NO3+P+pH+WHC+logMO+logSandCl
ay,data=Daou)
RDA.Daou
```

```
Call: rda(formula = Daou[, 3:6] ~ NO3 + P + pH + WHC + logM
O +
logSandClay, data = Daou)

              Inertia Proportion Rank
Total         596.7890     1.0000
Constrained   190.1467     0.3186    4
Unconstrained 406.6423     0.6814    4
Inertia is variance

Eigenvalues for constrained axes:
  RDA1   RDA2   RDA3   RDA4
174.36  11.44   3.09   1.26

Eigenvalues for unconstrained axes:
   PC1    PC2    PC3    PC4
219.12  87.21  58.18  42.13
```

[41] One could also modify the RDA analysis by replacing a linear regression model with a nonlinear regression model but this cannot be done in the `rda()` function of vegan.

We see that about 32% (i.e. "proportion constrained" = 0.3186) of the variance in the RGR values of the four species is captured using these six soil properties. The variance captured by each RDA axis is given next ("Eigenvalues for constrained axes"). If we plotted the eigenvalues for the constrained axes then this would result in a scree plot like we used in Chapter 2. The first RDA axis captures almost 92% of the constrained variance since the variance captured by it is 174.36 and the total variance captured by all of the RDA axes ("Total inertia") is 190.1467 (174.36/190.1467=0.9170). Since the drop is so precipitous when we look at the variance captured by the second RDA axis (11.44), this suggests that there might only be one important RDA axis. Looking at the variance (i.e. eigenvalues) accounted for by the unconstrained axes, we also notice that the first unconstrained axis accounts for almost 54% (219.12/406.6423=0.5389) of the unconstrained variation in the RGR values; that is, the variation in the RGR values that is not related to the measured soil properties. Since the drop is also quite precipitous when we look at the unconstrained variance accounted by second unconstrained axis, this suggests that there is only one additional pattern of covariation beyond that captured by the RDA axes. Of course, we can't know the source of this additional pattern of covariation but we can say that it is not related to the measured soil properties.

To show the flexibility of the `plot()` function, I will only show the scores of the predicted RGR values using type 1 (distance-preserving) scaling. Rather than plotting these values for the 76 soils as points, I will use the site numbers of the soils. This is done by first creating a blank plot using the `type=c("none")` argument in `plot()` and then adding the site numbers using the `text()` function. I will then show the full triplot for type 2 scaling. I have also used some basic plotting arguments (`cex=0.75, xlim=c(-6,7)`) that are common to all plots in **R**. Here is the code:

```
par(mfrow=c(1,2))
#type 1 plot; only scores
plot(RDA.Daou,choices=c(1,2),scaling=1,type=c("none")
,display="lc")
title(main="RDA  of  Daou  data,  type  1  scaling",
sub="distance triplot")
text(RDA.Daou,choices=c(1,2),display="lc",scaling=1,
col="black",cex=0.75,labels=as.character(1:76))

#type 2 plot; full triplot
out<-
plot(RDA.Daou,choices=c(1,2),scaling=2,type=c("none")
,
          display=c("sp","lc"),xlim=c(-6,7))
title(main="RDA of Daou data, type 2 scaling",
      sub="correlation triplot")
text(RDA.Daou,choices=c(1,2),display="bp",scaling=2,
col="blue",cex=0.75)
text(RDA.Daou,choices=c(1,2),display="lc",scaling=2,
col="black",
     cex=0.75,labels=as.character(1:76))
text(RDA.Daou,display="species",scaling=2,col="red",
cex=0.75)
for(i in 1:dim(out$species)[1]){

lines(x=c(0,out$species[i,1]),y=c(0,out$species[i,2])
,
        col="red",lty=2)
}
```

```
par(mfrow=c(1,1))
```

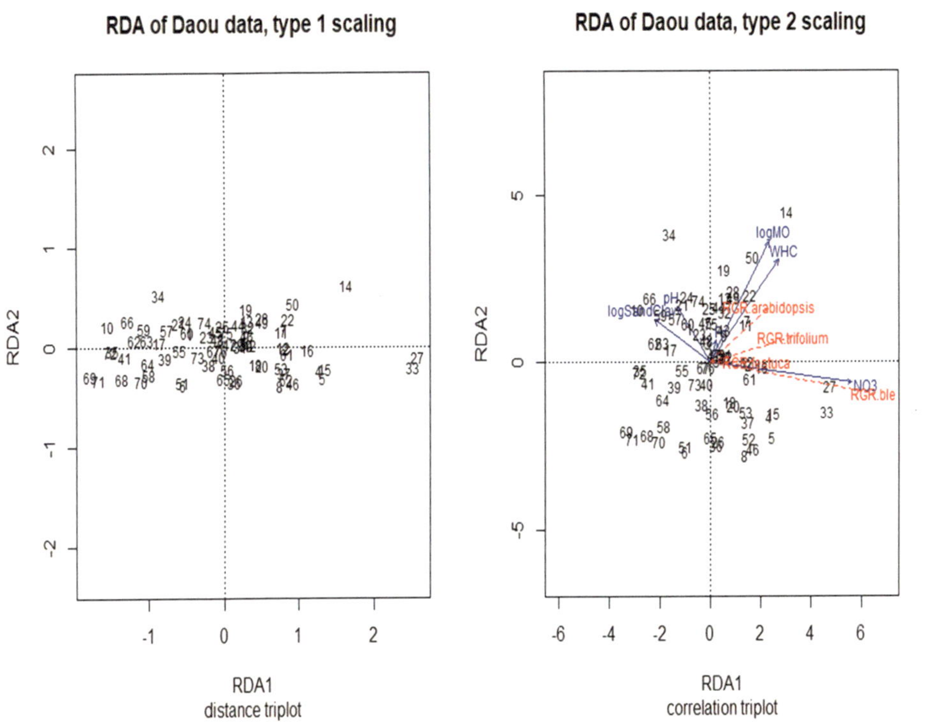

Figure 5.4. Triplots of the RGR values of four plant species as a function of six environmental variables based on type 1 scaling (left) and type 2 scaling (right).

The result is shown in Figure 5.4. Here is how to interpret it. The distance-preserving plot (type 1 scaling) shows where each of the sites occur on the first two constrained axes (Figure 5.4 left). The dominance of the first RDA axis relative to the second one is clear since the range of scores along the first RDA axis is so much greater than the range of scores along the second RDA axis. The soils in sites 27 and 33 have the largest values on the first constrained axis. Since these two soils are very close together, and since this is a type 1 (distance preserving) plot, this means that the four species had very similar RGR values in these two soils. We can see what is unusual about these two soils by looking at the correlation-preserving biplot (Figure 5.4, right). We see that the RGR values of all four species are positively correlated with the first constrained axis since the angles between each red line (species) and RDA1 are small (smaller angles mean a higher positive correlation). In other words, sites further to the right on RDA1 have larger values of RGR for all four species and so sites 27 and 33 have soils that produce the largest growth rates of all four species. Furthermore, the first constrained axis is strongly correlated with soils having high flux rates of NO_3 since the angle between its blue arrow and RDA1 is very small. The interpretation of the second constrained axis is done in the same way. Since the angles between the red lines are small, the correlations between the RGR values of each pair of species is large. Since the angle between soil organic matter (logMO) and soil water-holding capacity (WHC) is small, these two soil properties are strongly positively correlated. Similarly, soil sand:clay and soil pH are strongly positively correlated and they are somewhat negatively correlated with soil NO_3 flux because the angles between them are larger than 90°.

We also want to make sure that the predictor variables (the soil properties) are not so highly correlated that the RDA results become unstable and uninterpretable. To find out, we must calculate the variance inflation factors. The results are reassuring since all of the variance inflation factors are much lower than 10.

```
vif.cca(RDA.Daou)

        NO3              P             pH            WHC          logMO
logSandClay
   1.237323       1.069284       1.059294       1.629512       1.293236
1.359779
```

Next, we need to assure ourselves that the RDA axes actually account for real patterns of variation in the RGR values before continuing with our interpretation. This requires a permutation ANOVA. We will also test for the significance of each RDA axis.

```
anova(RDA.Daou)

Permutation test for rda under reduced model
Permutation: free
Number of permutations: 999

Model: rda(formula = Daou[, 3:6] ~ NO3 + P + pH + WHC + log
MO + logSandClay, data = Daou)
         Df Variance      F Pr(>F)
Model     6   190.15 5.3774  0.001 ***
Residual 69   406.64

anova(RDA.Daou,by="axis")

Permutation test for rda under reduced model
Forward tests for axes
Permutation: free
Number of permutations: 999

Model: rda(formula = Daou[, 3:6] ~ NO3 + P + pH + WHC + log
MO + logSandClay, data = Daou)
         Df Variance       F Pr(>F)
RDA1      1   174.36 30.4440  0.001 ***
RDA2      1    11.44  1.9971  0.823
RDA3      1     3.09  0.5388  1.000
RDA4      1     1.26  0.2197  1.000
Residual 71   406.64
```

The first ANOVA result confirms that the six RDA axes together account for non-random variation. There are six RDA axes because there were six soil properties in the model, which is why there are six model degrees of freedom. The second ANOVA result confirms that only the first RDA axis captures clearly non-random variation since the null probabilities associated with the remaining axes are all large. This means that we should not be making any biological conclusions about the patterns in Figure 5.4 concerning the second constrained axis.

How much non-random variation has been captured? The output from the RDA object told us that 31.86% of the variation was captured by the predictor variables (soil properties) but we know that this is a biased estimate since we included six predictor variables in the model. To correct for this bias, we calculate the adjusted R^2 value. We see that an unbiased estimate is about 26% of the variation.

```
RsquareAdj(RDA.Daou)
```

```
$r.squared
[1] 0.3186162

$adj.r.squared
[1] 0.2593655
```

We have established that the RDA axes account for a significant amount of the variance among the RGR values of our four species and that only one RDA axis is clearly capturing non-random variation. We also have some indication that not all of the soil properties are required in our model since the first RDA axis seems to be only strongly correlated with soil nitrate flux. Can we produce a more parsimonious RDA model, i.e. one with fewer soil properties, that still captures essentially the same proportion of the variance among the RGR values as our current RDA? To find out, we can use a multivariate equivalent of stepwise regression. Stepwise regression – at least, the version provided in the vegan package – uses the Akaike Information Criterion (AIC) (Akaike 1973) to search for subsets of an initial multiple regression that provide the best trade-off between maintaining the explained variance and minimizing the number of predictor variables (Burnham and Anderson 2002). There are different strategies for doing this; one can start with the full set of predictors and sequentially remove them (backwise selection), one can start with the simplest model (i.e. an intercept) and sequentially add predictors (forward selection) or one can do first backward selection and then forward selection. The equivalent method in RDA is found in the `ordistep()` function of the vegan package. By default, this function produces lots of output because it describes the result of each step but you can supress this output if you add the `trace=FALSE` argument. If you want to limit the combinations of predictor variables that can be considered for removal or addition then you can specify this via the `scope=` argument but the full range of possible models in used by default. Here is how to use the `ordistep()` function with these data:

ordistep()

```
parsimonious.RDA.Daou<-ordistep(RDA.Daou,direction="b
oth",trace=F)
parsimonious.RDA.Daou

Call: rda(formula = Daou[, 3:6] ~ NO3, data = Daou)

              Inertia Proportion Rank
Total         596.789      1.000
Constrained   156.954      0.263    1
Unconstrained 439.835      0.737    4
Inertia is variance

Eigenvalues for constrained axes:
  RDA1
156.95

Eigenvalues for unconstrained axes:
   PC1    PC2    PC3    PC4
238.61  95.49  62.11  43.61
```

The output tells us that the most parsimonious RDA model is one with only one predictor variable whose formula is returned: `rda(formula = Daou[, 3:6] ~ NO3, data = Daou)`. This model, with only one soil property (soil nitrate flux), accounts for 26.3% of the variance in the RGR values of the four species and this is slightly more that the unbiased variance that was captured by the initial model with six soil properties (i.e. 25.9%). When we use the `R2Adjust()` function on this parsimonious model we see that its unbiased estimate is 25.3%.

We can now look at the triplot of our new model but there is one complication. The `plot()` function cannot plot results when there is only one RDA axis since such a plot would be a 1-dimensional line rather than a 2-dimensional plane. I therefore include the soil phosphorus flux into the RDA since that was the variable that was last removed from the `ordiplot()` analysis and the variable whose null probability almost reached significance. I do this simply to get a model with two RDA axes. We must remember not to place much importance onto this second RDA axis in the interpretation since it is not significant. The result is shown in Figure 5.5. The main conclusions from this final RDA are:

1. The six measured soil properties generate patterns of variation in the RGR values of the four species that can be expressed along a single dimension (RDA axis).
2. Of these six measured soil properties, only the soil nitrate flux is required to account for such patterns of variation in the RGR values.
3. Soil nitrate flux alone captures about 26% of this variation in the RGR values and an unbiased estimate is 25%.
4. The remaining 76% of the variance is not accounted for by any of the other soil properties.

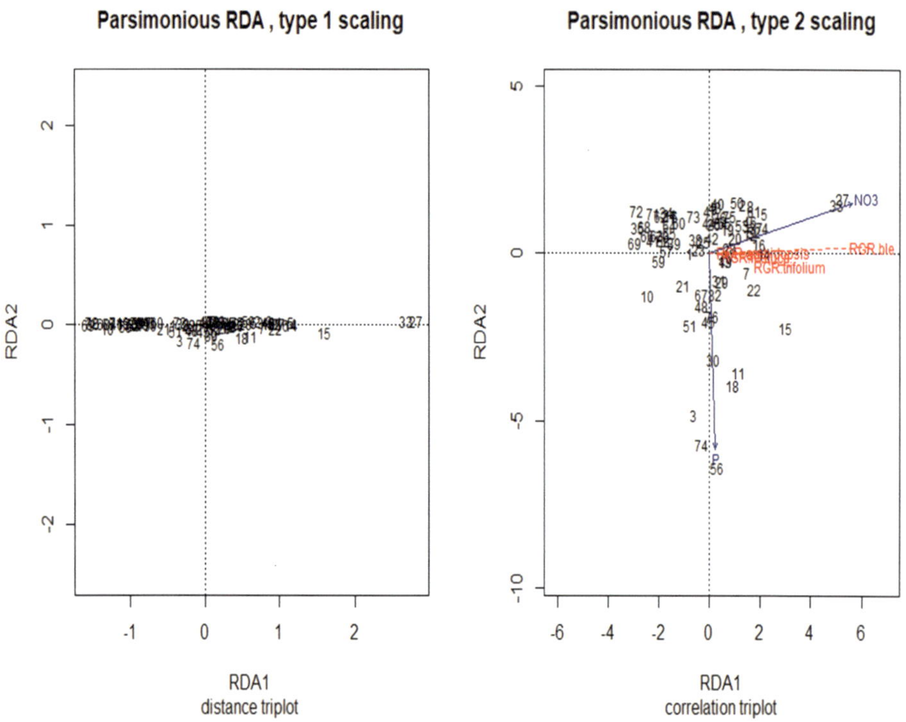

Figure 5.5. Triplots of the RGR values of four plant species as a function of nitrate (NO3) and phosphate (P) soil fluxes based on type 1 scaling (left) and type 2 scaling (right). The second RDA axis is not significant.

Of the remaining 76% of unexplained variance in the RGR values ("Eigenvalues for unconstrained axes"), a value of 238.61 is captured by the first unconstrained axis and this represents 54% of all of the unconstrained variance ($100 \cdot 238.61/439.835$). This strongly suggests that there is only one additional dimension whose origin we will eventually have to explain using additional predictor variables. These unconstrained axes will have to be explored and interpreted as we did in Chapter 2 when looking at unconstrained PCA. One way of generating hypotheses about what these additional predictor variables might be by searching for correlations between new potential predictor variables and the loadings and scores on this first unconstrained axis. The loadings on this first unconstrained axis can be found in the `parsimonious.RDA.Daou$CA$v[,1]` object and one could search for correlations between these loading and properties of the species. The scores can be found in the `parsimonious.RDA.Daou$CA$u[,1]` object and one could search for correlations between these scores and new properties of these sites beyond the measured soil properties.

5.8) Partial Redundancy Analysis

When we are using several predictor variables, we sometimes want to know how a subset of these predictor variables relate to the set of response variables while statistically holding constant the remaining variables. This is the goal of partial redundancy analysis. For instance, the above analysis has shown us that all of the patterns of variance and covariance among the RGR values of the four species can be associated with a single predictor variable (soil nitrate flux). In other words, the remaining predictor variables provide no additional information concerning the patterns of variance and covariance among the RGR values beyond that already contained in soil nitrate flux. This does not necessarily mean that the remaining variables are independent of the RGR values. Rather, it means that whatever dependencies they might have with the RGR values are removed by soil nitrate flux. However, we might want to know how the patterns of variance and covariance among the RGR values of the four species relate to the three remaining physical characteristics of the soils (organic matter concentration, water-holding capacity and soil texture) after controlling for the nutrient fluxes (nitrate and phosphate).

To start, we will do a RDA using only the three soil physical properties, determine if this set of three physical properties is significantly related to the set of four RGR values, and how many of the three resulting constrained axes are statistically significant:

```
soil.physical<-
rda(Daou[,3:6]~WHC+logMO+logSandClay,data=Daou)
anova(soil.physical)
Permutation test for rda under reduced model
Permutation: free
Number of permutations: 999

Model: rda(formula = Daou[, 3:6] ~ WHC + logMO + logSandClay,
data = Daou)
         Df Variance      F Pr(>F)
Model     3    57.88 2.5775  0.019 *
Residual 72   538.91

anova(soil.physical,by="axis")

Permutation test for rda under reduced model
Forward tests for axes
Permutation: free
Number of permutations: 999

Model: rda(formula = Daou[, 3:6] ~ WHC + logMO + logSandClay,
data = Daou)
```

```
         Df Variance      F Pr(>F)
RDA1      1    49.91 6.6683  0.024 *
RDA2      1     6.72 0.8973  0.698
RDA3      1     1.25 0.1668  0.921
Residual 72   538.91
```

The output of the second line (`anova(soil.physical)`) tells us that this set of three soil physical properties are significantly related to the set of four RGR values since the null probability of the model is only 0.02. The output of the third line (`anova(soil.physical,by="axis")`) tells us that this set of three soil physical properties relates to the set of four RGR values only through one dimension since only the first of the three constrained axes is significant. Figure 5.6 plots the first two constrained axes using type 2 scaling. Remember that only the first constrained axis is significant so don't attach much importance to the second axis. We see that increasing values of RGR for each of the four species are associated with increasing amounts of soil organic matter and of soil water-holding capacity, and lower amounts of sand versus clay.

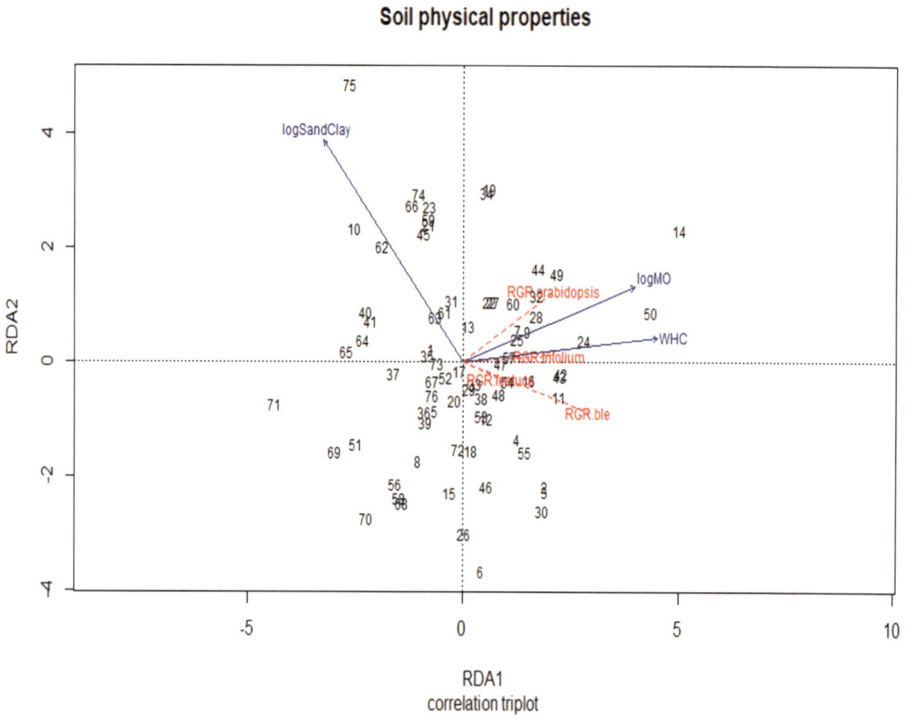

Figure 5.6. A correlation triplot of the RGR data of Daou and Shipley (2019) using only the three soil physical properties: organic matter content (MO), water-holding capacity (WHC) and the amount of sand versus clay in the soil. Only the first RDA axis is significant.

We know, from the above result, that the physical properties of the soil are associated with increasing growth rates of all four plant species through a single dimension, that is, through a single constrained axis of these soil physical properties. However, we also know that when we included the soil chemical properties in our first redundancy analysis of these data, there was also only a single significant RDA axis and that soil NO_3 flux seemed to be the most important. How can we reconcile these two conclusions? Since much of soil NO_3 is generated by the decomposition of soil organic matter, and since soil NO_3 must be dissolved in water in order for plant roots to absorb it, perhaps the causal chain is that these physical properties of the soil cause the soil NO_3 flux which then causes the increased RGR of each of the four plant species? If this hypothesis is true then these soil physical properties would be significantly related to the set of RGR values, but their causal effect would be blocked if we "held constant" the soil NO_3 flux.

How can we test this hypothesis? Clearly, we can't physically hold constant the NO_3 flux of these soils while still allowing the physical properties to vary, but we can statistically hold constant the NO_3 flux. This is what a partial RDA does. Here is how to test this hypothesis. First, we specify a partial RDA for these data, in which the three physical soil properties are related to the set of four RGR values while statistically controlling for (or statistically "holding constant") the soil NO_3 flux. Second, we test the null hypothesis, implied by this model that the soil physical properties are not associated with the variation in the RGR values of the set of four species once we statistically hold constant the soil NO_3 flux:

```
partialRDA.Daou<-
rda(Daou[,3:6]~WHC+logMO+logSandClay+Condition(NO3),d
ata=Daou)
```

The predictor variables of interest (`WHC`, `logMO` and `logSandClay`) are placed first on the right-hand part of the formula and the conditioning variables (the predictor variables whose effects we want to hold constant, i.e. `NO3`) are given next and enclosed in the `Condition()` argument. The resulting output changes slightly from RDA without conditioning variables:

```
partialRDA.Daou

Call: rda(formula = Daou[, 3:6] ~ WHC + logMO + logSandClay
+ Condition(NO3), data = Daou)

              Inertia Proportion Rank
Total         596.78898    1.00000
Conditional   156.95433    0.26300    1
Constrained    19.30709    0.03235    3
Unconstrained 420.52757    0.70465    4
```

The total variance ("Inertia") is listed first (596.78898). The table then decomposes this total variance into the variance captured by the conditioning variable `NO3` ("Conditional", which is 26.3% of the total variance, i.e. $100 \cdot 156.95433 / 596.78898$), the variance captured by the chosen predictor variables `WHC`, `logMO`, and `logSandClay` ("Constrained", which is 3.2% of the total variance), and the remaining residual variance (Unconstrained, which is 70.5% of the total variance). However, is this constrained variance significantly more than one would expect by chance? According to our hypothesis, the three soil physical properties (the variables doing the constraining) should not be associated with the patterns of correlation and covariation in the set of four RGR values once we statistically hold constant the soil NO_3 flux (the conditional variable). If our hypothesis is right then this constrained variance should not be statistically significant. To determine this, we need to test this using the `anova()` function:

```
anova(partialRDA.Daou)
```

```
Permutation test for rda under reduced model
Permutation: free
Number of permutations: 999

Model: rda(formula = Daou[, 3:6] ~ WHC + logMO + logSandClay
+ Condition(NO3), data = Daou)
         Df Variance      F Pr(>F)
Model     3    19.31 1.0866  0.365
Residual 71   420.53
```

The null probability is 0.365. We had already concluded that these three physical properties of the soil were significantly related to the joint RGR responses of the plants when we ignored `NO3` since the null probability on the first RDA axis was 0.024. However, when we hold constant, or "condition on", `NO3`, the null probability increases to 0.365. This agrees with our biological hypothesis: these three physical properties of the soil do affect the joint RGR responses of the plants along a single dimension, but this affect on RGR is blocked when we hold constant NO3.

5.9) Canonical Correspondence Analysis (CCA)

If you have understood the section on redundancy analysis then you already understand most of canonical correspondence analysis, which is the constrained version of correspondence analysis. As you know, correspondence analysis can be described both from the viewpoint of the reciprocal averaging method that we studied in Chapter 3 or as an eigen technique like PCA. This is also true for canonical correspondence analysis. The best way of visualizing CCA is by revisiting the reciprocal averaging method that you learned in Chapter 3. Here is the same text that I gave in Chapter 3 except that I have added one additional step in **bold** font.

Initialization step. Choose initial estimates (x_i) of the unknown environmental gradient (i.e. the row scores).

3. Do the next four steps.
    - E. Given x_i, calculate new species scores (μ_j) by weighted averaging using Equation 3.4
    - F. Given the new μ_j, calculate new estimates of the unknown environmental gradient x_i^* (i.e. the sampling unit, or row, scores) using Equation 3.5.
    - **G. Regress the x_i^* onto the measured predictor variables and replace the x_i^* by their predicted values.**
    - H. Rescale the estimates of the environmental gradient to give x_i using Equations 3.6 to 3.8.
    - I. Calculate the difference between the previous estimates of the environmental gradient with these new estimates.
4. If the difference between the previous and current estimates of the environmental gradient are greater than some small amount then repeat steps 1A to 1D; otherwise stop.

In other words, the only difference is that the constrained version of correspondence analysis (CCA) uses the *predicted* site scores (row, or sampling unit, scores) given the predictor variables rather than the actual site scores themselves. This is similar to RDA, which uses the *predicted* values of the response variables given the predictor variables rather than the response variables themselves. You might notice that this is also very similar to how we analysed the Axe Lake data at the beginning of this chapter. Everything else that you learned in redundancy analysis applies to CCA and so there is no need to repeat everything. Instead, we will use CCA to again analyze the Wisconsin understory plant data.

The data on understory plant communities of the forested regions of Wisconsin consist of 253 rows (observational units that we will call sites) and 575 columns of species' abundances (response variables). These data are held in a data frame called `siteXspecies`. We also have a second data set in which the values of 11 environmental variables, transformed to normality as explained in Chapter 3, were recorded for each site. These environmental data are held in a data frame called `siteXenv`. We will first do a CCA of the species' abundances constrained by the 11 environmental variables, even though we suspect that this is generally not a good idea, for two reasons. First, we know from Chapter 3 that several of the environmental variables are highly correlated and this might lead to problems of collinearity. Second, it is often very difficult to biologically interpret canonical axes that are composed of so

cca()

many environmental variables, especially when these environmental variables are correlated. Here is the **R** code:

```
wisconsin.CCA1<-
cca(siteXspecies~pH+sand+silt+clay+OM+Ca+K+Mg+basal.a
rea+pet+ai,
data=siteXenv)
```

The left-hand part of the formula in the `cca()` function of the vegan package gives the matrix or data frame containing the response variables (here, the abundances of the 575 understory plant species that are held in `siteXspecies`). The right-hand part of the formula gives the eleven environmental variables that will constrain the axes of the correspondence analysis. These variables are found in the data frame called `siteXenv`. Here are the first few lines of the basic output:

```
wisconsin.CCA1

Call: cca(formula = siteXspecies ~ pH + sand + silt + clay
+ OM +
Ca + K + Mg + basal.area + pet + ai, data = siteXenv)

              Inertia Proportion Rank
Total         10.9707     1.0000
Constrained    1.7386     0.1585   10
Unconstrained  9.2321     0.8415  242
Inertia is scaled Chi-square
Some constraints were aliased because they were collinear (
redundant)
```

The first part of the output, the decomposition of the variance, is the same as with RDA except that we are no longer measuring "variance" but rather "inertia". The combination of the 11 environmental variables (i.e. their combination into 10 canonical axes) accounts for 15.85% of the total inertia ("constrained inertia") and the remaining 242 unconstrained correspondence analysis axes accounts for the remaining 84.2% of the inertia. You might wonder why there are only 10 canonical axes even though there are 11 environmental variables? After all, there are supposed to be as many constrained axes as there are constraining variables. The reason is given rather cryptically in the last line of output ("Some constraints were aliased because they were collinear (redundant)"). Three of the environmental variables were the percentages of sand, silt and clay in the soil. Since these three values must always sum to 100%, this means that we can always perfectly predict the third amount if we know the other two amounts. Another way of saying the same thing is that these three variables are perfectly "collinear". Although I will show you a better way of dealing with this problem in terms of biological interpretability, one simple way of removing collinear variables is to remove one from the analysis. If we remove silt as one of the predictor variables and redo the analysis we get the following basic output:

```
Call: cca(formula = sitexspecies ~ pH + sand + clay + OM +
Ca + K +
Mg + basal.area + pet + ai, data = sitexenv)

              Inertia Proportion Rank
Total         10.9707     1.0000
Constrained    1.7386     0.1585   10
Unconstrained  9.2321     0.8415  242
Inertia is scaled Chi-square

Eigenvalues for constrained axes:
  CCA1   CCA2   CCA3   CCA4   CCA5   CCA6   CCA7   CCA8   C
CA9  CCA10
```

```
0.6546 0.3053 0.2544 0.1365 0.1074 0.0720 0.0671 0.0573 0.0
470 0.0369

Eigenvalues for unconstrained axes:
   CA1    CA2    CA3    CA4    CA5    CA6    CA7    CA8
0.4011 0.3414 0.2480 0.2269 0.2148 0.1964 0.1782 0.1665
(Showing 8 of 242 unconstrained eigenvalues)
```

We no longer have any perfectly collinear predictor variables but this does not mean that we have removed any problems of highly, but not perfectly, collinear, predictor variables that would render unstable the scores of the constrained axes. We must check for this possibility before proceeding with the analysis by looking at the variance inflation factors via the `vif.cca()` function:

```
vif.cca(wisconsin.CCA1.2)
```

```
         pH        sand        clay          OM          Ca
K
  6.274601   2.020565   1.164493   2.739770   9.358039   2.
685078
         Mg basal.area        pet          ai
 16.963677   1.339184   5.937888   5.478869
```

Large variance inflation factors, approximately 10 or more, will cause problems. We see that soil Mg and Ca concentrations are definitely suspect but even soil pH, as well as potential evapotranspiration (pet) and the aridity index (ai), have rather large variance inflation factors. We could simply remove one or more of these highly correlated predictor variables but I will instead use principal components analysis to combine the highly correlated sets of predictors as we did in Chapter 3. In Chapter 3 we constructed composite variables using the site scores of subsets of the environmental variables on the first axis of three principal component analyses. The first composite variable, which I called "NPP" was the first PCA axis (93% of the total variance) of potential evapotranspiration and the aridity index. Higher values of this first PCA axis indicated sites that were colder and/or with less precipitation. The second composite variable, which I called "soil texture", was the first PCA axis (87% of the total variance) of the three soil texture variables (sand, silt and clay). The third composite variable, which I called "cation exchange", was the first axis (83% of the total variance) of a PCA of the cation concentrations (Mg, Ca and K) plus soil pH. I first construct a new data frame of predictor variables called `siteXenv2` containing these three new composite variables plus soil organic matter content (OM) and basal area and then do a canonical correspondence analysis of the `siteXspecies` data frame of understory species abundances constrained by these new predictor variables. Here is the call to `cca()` and the resulting basic output:

```
wisconsin.CCA3<-
cca(siteXspecies~soil.texture+OM+cation.exchange+
basal.area+NPP,data=siteXenv2)
wisconsin.CCA3

Call: cca(formula = siteXspecies ~ soil.texture + OM +
cation.exchange + basal.area + NPP, data =
siteXenv2)

              Inertia Proportion Rank
Total         10.9707     1.0000
Constrained    1.2849     0.1171    5
Unconstrained  9.6858     0.8829  247
Inertia is scaled Chi-square

Eigenvalues for constrained axes:
  CCA1   CCA2   CCA3   CCA4   CCA5
```

```
0.6315 0.2551 0.2121 0.1072 0.0790
```

In our first model, including each of the original predictor variables, the constrained axes accounted for 15.85% of the total inertia while the constrained axes of this new model, using the new predictor variables, accounted for 11.71% of the total inertia. However, remember that these are biased estimates. We can use the `RsquareAdj()` function to get the unbiased estimates for the two models. These unbiased estimates are 12.4% for the first model and 9.9% for the new model. Although the constrained axes of the second model accounted for slightly less (2.5%) of the total variance, the biological interpretability of the second model is better since the variance inflation factors of the new model are all less than 4.0 and the new composite variables are clearly interpretable.

At this stage, and before we look at the detailed output and plots, we must first make sure that the constrained axes actually account for real patterns in the data. We determine this by doing a permutation ANOVA (`anova(wisconsin.CCA3,by="axis")`) just as we did in redundancy analysis. Since the Wisconsin data set is much larger, this will take a few minutes. In fact, all six constrained axes are highly significant.

Typing `summary(wisconsin.CCA3)` produces detailed information on the resulting analysis. The `summary()` output is exactly the same as with an RDA object but can be quite voluminous since we now have 575 response variables and five predictor variables. There are 575 response variables because there are 575 columns in the siteXspecies matrix contained the abundance of a given species and the abundance of each species is a response variable. There are five predictor variables (cation.exchange, soil.texture, NPP, basal.area and OM) because we are using five columns of the siteXenv2 matrix. Why does the output tell us that there are only 247 unconstrained axes (the rank of the unconstrained variables) if we have 575 response variables? The reason is because we only have 247 rows (i.e. sites) in the siteXspecies matrix; that is, we have fewer observational units than variables.

Just as when we used summary() on an RDA object, the summary() output using a CCA object can be divided into seven parts but you can suppress parts 4 to 6 by adding the axes=0 argument to the summary() function:

(1) The partitioning of the total inertia between the constrained and unconstrained axes;

(2) The proportion of the total inertia accounted for by each individual axis;

(3) The proportion of the total inertia assigned to each constrained axis as a percentage of the total inertia accounted for by all of the constrained axes;

(4) The species scores along each axis;

(5) The site scores along each axis;

(6) The site constraints (i.e. the values of the constrained axes for each site);

(7) The biplot scores for each constraining variable.

The first important information is the partitioning of the inertia between the constrained axes (determined entirely by the five predictor variables) and the unconstrained axes:

```
Partitioning of scaled Chi-square:
              Inertia Proportion
Total          10.971     1.0000
Constrained     1.285     0.1171
Unconstrained   9.686     0.8829
```

In interpreting these values, remember that there are five constrained axes because we used five predictor variables and that there will be 247 unconstrained axes because we had 247 rows (i.e. sites) in the `siteXspecies` data frame. Since the sum of the inertia associated with each of these 247+5 = 252 axes must account for 100% of the inertia, it is quite normal for each axis, including the five constrained axes, to each account for only a small proportion of the total inertia. Even though the five constrained axes only account for 11.71% of the total, you must not conclude that this is an insignificant percentage since every axis is only accounting for a small percentage of the total.

The next important information that is output is the proportion of the total inertia accounted for by each individual axis. This section is rather long since there are 252 axes, but the first few lines are as follows:

```
Eigenvalues, and their contribution to the scaled Chi-squar
e

Importance of components:
```

```
                              CCA1    CCA2    CCA3     CCA4      C
CA5      CA1
Eigenvalue               0.63147 0.25513 0.21210 0.107191 0.07
9002 0.45675
Proportion Explained   0.05756 0.02326 0.01933 0.009771 0.00
7201 0.04163
Cumulative Proportion 0.05756 0.08082 0.10015 0.109920 0.11
7121 0.15876
```

The five constrained axes are indicated as CCA1 to CCA5. CA1 is the first unconstrained axis and there are 246 more axes although I have not shown them here. The first constrained axis (CCA1) accounts for 5.8% of the total inertia and the second constrained axis (CCA2) accounts for an additional 2.3%. The fist unconstrained axis (CA1) accounts for 4.1% of the total inertia. What does this mean? The constrained axes are those that are completely determined by the predictor variables (here, the five environmental variables) and the unconstrained axes are those that are independent of the constrained axes. These numbers are therefore telling us that the first constrained axis (CCA1) is capturing slightly more of the overall patterns of association in the species composition ("proportion explained" = 0.05756, or ~5.8%) in the Wisconsin understory plant communities that the most important unconstrained axis ("proportion explained for CA1 = 0.04163, or ~4.2%). However, the second constrained axis (CCA2), while statistically significant, is capturing slightly less of the overall patterns of association in the species composition than the most important unconstrained axis. It is therefore likely that there are additional variables, beyond the five environmental variables that we used to constrain the ordination, that are structuring these understory plant communities.

The next bit of important information is the following:

```
Accumulated constrained eigenvalues
Importance of components:
                        CCA1   CCA2   CCA3    CCA4    CCA5
Eigenvalue            0.6315 0.2551 0.2121 0.10719 0.07900
Proportion Explained  0.4915 0.1986 0.1651 0.08342 0.06148
Cumulative Proportion 0.4915 0.6900 0.8551 0.93852 1.00000
```

As with the summary output of the `rda()` function, this information is often misinterpreted by users. We read that the first constrained axis (CCA1) accounts for 49.2% of the "constrained eigenvalues". This does not mean that this axis accounts for 49.2% of the total inertia; we already know that it only accounts for 5.8%. Rather, it means that the first constrained axis accounts for 49.2% of the 5.8% of the total inertia that is attributed to the five constrained axes. None the less, this is important information because it means that the first two constrained axes (CCA1 and CCA2) together represent 69% of that inertia that is associated with our five predictor variables, and so we can reasonably use these two constrained axes to represent the patterns associated with our five predictor variables. Since we know that the remaining three constrained axes were statistically significant, this means that using only CCA1 and CCA2 will ignore some of the real patterns in the data but it does allow us to visually see the main patterns.

The `summary()` object next gives us the species scores along each of the constrained axes, which are held in `summary(X)$species`, where

X is your chosen name for the cca object. These are (approximately) the points along each axis at which each of the species are most abundant. If type 2 scaling is chosen (the default), then remember that these have been scaled, i.e. multiplied, by a constant to make them proportional to the eigenvalues. The unscaled values are also stored in `X$CCA$v` of the cca object, where X is again your chosen name for the cca object. Each species has a score on each of the five constrained axes and on each of the unconstrained axes, but `summary()` only outputs the species scores on the constrained axes and the first few unconstrained axes. Since there are many species in the data set, there are many lines of output but here are the first few species and for the first four constrained axes.

```
Scaling 2 for species and site scores
* Species are scaled proportional to eigenvalues
* Sites are unscaled: weighted dispersion equal on all dime
nsions


Species scores

                                        CCA1      CCA2       C
CA3        CCA4
Abies_balsamea                      1.004116  0.7723650  0.0
10317  0.0165725
Acalypha_virginica                 -1.006036  0.4410819  0.6
45119  0.4145647
Acer_negundo                       -0.858996 -0.2094229 -0.1
13225 -0.2894679
```

By default, the scaling if Type 2 but I could have chosen Type 1 scaling by including the `scale= "site"` argument in the `summary()`

function. This output is telling us that the seedlings[42] of *Abies balsamea* are (approximately) most abundant in sites located at a value of 1.0 and 0.77 on the first and second constrained axes while seedlings of *Acer negrundo* are most abundant in sites located at -0.86 and -0.21 on the first two constrained axes. Later, we will see how each of the predictor variables load onto each of these constrained axes and this will inform us about the environmental conditions of sites located at these points on the axes.

Next, we look at the site scores, which are also available in `summary(X)$CCA$sites` where X is your chosen name for the cca object. As the output states, these site scores are computed as the weighted averages of the species scores. These site scores are a compromise between the scores of an unconstrained ordination (i.e. a correspondence analysis) and the site scores of the constrained axes, which are given later. Again, `summary()` only outputs the scores for the constrained axes and the first few unconstrained axes. Here are the first few lines of output:

```
Site scores (weighted averages of species scores)

          CCA1      CCA2      CCA3       CCA4      CCA5       C
A1
1     -0.68353  0.264431 -0.302491   0.120417  1.444767    1.0
57701
2     -0.93406  0.219126 -0.375020  -0.351216  1.679999    1.0
49069
3     -0.78679  0.231248 -0.550371   0.401315  1.767864    1.7
26879
```

[42] Because this data set only records the understory vegetation and so only the seedlings of the canopy trees are included.

The next information that is output by the `summary()` function is called "site constraints" or "linear combinations of constraining variables". These are similar to the site scores that we just looked at, but are site scores based on the predicted species abundances rather than on the observed species abundances. One could use either the "site scores" or the "site constraints" to situate each site on each of the ordination axes. There is some disagreement between authors about which is more appropriate. The consensus seems to be that the site scores are better when looking as siteXspecies data because the site constraints assume no error in the environmental variables. These values can also be obtained in `summary(X)$constraints`, where X is your chosen name of the cca object. Here are the first few lines of the output:

```
Site constraints (linear combinations of constraining varia
bles)

          CCA1      CCA2      CCA3       CCA4      CCA5      C
A1
1    -1.22229 -1.340631  0.223504 -0.2197590  2.004258    1.0
57701
2    -0.94123  0.010693  0.260482 -0.3505410  1.877497    1.0
49069
3    -1.02040 -0.161158  1.435973 -1.1731896  0.654854    1.7
26879
```

The final bit of information provided by the `summary()` function are the biplot scores for the predictor variables. These are the loadings of each predictor variable on each of the constrained axes. The values are held in `summary(X)$biplot`, where X is your chosen name for the cca object. Here is this information for our example:

```
Biplot scores for constraining variables
```

```
                    CCA1    CCA2    CCA3      CCA4     C
CA5 CA1
soil.texture      0.7532 -0.3630 0.26880 -0.478224 -0.
004802    0
OM               -0.3190  0.4039 0.72165  0.424883 -0.
183836    0
cation.exchange  -0.8875  0.2816 0.20527 -0.060221 -0.
295540    0
basal.area       -0.3674  0.5227 0.19796 -0.389919  0.
632917    0
NPP              -0.8552 -0.5047 0.06886 -0.004012  0.
095643    0
```

We see that the first constrained axis (CCA1) is more strongly determined by soil texture (0.7532), soil cation exchange (-0.8875) and NPP (-0.8552) and less by soil organic matter (-0.3190) and the basal area of the canopy trees (-0.3674). Sites high on this first axis have large values of soil texture (more sand, less silt), low values of soil cation exchange (higher concentrations of Mg, Ca and K and less acidic soils) and lower values of NPP (more potential evapotranspiration and lower aridity). The second constrained axis (CCA2) reflects more equitable contributions of all of the environmental variables since the biplot scores only vary in absolute value from 0.52 to 0.28. The third constrained axis (CCA3) is almost entirely reflecting variation in soil organic matter content.

You can graphically display the information in the `summary()` function using `plot()` just as we did for RDA objects. However, since there are so many more response variables (i.e. the abundances of each species), it is not a good idea to display these response variables as red lines as we did previously. Instead, we will display these as red crosses. As well, since there are so many more sites, it is not a good idea to display these as site numbers as we did previously since the numbers will be superimposed and unreadable. Instead, we will display these as black dots. In fact, these are the default settings of the `plot()` function and so we can create such a triplot with fewer lines of code. Alternatively, you could use black dots for most of the sites and use site numbers for a few key sites. However, since the sites in the Wisconsin data had also been classified into five vegetation types, we can use different point symbols to represent each type. The site classification into vegetation types is in a vector (since there are 253 sites) called `site.classification`. There are five such vegetation types whose names are "SUF", "BEECH", "NUF", "PB", and "SLF"; these names are acronyms meaning "southern upland forests", "beech forests", "northern upland forests", "Pine barrens" and "southern lowland forests". The result, using type 2 scaling, is shown in Figure 5.7. Here is how to produce it.

First, we specify that we only want to plot the observed species scores (not the predicted ones) by adding the argument `display=c("sp")` and change the limits of the plot to allow the output to be easily visible:

```
plot(wisconsin.CCA3,display=c("sp"),xlim=c(-
4,4),ylim=c(-5,3))
```

Next, we create a vector containing the five different vegetation types and a vector specifying the colours that we want to use for each vegetation type:

```
unique.veg.types<-c("SUF","BEECH","NUF","PB","SLF")

veg.type.colours<-
c("black","blue","brown","green","yellow")
```

Finally, we plot the site scores on the first two constrained axes and add the legend:

```
for(i in 1:5){
  sel<-site.classification==unique.communities[i]
  points(summary(wisconsin.CCA3)$sites[sel,2]~

summary(wisconsin.CCA3)$sites[sel,1],col=veg.type.col
ours[i],pch=16)
}
legend("topright",legend=unique.communities,col=veg.t
ype.colours,pch=16)
```

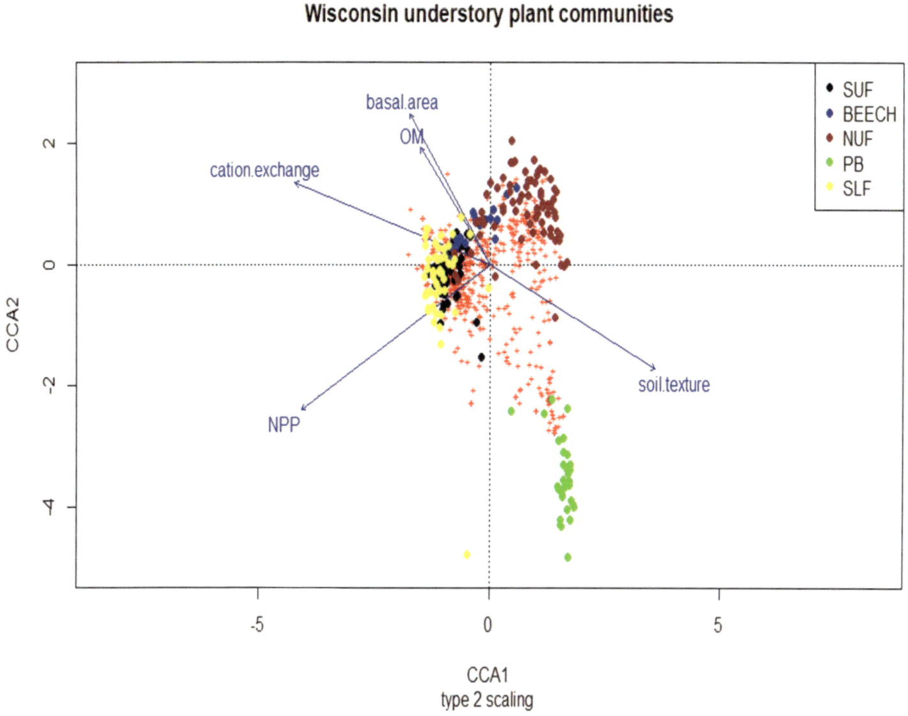

Figure 5.7. Canonical correspondence analysis of the Wisconsin understory forest vegetation.

You interpret this figure as always. Drawing a perpendicular line from a site (point) to an environmental arrow indicates the environmental value at that site. Drawing a perpendicular line from a species (red cross) to an environmental arrow indicates the environmental value in those sites at which that species reaches maximum abundance. We see a sequence of vegetation types in Figure 5.7 that are arranged along the NPP arrow going from southern lowland forests (SLF) to southern upland forests (SUF) to

Beech forests (BEECH) to northern upland forests (NUF). Since NPP is a combination of potential evapotranspiration and aridity, and since these environmental variables increase from north to south in Wisconsin, this shows the dominant effect of latitude and the increasing net primary productivity as one moves south. The Pine Barrens (PB) are rather far from the other groups of points, indicating that the understory vegetation in these sites are quite different from the others. The Pine Barrens sites are about midway along the NPP axis and so these sites are not unusual in terms of NPP. However, they are associates with very high levels of soil texture (i.e. sandy soil) and very low levels of cation exchange, organic matter content (OM) and basal area (i.e. fewer or smaller canopy trees).

5.10) Partial canonical correspondence analysis

The results in Figure 5.7 are not surprising since they mostly mirror the same conclusions that we drew from the unconstrained ordinations in Chapters three and four. However, with canonical correspondence analysis we can go further. Since the site scores in Figure 5.7 are well separated in terms of the five vegetation types, we can use this variable (`vegetation.type`) to remove this overall trend. This will allow us to see if the predictor variables (the five environmental variables) also explain significant variation in the understory plant communities *within* each vegetation type. We can do this by using `vegetation.type` as a conditioning variable. In other words, we first conduct a CCA using only the conditioning variables whose effect we want to remove (here, *vegetation.type*) and then conduct a second CCA on the residuals of the first CCA. This is called a *partial* canonical correspondence analysis.

```
wisconsin.CCA4<-
cca(siteXspecies~soil.texture+OM+cation.exchange+basa
l.area+
plant.available.water+Condition(site.classification),
data=siteXenv2)
anova(wisconsin.CCA4,by="axis")
```

The resulting ANOVA tells us that each of the five constrained axes are significant even after controlling for the larger vegetation types. This is important information because it means that our chosen environmental variables are also affecting variation in the understory plant communities within each of the larger vegetation types. Plotting the first two constrained axes in the same way as for Figure 5.7 gives the triplot shown in Figure 5.8.

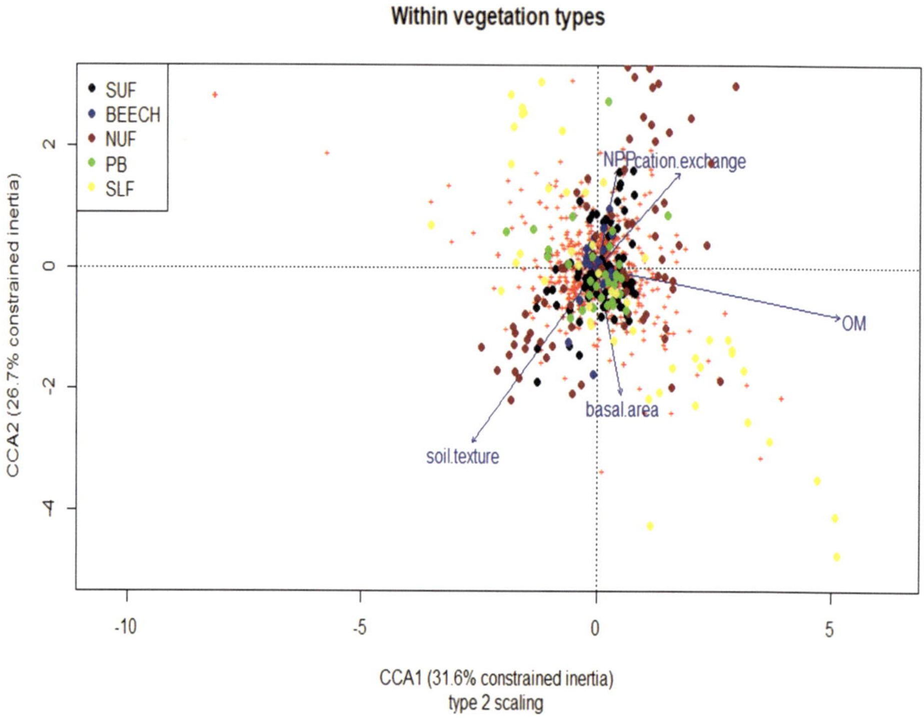

Figure 5.8. A canonical correspondence analysis of the Wisconsin understory vegetation after removing the main differences between vegetation types.

Notice that the colours, representing the different vegetation types, are completely mixed. This is because, by conditioning on these vegetation types, we have removed the differences between them. This allows us to look at how the predictor variables relate to the trends within the vegetation types. The first two constrained canonical axes account for ~ 58.3% of the inertia that is captured by all five constrained axes, although these constrained axes now only account for 4% of the total inertia while the conditioning variable (the vegetation type) accounts for 14% of the total inertia. Said differently, the trends between the vegetation types account for about 3.5 times more inertia than the trends within each vegetation type. We can see that different combinations of environmental variables appear to control the variation between sites within each of the larger vegetation types. For instance, variation within the southern lowland forest group appears to be mostly related to differences in soil organic matter and basal area while variation within the northern upland forests appears to be mostly related to differences in soil texture, cation exchange and NPP.

5.11) Hypothesis testing with CCA

The next empirical example shows how canonical correspondence analysis can be used to test between different competing hypotheses in conservation ecology. The last glaciation removed all earthworms from the glaciated regions of North America and these regions remain, to this day, almost entirely without native earthworm species. European earthworms were brought to North America with European colonists and these earthworms are primarily found in agricultural soils or other places heavily affected by human activity. In particular, many forested areas in Canada remain largely free of earthworms and this means that the plant communities in the forest understory of these areas have developed without earthworms for approximately the last 10,000 years. It is easy to imagine that the introduction of earthworms into such forests could change their understory plant communities by removing the thick layer of organic matter, often 5 – 10 cm thick that lays over the mineral soil. For instance, the roots of some understory herbs are restricted to this organic matter while the roots of other understory species grow down into the mineral layer.

Ndabarushimana (2019) wished to test the hypothesis that the introduction of invasive earthworms into a forested ecosystem can cause changes in the structure and composition of understory plant communities. To do this, he measured the abundance of understory species in many quadrats placed in the Johnville Parc Écoforestier, which is a nature reserve in southern Quebec (Canada) that varies in the abundance of invasive earthworms; some areas have almost no earthworms while other areas have earthworms in abundance. This nature reserve is now a mixed conifer-deciduous forest but different parts of it have different past histories of land use. One zone of the park consisted of cultivated fields until about 100 years ago and has subsequently developed a forest through secondary succession. This area is open for recreational activities and is coded as "R" in the data base. The second zone has never been used for agriculture although it had been lightly logged until about 50 years previously. The public has limited access to this second area, is reserved for conservation activities, and is coded "C" in the data base.

Ndabarushimana (2019) tested between two competing scenarios, only one of which implicates earthworms as causal agents of changes in the structure and composition of the understory plant communities. In the first scenario, the previously agricultural areas introduced earthworms and, since earthworms spread very slowly, this has caused differences in contemporary variation in earthworm abundance. The earthworms then cause changes in the understory forest vegetation: the causal scenario is past land-use → earthworm abundance → understory forest plant composition. This first scenario therefore hypothesizes that earthworms are the direct cause, and past land-use the indirect cause, of changes in the contemporary structure and composition of the understory plants. The second scenario assumes that the past land-use history of the three areas has both caused the variation in earthworm abundance and has caused changes in the contemporary structure and composition of the understory plants: earthworm abundance ← past land-use → understory forest plant composition. This second scenario therefore hypothesizes that earthworms do not cause changes in the contemporary structure and composition of the understory plants and so the current association of earthworms and understory plant communities is spurious.

These two scenarios can be tested using partial canonical correspondence analysis. In the first scenario, a partial CCA using earthworm abundance as a predictor variable and the three zones of the park as a conditional factor variable would reveal that earthworms are significant but a partial CCA using the three zones as a predictor variable with earthworm abundance as a conditional variable would reveal that the park zones are not significant. The second scenario would produce the opposite results.

The siteXspecies data set, called `johnville.understory`, consists of 82 rows (quadrats or sites) and 49 columns (species). Each column gives the number of rooted plants of a given species in each of the quadrats. The data frame `johnville.env` contains the earthworm abundances in each quadrat and its classification zone. Remember that the reciprocal averaging algorithm of CCA consists of a multiple linear regression of the site scores on the predictor variables. Since the earthworm abundances are counts (0, 1, …), the first step is to verify that it is approximately normally distributed, or at least that the distribution is approximately symmetrical. The permutation ANOVA of CCA does not require normality but an approximately symmetrical distribution is better for the linearity assumption. Plotting the histogram of earthworm abundance (Figure 5.9) shows a highly asymmetric distribution that is approximately symmetrical after transforming the earthworm counts to ln(abundance + 1); we must add 1 to the abundances because some quadrats had no earthworms.

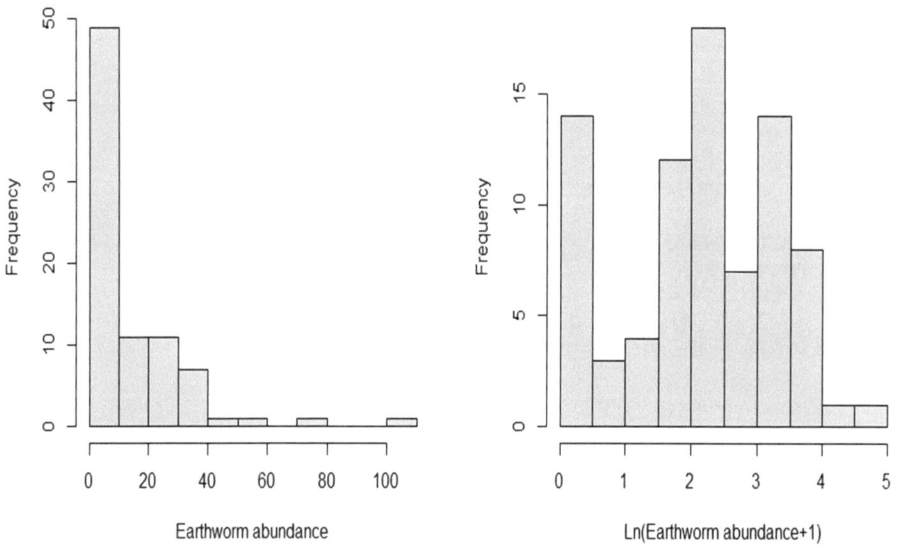

Figure 5.9. Empirical distribution of the number of earthworms per quadrat.

Since both scenarios require that the understory vegetation is associated both with the park zones and with the earthworm abundances, we first do two CCAs using each environmental variable in turn:

```
johnville.CCA1<-
cca(johnville.understory~log(earthworms+1),data=johnv
ille.env)
anova(johnville.CCA1)
Permutation test for cca under reduced model
Permutation: free
Number of permutations: 999

Model: cca(formula = johnville.understory ~ log(earthworms
+ 1), data = johnville.env)
```

```
          Df ChiSquare       F Pr(>F)
Model      1    0.3683 2.3594  0.002 **
Residual 80   12.4868
```

```
johnville.CCA2<-
cca(johnville.understory~zone,data=johnville.env)
anova(johnville.CCA2)
Permutation test for cca under reduced model
Permutation: free
Number of permutations: 999

Model:  cca(formula  =  johnville.understory  ~  zone,  data  =
johnville.env)
          Df ChiSquare       F Pr(>F)
Model      1    0.4467 2.8797  0.001 ***
Residual 80   12.4084
```

These first two CCAs confirm, as assumed by both scenarios, that there are systematic relationships between the understory vegetation in the park and both (i) earthworm abundances and (ii) the zone in which the quadrat resided. Next, to differentiate between the two scenarios, we perform the two partial CCAs. Here is the first partial CCA, with `earthworms` as the predictor variable and `zone` as the conditioning variable:

```
johnville.CCA3<-
cca(johnville.understory~log(earthworms+1)+
Condition(zone),data=johnville.env)
anova(johnville.CCA3)

Permutation test for cca under reduced model
Permutation: free
Number of permutations: 999

Model: cca(formula = johnville.understory ~ log(earthworms
+ 1) + Condition(zone), data = johnville.env)
          Df ChiSquare       F Pr(>F)
Model      1    0.1883 1.2176  0.211
Residual 79   12.2200
```

This first result tells us that there is not a significant association between differences in the understory vegetation and earthworm abundance after we have controlled for site history (i.e. `zone`). Here is the second partial CCA, with `zone` as the predictor variable and `earthworms` as the conditioning variable:

```
johnville.CCA4<-cca(johnville.understory~zone+
Condition(log(earthworms+1)),data=johnville.env)
anova(johnville.CCA4)
Permutation test for cca under reduced model
Permutation: free
Number of permutations: 999

Model:  cca(formula  =  johnville.understory  ~  zone  +
Condition(log(earthworms + 1)), data = johnville.env)
         Df ChiSquare      F Pr(>F)
Model     1    0.2667 1.7244  0.013 *
Residual 79   12.2200
```

This second result tells us that there is still a significant association between differences in the understory vegetation and site history (i.e. zone) even after we have controlled for earthworm abundance. This means that we must reject the first scenario but cannot reject the second scenario. It appears that relationship between the understory vegetation and earthworms, while real, is spurious and that both the understory vegetation and the earthworms are each caused by the land-use history of the park.

5.12) Distance-based redundancy analysis (db-RDA)

You know, from Chapter 4, that principal coordinates analysis (PCoA) is a generalization of both PCA and CA. If you calculate a dissimilarity matrix from your original data using Euclidean distances and then perform a PCoA on this Euclidean dissimilarity matrix then you will get the same row scores as if you did a PCA directly on the original data. If you calculate a dissimilarity matrix from your original data using Chi-square distances and then perform a PCoA on this Chi-Squared dissimilarity matrix then you will get the same row scores as if you did a CA directly on the original data. The advantage of PCoA is that one can use many other types of dissimilarity measures, including the Bray-Curtis dissimilarity that has advantages for data involving species' abundances, as will be further explained in the next chapter. It will therefore come as no surprise that there also exists a constrained version of PCoA, which is called distance-based redundancy analysis (db-RDA). As an aside, there also exists a constrained version of non-metric multidimensional scaling (de Leeuw and Mair 2009) that is implemented in the smacof package. However, since the whole point of using constrained ordination is to perform inferential tests, and since the constrained version of NMDS does not (yet) have an inferential test associated with it, I will not discuss constrained NMDS further.

If you have already understood both PCoA (Chapter 4) and RDA then constrained PCoA will be easy to understand. The method consists of only three steps:

1. Compute a dissimilarity matrix of the response variables using whatever dissimilarity measure that is appropriate; these choices were discussed in Chapter 4.
2. Perform a principal coordinates analysis on this dissimilarity matrix. All of the principal coordinates (i.e. all of the PCoA axes) are stored in a matrix or data frame. This produces a square matrix with as many rows and columns as there are rows (observational units) in data matrix of the original response variables. Remember that the full set of axes perfectly reproduces the dissimilarity matrix based on the original response variables. If the variances of some of the PCoA axes (i.e. eigenvalues) are negative then you will either have to transform the dissimilarities or else use a Lingoes correction to correct these axes having negative variances. These points were explained in Chapter 4.
3. Use RDA to relate the matrix of principal coordinates (step 2), treated as the response variables, to a set of predictor variables.

Since everything that you learnt about RDA applies with db-RDA, I will do both a db-RDA and the conditional version of db-RDA and reproduce the same analysis as shown using CCA in Figures 5.7 and 5.8. Figure 5.7 ordinated the Wisconsin understory vegetation in relation to the environmental variables and Figure 5.8 did the same thing after removing the influence of the five vegetation types to look at the within-vegetation type relationships between the environmental variables. Here is how to conduct a constrained principal coordinates analysis (i.e. a constrained db-RDA) of the Wisconsin understory vegetation using the Bray-Curtis dissimilarity. First, we calculate the dissimilarity matrix using the `vegdist()` function that we saw in Chapter 4.

```
wisconsin.bray<-
sqrt(vegdist(siteXspecies,method="bray"))
```

Next, we use the `capscale()` function that does the work of combining the PCoA (step 2) with the RDA (step3). The formula inside the `capscale()` function is identical to the one used in RDA. I added the argument `sqrt.dist=TRUE` in order to avoid obtaining principal coordinate axes with negative values, as explained in Chapter 4. Since the conditioning variable, i.e. `Condition(site.classification)`, is a factor variable indexing to which of the five community classifications each site belongs, this conditional constrained principal coordinates analysis will test for the significance of the environmental variables after removing the average differences between the community classifications.

capscale()

```
wisconsin.dbRDA<-
capscale(wisconsin.bray~soil.texture+OM+cation.exchan
ge+basal.area+
plant.available.water,data=siteXenv2,sqrt.dist=TRUE)
wisconsin.cond.dbRDA<-
capscale(wisconsin.bray~soil.texture+OM+cation.exchan
ge+basal.area+
plant.available.water+Condition(site.classification),
data=siteXenv2,
sqrt.dist=TRUE)
wisconsin.dbRDA
```

```
wisconsin.cond.dbRDA
anova(wisconsin.dbRDA,by="axis")
anova(wisconsin.cond.dbRDA,by="axis")
```

The output of `wisconsin.dbRDA` informs us that the constrained axes, and thus the five environmental variables, account for 11.63% of the inertia. The output of Wisconsin.cond.dbRDA informs us that the constrained axes account for 3.48% of the inertia within the five vegetation types. The two permutation ANOVAs tell us that all of the constrained axes capture significant amounts of the inertia. Figure 5.10 plots the resulting triplots. The code for doing this is identical to what you have already learned. Note that there are no red crosses in Figure 5.10. This is because db-RDA, being based on a dissimilarity matrix, only ordinates the rows (i.e. the sites in the `siteXspecies` matrix) and therefore looses the information about the columns (i.e. the species in the `siteXspecies` matrix). If you want to recover the positions of the species along the constrained axes then you would have to calculate the weighted means of each species yourself. In comparing Figure 5.10 with Figures 5.7 and 5.8, remember that ordination axes have no natural direction and so we can rotate then 180 without changing anything.

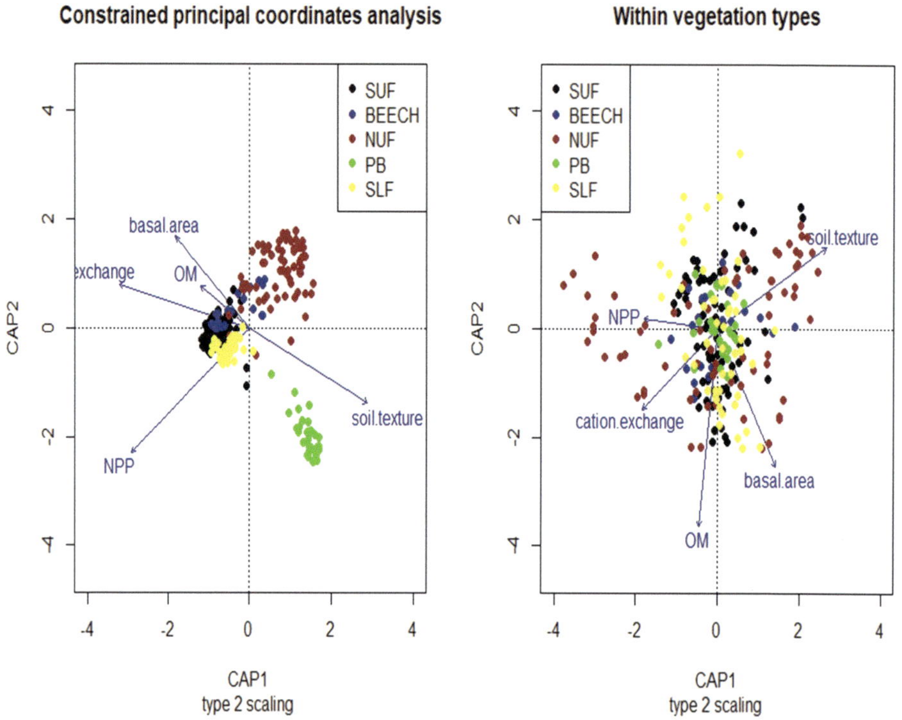

Figure 5.10. Constrained principal coordinates analysis of the Wisconsin data based on Bray-Curtis dissimilarities. The unconditional triplot (left) and a conditional triplot after controlling for differences in the five vegetation types (right).

Chapter 6 Gradient analysis using siteXspecies data

6.1) What is an environmental gradient?

This chapter deals specifically with the ordination of siteXspecies data. Such siteXspecies data concern the abundance of all species occurring in different sites. Each row represents a single observational unit (a "site") and each column represents the abundance of a given species in each of these observational units. Of course, it is up to the ecologist to decide the appropriate scale of a single observational unit, or "site". A single site might be a single 50cm square quadrat in herbaceous vegetation or a single hectare in a mature forest. Such analyses are very common in community ecology but they present important challenges that will require appropriate intermediate steps in the ordination. You have already encountered such data several times in this book and have learned several different ordination methods that could potentially be used with such data. At this point, you are probably wondering which methods are most appropriate for the ordination of siteXspecies data? The goal of this chapter is to provide some answers to this question.

The most common reason for performing constrained or unconstrained ordinations of siteXspecies abundance data is to discover (via unconstrained ordinations), or to test hypotheses (via constrained ordinations), concerning the underlying environmental variables that structure such ecological gradients. The previous sentence referred two concepts – ecological gradients and environmental variables – that are often incorrectly conflated. An "environmental variable" is not the same thing as an "environmental gradient" and understanding the difference between the two goes a long way towards understanding why ordinating siteXspecies abundance data is challenging. Ecologists often use shorthand words like "a gradient of water availability" and it is easy to think that this means the same thing as "increasing (or decreasing) values of water availability" but the two do not mean the same thing! An environmental gradient is a systematic relationship *between* environmental variables and some ecological response. Both components (environmental variables and ecological responses) are required to define an environmental gradient. Both components are explicitly used in direct gradient analysis, like in the response curves of the shoreline plants in Chapter 3 or the methods of constrained ordination in Chapter 5. However, the confusion between environmental variables and environmental gradients probably arises because the various methods of unconstrained ordinations only use one of the two components (the ecological response) to *infer* the full environmental gradient.

To see why an environmental gradient requires both a set of environmental variables and a set of ecological responses, consider the two panels of Figure 6.1. This figure plots the changes in abundance of a series of hypothetical species as a function of mean annual temperature. The environmental variable (mean annual temperature) is the same in both panels, as is the range of this environmental variable. However, the species in the left panel are much more sensitive to changes in temperature than the species in the right panel, resulting in much more differentiated communities occurring over this range of temperature in the left panel. The environmental gradients are different in the two panels even though the environmental variable is the same because the species experience the same environmental variable differently. An environmental gradient describes the relationship *between* the environmental variable (here, mean annual temperature) and the ecological response (here, the changes in abundance of the species). Using ordination techniques to reveal environmental gradients is challenging because the relationship between environmental variables and species' abundances is inherently nonlinear.

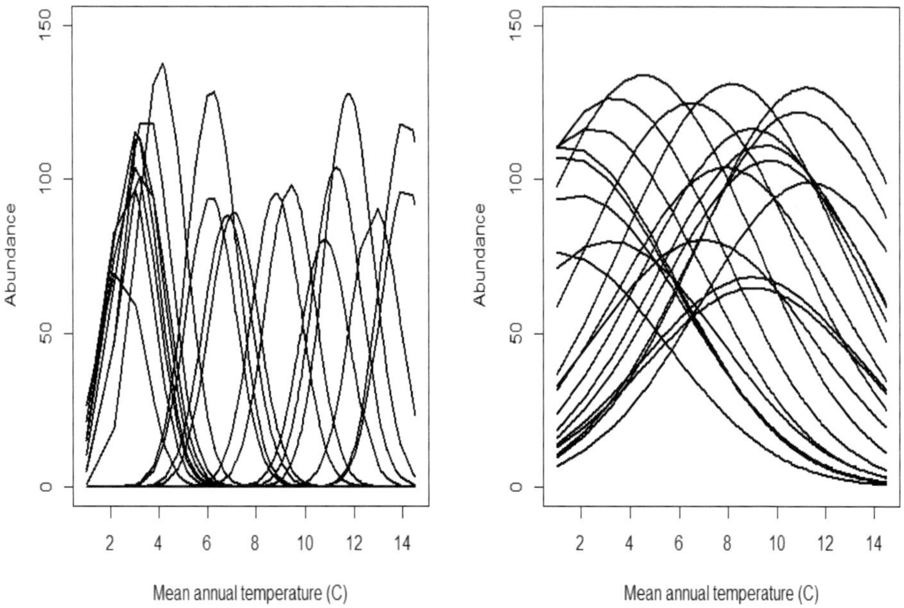

Figure 6.1. Two different ecological gradients based on the same environmental variable.

The ecological justification for using unconstrained ordinations of siteXspecies data to identify underlying environmental gradients rests on the following two-part assumption.

1. *Sites that are very similar in terms of the composition and abundance of species will also be very similar in terms of the values of the combination of environmental variables that control the assembly of such species.*

2. *The more different the environmental conditions between sites that control the composition and abundance of species, the more different will be the communities of species found in them.*

There are three important points to this assumption that require consideration when we choose an unconstrained ordination technique. First, this assumption does not require the rate of change of environmental similarity between sites to match the rate of change of compositional similarity between sites. Second, the assumption only applies to those environmental variables that actually control the assembly of the ecological communities. This means that if you include irrelevant environmental variables then this will reduce (or even prevent) the ordination technique to identify environmental gradients. Finally, taken to an extreme, the assumption allowing ordination methods to reveal environmental gradients is based on a rather strict deterministic view of community assembly that is rarely true. Remember that all unconstrained ordination methods, when applied to siteXspecies data, describe the patterns of association between the species across the sites without reference to any external information about environmental conditions. However, some patterns of association between the species might be generated by factors other than the present environment such as past biogeographical processes or past land use. The example in of earthworms and understory forest vegetation in Chapter 5 is one example since that association appears to have arisen due to the reversion of part of the park from agricultural land to forest a hundred years previously.

Since the goal of this chapter is to compare the abilities of the different ordination methods to accurately reveal environmental gradients in community ecology, what is the best way of doing this? The best way is to compare the different methods using data for which we already know the relationship between changes in species' abundances and the underlying environmental gradients that control these changes in abundance. We can obtain such data by generating them ourselves using numerical simulations.

6.2) Simulating environmental gradients

There are many different ways of simulating community assembly along environmental gradients. These different ways vary in numerical complexity and in the degree of realism of the simulation. In this chapter, we will use a method of intermediate complexity that allows us to explore the relative ability of different ordination methods to capture environmental gradients without getting lost in the complexity of coding in **R**. We start with Hutchinson's (1978) concept of a realized niche. The realized niche of a species is inherently multivariate, since there are many different environmental variables that control the abundance of a species, but the concept of a realized niche can be most easily visualized and understood using a single environmental variable (Figure 6.2). The combination of environmental conditions to which a species is best adapted defines its realized niche optimum. In Figure 6.1 this optimal environmental value is 8. The fitness, and so the abundance, of this species decreases as the environment deviates from this optimum and as the species becomes increasingly less adapted to the environmental conditions. The amount by which the fitness of a species decreases with increasing deviations from the optimum is called its ecological tolerance. A species with a wide tolerance can perform quite well even when the environment deviates far from its optimum. A species with a narrow tolerance will have its performance decrease rapidly even when the environment deviates just a small amount from its optimum. The mathematical function that describes the way that fitness changes with the environment is the "response" function. There is nothing in the niche concept that specifies this response function but we will use a Gaussian response function, as described in Chapter 3, for two reasons. First, the way that many (but certainly not all) species actually change along environmental gradients can be approximated by this function. Second, it is easy to simulate; we can completely define a Gaussian response function by giving its optimum value (the mean) and its tolerance (the standard deviation). If the realized niche of a given species is defined along *n* environmental variables then you would give *n* means and *n* standard deviations[43]. Finally, we want to specify a minimum abundance (i.e. a minimum population size) below

[43] A more complete description would also include the covariance matrix for the *n* environmental variables, but we will allow the realized niche to be mutually independent along these axes, and generate correlations between the environmental variables themselves in the `gen.env()` function.

which the species cannot exist. This is the broken horizontal line in Figure 6.2. The response functions of each of our simulated species along our chosen environmental variables will be specified in the `gen.abund()` function and the output of this function will give the observed abundances of each species across sites. You are invited to modify the `gen.abund()` function that is given at the end of this chapter if you want to try other types of response functions.

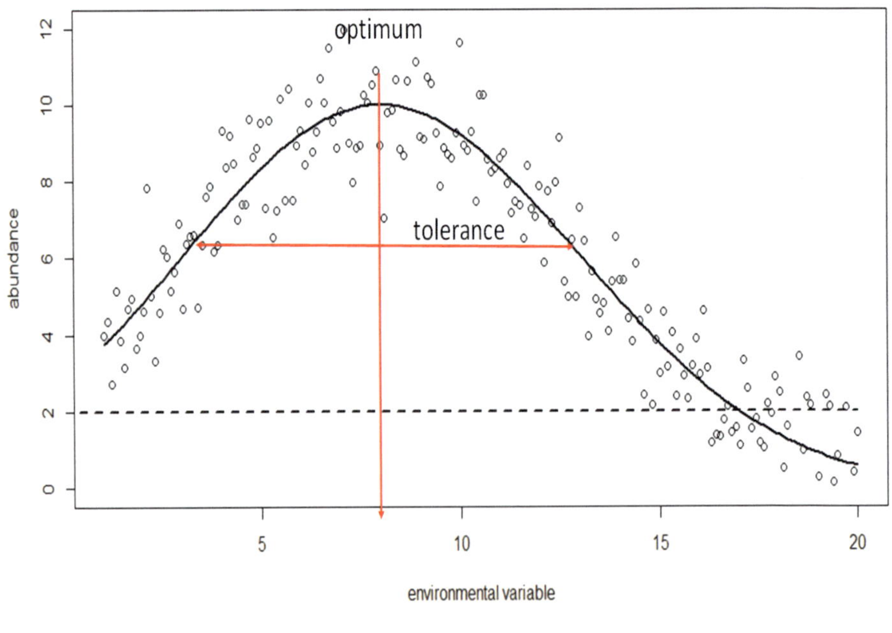

Figure 6.2. A one-dimensional realized niche showing how the fitness, or abundance, of a species changes as a function of some environmental variable.

The `gen.abund()` function will be called separately for each species, given the values of the environmental variables in each site, in order to simulate how its abundance will change across sites. The output of `gen.abund()` will be the abundance of a single species having a specified set of means and tolerances for the underlying environmental variables, in each site. Each site is specified by the values that each of the underlying environmental variables take in that site. The arguments of `gen.abund()` are:

1. `env`: a matrix holding the values of each of the environmental variables (columns) in each of the sites (rows).
2. `means`: a numerical vector holding the optima of the species for each of the environmental variables.
3. `tolerances`: a numerical vector holding the tolerances of the species for each of the environmental variables.
4. `max.abund`: a single numerical value specifying the maximum predicted abundance of the species.
5. `random.sd`: a single numerical value specifying by how much the actual observed abundance of the species in a site can randomly vary from the Gaussian response curve.

Next, we need to simulate the pattern of correlation between the environmental variables. We will want to explore how the different ordination methods perform when there are different numbers of environmental variables included in the environmental gradient, when these environmental variables are correlated to different degrees, and when the relationships between the environmental gradients are nonlinear. Not all of these combinations of simulations will be presented in this chapter but you are invited to try the ones that interest you. We will do this using the `gen.env()` function, which will generate normally distributed values for each environmental variable in each site. The output of `gen.env()` is a numerical matrix with `n.sites` rows and `n.env` columns and this output matrix will be input into the `gen.abund()` function. The arguments of `gen.env()` are:

1. `n.sites`: a single numerical value specifying the number of sites that you want to simulate.
2. `n.env`: a single numerical value specifying the number of environmental variables whose values will be assigned to each site.
3. `r`: a single numerical value. If r=0 then the values of the different environmental variables will be completely independent. If r is between 0 and 1 then this will specify the positive correlation between each environmental variable. If r is between 0 and -1 then this will specify the negative correlation between non-sequential environmental variables.
4. `quad`: a single logical value (TRUE/FALSE). If quad=TRUE then this will simulate a quadratic relationship between environmental variables.

Finally, the `gen.community()` function calls the `gen.env()` function to simulate the environmental values of each site and then calls the `gen.abund()` function many (`n.species`) times in order to simulate the abundance of each species in each site, producing a siteXspecies matrix. The arguments are:

1. `env`: an input numerical matrix that is produced from the `gen.env()` function.

2. `n.species`: a single numerical value specifying the number of species to simulate.
3. `min.abund`: the minimum abundance that a species can have in a given site and still be part of the simulation. Values below `min.abund` are replaced with zero abundance values. Smaller values, in combination with lower tolerance and lower maximum abundance (see below) result in more species having zero abundance in sites whose environmental conditions are far from their optima. The default value is 5.
4. `max.abund`: the *average* maximum abundance of the species. The maximum abundance of each species is a random draw from a normal distribution whose mean is this value and whose standard deviation is 20% of this value, with a lower bound at 1. The default value is 100.
5. `tolerance`: the *average* tolerance value of the species. The tolerance value of each species is a random draw from a normal distribution whose mean is this value and whose standard deviation is 20% of this value, with a lower bound of 0.1. The default value is 5.
6. `random.sd`: the standard deviation of the actual abundance of each species in each site around the theoretical value from the response function. The actual abundance of each species in each site is a normally distributed random value as produced by the `gen.abund()` function (see above) and this is the argument that is passed to that function. The default value is 1.

A single simulated siteXspecies data set is generated by first calling the `gen.env()` function to produce the environmental values of each site and then calling the `gen.community()` function to produce the abundance of each species in each site. The output of `gen.env()` is a matrix with `n.sites` rows and `n.env` columns. Each column gives the value of the environmental variable in each of the sites. The output of `gen.community()` is a list. The first element of the list is a numeric vector (`site`) containing the number of each site. The second element of the list is a matrix (`community`) containing the siteXspecies abundances of each species (columns of the matrix) in each site (rows of the matrix). The third element of the list is a matrix (`predicted.community`) containing the siteXspecies predicted abundances (i.e. without the random variation) of each species (columns) in each site (rows). Figure 6.3 gives an example of how this works. The abundance response function for a species with optimal fitness of 8 and tolerances of 10 along each of two environmental variables is shown in the left panel[44]. This response surface is plotted in "environmental" space since it shows how the abundance changes over all combinations of possible values of the two environmental variables. This response surface is generated by a call to `gen.abund()` by providing a matrix (`env.all`) in which all combinations of values for the environmental variables e1 and e2 are represented.

```
gen.abund(env.all,means=c(8,8),tolerances=c(10,10),ma
x.abund=10,
random.sd=2)
```

[44] This response surface was drawn using the `persp()` function in **R**.

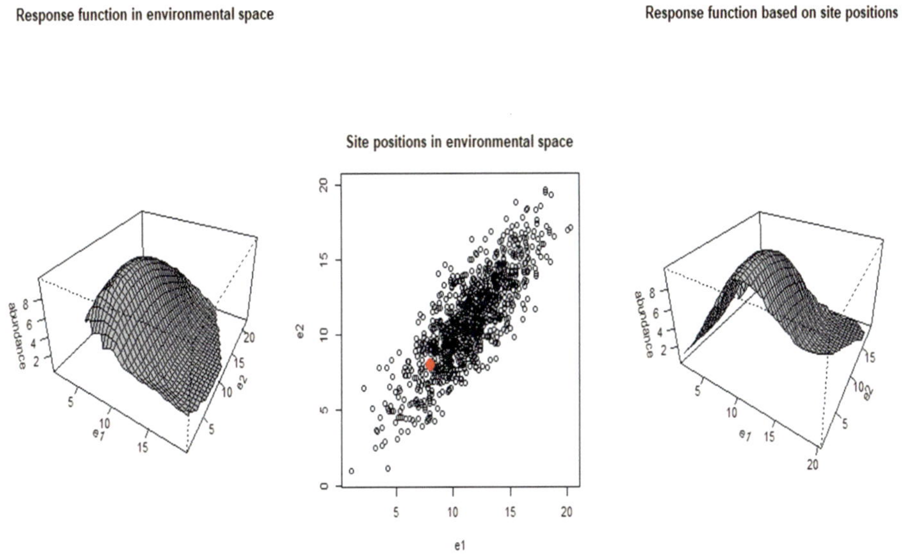

Figure 6.3. (Left panel) A Gaussian response surface of a particular species in environmental space along two uncorrelated environmental variables (e1 and e2). (Middle panel) The actual combinations of these two environmental variables in the "field" over many different sites (points), in which the two environmental variables are positively correlated. The position of the optimum of that particular species in the field is shown by the red dot. (Right panel) The theoretical abundance of this species that is actually encountered in the field along these two correlated environmental variables.

However, this full response surface will not be realized if some combinations of the two environmental variables are never encountered in the sampled sites. This would happen if, for instance, the two environmental variables were correlated in the field. The middle panel of Figure 6.3 shows the actual combinations of the two environmental variables that are found in a set of sites, as generated by `gen.env()`. Each point represents a site and the position of the site in the graph indicates the values of the two environmental variables (e1 and e2) in this site. The red dot indicates the values of the two environmental variables to which our species is optimally adapted. Not all of the environmental space is included in our sites; for instance, there are no sites in the top left or bottom right corners of this middle panel. Here is how the values in the middle panel were generated for 1000 sites and two positively and linearly correlated (r=0.9) environmental variables:

```
gen.env(n.sites=1000,r=0.9,n.env=2,quad=F)
```

The predicted abundance of this species across these 1000 sites is shown in the right panel of Figure 6.3. Note that part of the response surface in the left panel of Figure 6.3 is missing in the right panel because the combinations of environmental values in the missing part do not exist in the simulated environment. If we did this for each species that was encountered somewhere in these 1000 sites then we would have the output of the `gen.community()` function.

6.3) Using simulations to compare ordination methods

We can simulate different patterns of correlation between different numbers of underlying environmental variables using `gen.env()` while changing the input values of its arguments. We can then simulate how entire communities of species will response to these underlying environmental variables using `gen.community()`. By modifying the arguments of `gen.community()`, we can see how changing the number of species, the number of rare/common species, or the tolerances of these species will affect the patterns of association in the siteXspecies matrix. By ordinating these different simulated communities using the different methods and measures of similarity, we can then see how well the site scores of these different ordinations reflect the underlying patterns of correlation in the environmental variables. In other words, this will allow us to see how well these different ordination methods correctly reveal environmental gradients in our simulated ecological communities. We will evaluate the abilities of each ordination method both visually and numerically. First, we will plot the site scores on the first two axes and visually compare the results with the known position of each site (i.e. its environmental values) in physical space just like the middle panel of Figure 6.3. The more closely the positions of the site scores on the ordination axes agree with the positions of the sites in environmental space, the better the ordination will reveal the environmental gradients. Second, we will use the `procrustes()` function (Chapter 4) to compare the dissimilarity matrix between sites based on the first few ordination axes to the dissimilarity matrix between sites calculated directly from the environmental values of each site. Values closer to unity mean that the relative positions of the sites in the ordination space more closely reflect the positions of the sites along the environmental gradients. Finally, we will calculate the rank-order (Spearman) correlation relating the distances or dissimilarities between the sites based on their known positions along the environmental variables to the distances or dissimilarities between the sites based on the positions on the first few ordination axes.

We begin with a very simple simulation: 200 species in 100 sites distributed along a single environmental gradient. The optimum position of each species is randomly chosen and each species has a Gaussian response with a tolerance of $\sqrt{5}$ along a single environmental variable (Figure 6.3). The positions of each of the 100 sites on this single environmental gradient are shown by grey vertical lines and three sites (56, 33 and 3) have been highlighted by vertical black lines. Site 56 occurred at the lowest value of the environmental variable (1.0), site 3 occurred at the highest value of the environmental variable (15.8) and site 33 occurred close to the median value of the environmental variable (9.0). The theoretical Gaussian response curves of each species are shown by bell-shaped black curves and the response curves of the first two randomly generated species are shown by the red and blue curves, respectively. Because these are Gaussian (i.e. a normal distribution) response curves, approximately 95% of the theoretical abundance of a given species will be found within two times its tolerance value (i.e. within two times the standard deviation). For instance, the first species (red curve) has an optimum at 7.9 and a tolerance of $\sqrt{5}$ and so approximately 95% of its abundance occurs between 2.9 and 12.9. That means that species 1 will usually be found in site 33 but will almost always be absent from sites 56 and 3. The actual abundance of a species in a site is determined from its theoretical Gaussian response but will vary randomly around this curve, as seen in the bottom panel of Figure 6.4. Here is how I generated these data:

```
set.seed(101)
env1<-gen.env(n.sites=100,r=0,n.env=1,quad=F)
community1<-
gen.community(env=env1,n.species=200,min.abund=5,max.
abund=100,tolerance=sqrt(5),random.sd=0.25)
```

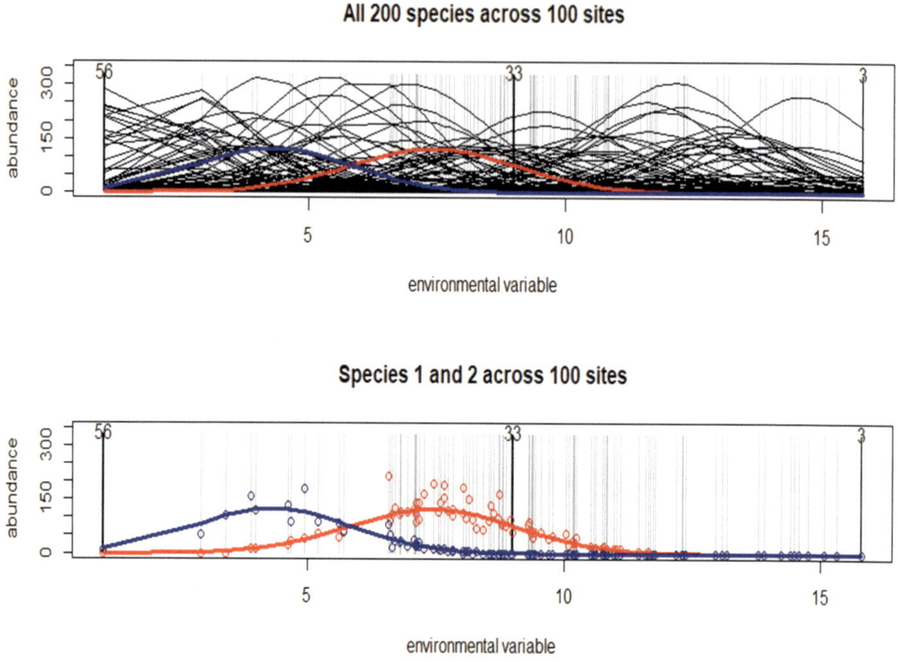

Figure 6.4. The top panel shows the theoretical distributions of 200 species along a single environmental variable. One hundred sites (the grey vertical lines) are also shown. The bottom panel shows the theoretical (curves) and actual (points) abundances of the first two simulated species (red and blue) among the 100 sites.

6.4) The "horseshoe" or "arch" effect

Now we can see how well the different unconstrained ordination methods can recover the single environmental gradient in these simulated data. If an ordination method performs perfectly using only two axes (dimensions) then we would see the positions of each of the 100 sites aligned along the first axis in the same order, and with the same spacing, as shown in Figure 6.4. The positions of each site on the second ordination axis would be random with respect to the underlying environmental variable. In other words, site 56 would be at one extreme on the first ordination axis, site 3 would be at the other extreme, and the remaining sites would be arranged between them in the correct order. Unfortunately, this is not what happens when we use the different ordination methods and dissimilarity measures that we studied in Chapter 4.

This problem is best seen using principal components analysis. We will (i) perform a PCA on the siteXspecies matrix of our 200 species in 100 sites, (ii) calculate the Euclidean distances between the sites based on their actual positions in environmental space and on their Euclidean distances based on the species' abundances on the first two PCA axes, and (iii) compare the positions of the sites in environmental space and in the two-dimensional ordination space using Procrustes analysis. Since there is actually only a single environmental variable, and the Procrustes analysis requires that the dimensions of the two distance matrices be equal, I trick the Procrustes analysis by adding a second environmental "variable" that is a constant (i.e. 1). Figure 6.5 summarizes the results; here is the relevant R code:

```
pca1.obs<-prcomp(community1$community,scale=FALSE)
pca1.dist<-
vegdist(pca1.obs$x[,1:2],method="euclidean")
```

```
pro.R2<-
Procrustes(Y=pca1.pred$x[,1:2],X=cbind(env1,rep(1,100
)))$congcoef^2
```

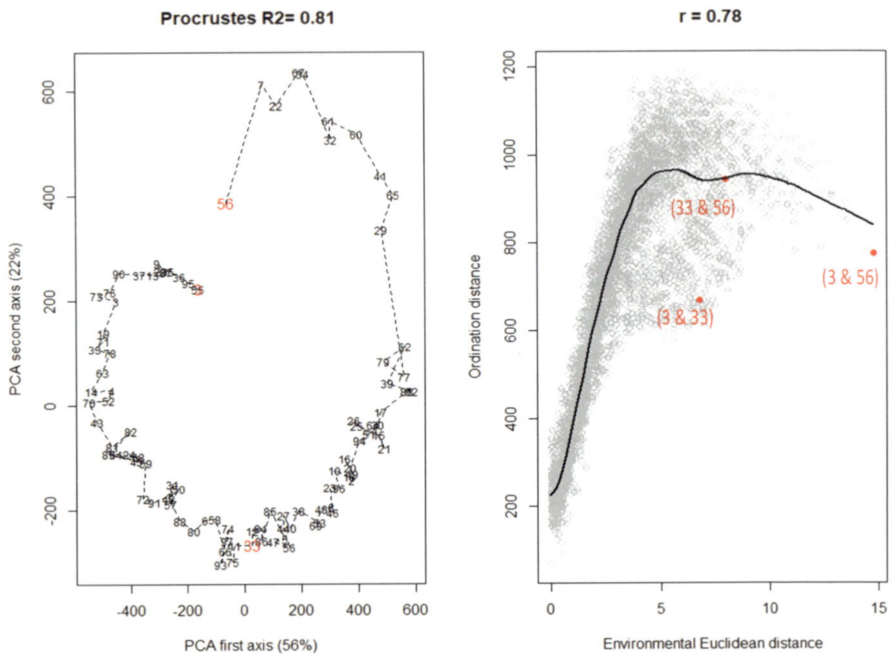

Figure 6.5. A principal component analysis of a siteXspecies matrix generated along a single environmental gradient. The left panel shows the position of each site calculated in the actual environmental space and in the ordination space. The physical distances between sites is proportional to the Euclidean distances between them. The positions of sites #3, 33 and 56 in the two spaces are shown in red. The right panel shows the relationship between the actual Euclidean distances in the environmental space and the Euclidean distances in the PCA ordination space. The positions of the sites with the lowest (#56), highest (#3) and median (#33) environments are shown in red.

Although all of the sites are arranged along a single environmental gradient, the arrangement of the sites in the PCA ordination (Figure 6.5, left) does not occur along a single ordination axis. I have joined sites that occur sequentially along the environmental gradient by a broken line. Instead, we see this single environmental gradient curled up in an arch spanning two dimensions and the range of the site scores on the second axis is almost as large as on the first axis. Notice that the straight-line (Euclidean) distance between site 56 (which had the lowest environmental value) and site 3 (which had the highest environmental value) is smaller than the distance between each and site 33 even though site 33 was half way between the other two along the environmental gradient! This means that neither the first or the second PCA axes properly captures the single underlying environmental gradient.

We see a similar phenomenon (Figure 6.5, right) when we plot the Euclidean distance between each unique pair of sites based on environmental space and ordination space: the two types of Euclidean distances increase together (as they should) for sites that are not too far apart in environmental space until they are about 9 units apart. However, the Euclidean distance between sites in the ordination space then begins to decrease even as the environmental distance between sites continues to increase. The pattern is known as the "horseshoe" effect because the linear environmental gradient has been both curved (the "arch effect") and the ends pulled together in the shape of a horseshoe. Since the common tolerance value in this simulation was $\sqrt{5}$, two tolerances in one direction is ~4.5 units and is ~9 units in both directions. This means that the average species will first become present, reach its optimum, and then disappear again over ~ 9 units along the environmental gradient. In other words, the distortion using PCA becomes severe once the environment changes enough so that there is a complete turnover of species. This simulation illustrates why PCA is not a particularly good ordination technique for siteXspecies data since, in a real empirical study, we would not know that the two dimensional pattern in the PCA is really a one dimensional environmental gradient that has been twisted.

Why does this distortion occur? After all, the PCA is doing exactly as advertised: it represents the linear patterns of association between many variables in a low dimensional Euclidean space. The reason is twofold. First, because the pattern of association between the variables in such siteXspecies data is not linear. Second, because a Euclidean distance is not appropriate to represent degrees of "similarity" in terms of the abundance of species across sites arranged along environmental variables. We will look at each reason in turn.

The first reason is that the pattern of association between the variables in such siteXspecies data is not linear. PCA ordinates observational units (rows) given a set of variables (columns). In siteXspecies data, the observational units are sites and the variables are the abundances of each species across these sites. However, the abundances of species do not change linearly along most environmental gradients.

Figure 6.6 (left) shows again the distribution of the first two species along the single environmental variable. I have chosen four points, marked as "a" to "d", along the resulting environmental gradient. Both species have low abundance at point "a" and so the graph showing how the abundance of each species changes (Figure 6.6, right) has a point close to the origin. A site at point "b" has an environment that is at the optimum for species 2 (blue). The abundance of species 2 is maximal at point "b" even though the abundance of species 1 (red) is still very low since the site at point "b" is still far from its optimum. As we move to point "c", the blue species has decreased its abundance substantially since point "c" is far from its optimum environment. However, point "c" has an environment that is at the optimum for species 1 (red) and so species 1 has high abundance. Moving again to a site at point "d", such a site is far from the optimum of both species and so both species have low abundance. The result is a very nonlinear pattern of changes in abundance for the two species as we move along the environmental gradient and this occurs for each pair of species. However, PCA assumes a linear pattern between the variables (species) across the observational units (sites) and so what is a linear pattern between sites in environmental space becomes a very complicated nonlinear pattern between sites in the space of the PCA ordination.

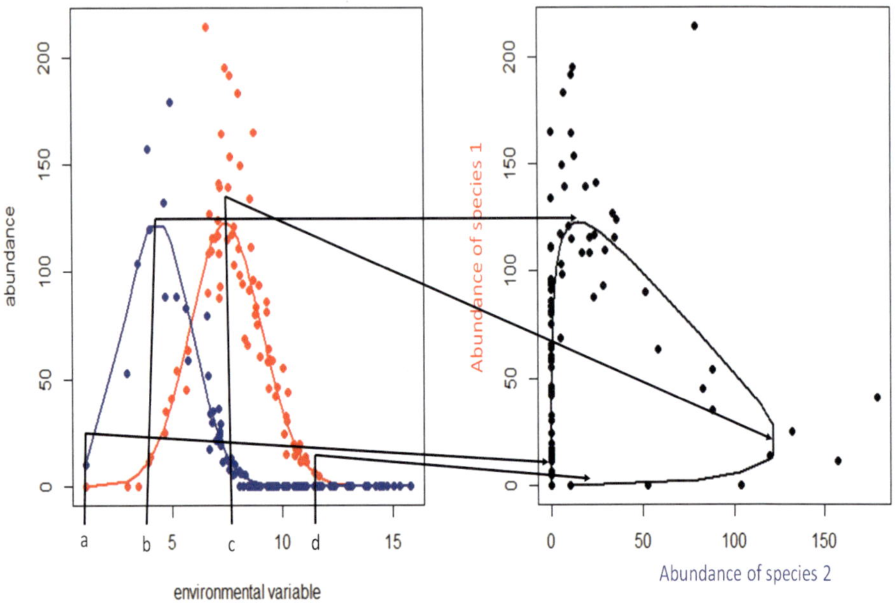

Figure 6.6. The theoretical (lines) and observed (points) abundances of the first two species are shown in the left panel. The theoretical (lines) and observed (points) relationship between the abundances of these two species are shown in the right panel.

The second reason why PCA distorts the underlying linear environmental gradient is because PCA is based on Euclidean distances. A Euclidean distance is not appropriate to represent degrees of "similarity" in terms of the abundance of species across sites arranged along environmental variables. To understand why, we must go back to the fundamental two-part assumption justifying the use of ordinations to reveal environmental gradients: "*(1) Sites that are very similar in terms of the composition and abundance of species will also be very similar in terms of the values of the combination of environmental variables that control the assembly of such species.* (2) *The more different the controlling environmental conditions between sites, the more different the communities of species found in them*". If the Euclidean distance agrees with this assumption then the environmental Euclidean distances between our sites based on the environments of the sites should be proportional to the Euclidean distances between our sites when calculated using the abundance of the species in each site. As you can see in Figure 6.5 (right), this is not true. The two types of Euclidean distances in Figure 6.5 do approximately increase together until the environmental Euclidean distance is about 9 (i.e. about half way along the environmental gradient) but then the pattern changes: the environmental Euclidean distance continues to increase while the ordination Euclidean distance begins to decrease. In particular, when looking at the ordination Euclidean distance, the two extreme sites (3 and 56) are closer together that either is to the site that is half way between them in terms of the environment (site 33). This is why the unidimensional environmental gradient (a line) becomes a two dimensional curve in the first two PCA axes. The extremes of the environmental gradient are pulled together in the Euclidean space. This distortion is more pronounced in PCA than in the other ordination methods because PCA is based on Euclidean distances but all of them suffer from such distortions to a certain extent using any of the distance or dissimilarity measures that we studied in Chapter 4 and we will need to find a solution to this problem. Why does this happen? The problem is that the Euclidean distance, and the other dissimilarity measures, violate the second part of our fundamental assumption: *the more different the controlling environmental conditions between sites, the more different the communities of species found in them.* This occurs for three reasons.

The first reason is that the Euclidean distance, like several of the other distance or dissimilarity measures described in Chapter 4, is based on the abundances of each species rather than on its relative abundance. This is why Legendre and Legendre (2012) recommend that a siteXspecies data matrix based on abundances be transformed via a Hellinger transformation into a siteXspecies data matrix based on relative abundances per site. Remember that a Hellinger transformation simply means that you divide the absolute abundance of each species in a given site by the total abundance of all species in that site. You can do this using the `decostand()` function of vegan and specifying `method="hellinger"`.

The second reason is that the Euclidean distance, like several of the other distance or dissimilarity measures described in Chapter 4, considers "double zeros" to be valid comparisons in terms of similarity. If the abundance (absolute or relative) of species *j* in site *i* is x_{ij} then the Euclidean distance, d(a, b), between two sites (a and b) is $d(a,b)=\sqrt{\sum_j \left(x_{aj}-x_{bj}\right)^2}$.

Every time a species is absent from both sites (a double zero) then this species counts as $(0-0)^2=0$ meaning that the Euclidean distance considers this an instance of perfect similarity (i.e. zero distance). The more species that are absent from both sites, the more such instances of perfect similarity are counted. This is certainly true in a purely mathematical sense. After all, the abundance of the species *is* the same in both sites (it is zero). However, this is not true in an ecological sense. A species can have the identical (zero) abundance in sites whose environments are very different. A species can have zero abundance either because the site has environmental conditions that are too low or too high relative to its optimum. For instance, the red species in Figure 6.6 (left) has the same low abundance at sites b and d even though site b has an environmental value of ~4 while site d has an environmental value of ~12. This "double-zero" phenomenon violates the second part of our fundamental assumption: *the more different the controlling environmental conditions between sites, the more different the communities of species found in them.* We should expect this violation of our assumption whenever the range of the controlling environmental variables is large relative to the tolerances of the species and this suggests that we should prefer dissimilarity measures that don't count double zeros. This last sentence is of limited practical use since the whole purpose of these unconstrained ordination methods is to use the patterns of abundance across sites in order to *identify* the environmental variables that are controlling their abundances. Since we haven't yet identified these environmental variables, we are in no position to compare their ranges. A more practical recommendation is to expect greater violations of our assumption as the number of species increases in the data set that have zero abundances in a site.

There is a third reason why even dissimilarity measures that don't count double zeros, like the Bray-Curtis or Chord distances, can violate our fundamental assumption. This third reason is related to, but conceptually different from, the double-zero problem. I'll call it the "upper boundary" problem. The upper boundary problem arises once we begin comparing sites whose underlying environmental values are sufficiently different that the sites no longer have *any* species in common. The dissimilarity measure then reaches a maximal value and can't increase any further even if the underlying environmental conditions continue to diverge. This phenomenon violates the second part of our fundamental assumption: *the more different the controlling environmental conditions between sites, the more different the communities of species found in them*. Imagine that we compare a series of sites taken at different latitudes in North America. One site is from a boreal forest in northern Canada. One site is from a deciduous forest in New England. One site is from a tropical forest in Mexico. If there is complete species' turnover from the boreal to the deciduous forest then the similarity between the boreal and deciduous site is zero and the dissimilarity between the two sites is maximal. However, this is also true of the similarity between the boreal and tropical site. The underlying environmental variable determining community structure continued to change from New England to Mexico even though the amount of dissimilarity in terms of the species' abundances between each of these two sites and the boreal site has not increased.

6.5) Flexible shortest path adjustments

Our imaginary trip from the Canada to Mexico suggests a method of removing the upper boundary problem. Imagine sampling forest vegetation at the following three "sites": (a) a boreal site in Northern Quebec, (b) a site in southern Quebec having a mixture of coniferous and deciduous plant species, (c) a site in southern New England having a deciduous forest and (d) a site in northern Florida. If we use a Bray-Curtis dissimilarity (Chapter 4), which varies from 0 (complete similarity) to 1 (complete dissimilarity) then the (imaginary) dissimilarity matrix might look like the one in Table 6.1.

	Boreal forest(1)	Mixed forest (2)	Deciduous forest (3)	Florida (4)
Boreal forest (1)	0	0.6	**1** (1.2)	**1** (2.1)
Mixed forest (2)	0.6	0	0.6	**1** (1.5)
Deciduous forest (3)	**1** (1.2)	0.6	0	0.9
Florida (4)	**1** (2.1)	**1** (1.5)	0.9	0

Table 5.1. Bray-Curtis dissimilarity values between pairs of four sites along an environmental gradient extending from northern Quebec (Boreal forest) to northern Florida.

Since we are using Bray-Curtis dissimilarities, we will not run into the double-zero problem. However, some of the pairs of sites in Table 6.1 have a Bray-Curtis dissimilarity of 1.0 and so these pairs of sites do not share any species in common. We have still run into the upper boundary problem. Note that the dissimilarity between the boreal and deciduous forest is 1.0, which is the same dissimilarity between the boreal forest and the forest in Florida. In a purely mathematical sense this information is correct: the boreal forest is completely different from both the deciduous and Florida forests. However, this is not true in an ecological sense because the underlying environmental variables are more different between the boreal and Florida sites than between the boreal and deciduous sites. In principal, this could be solved by increasing the largest dissimilarities to better reflect a linear relationship between underlying environmental distances and observed community distances. This solution is called a " flexible shortest path adjustment" by Bradfield and Kenkel (1987) or an "extended dissimilarity" by De'ath (1999). Both solutions rely on a "minimum spanning tree" and on a "cophenetic distance" (Legendre and Legendre 2012).

Minimum spanning trees
Cophenetic distances

We will start by constructing a "minimum spanning tree" of our dissimilarity matrix and we will use the minimum spanning tree to convert our dissimilarity matrix into a matrix of "cophenetic" distances. Here is how to construct a minimum spanning tree from a distance or dissimilarity matrix and then obtain the resulting cophenetic distances:

The pairs of sites (observational units) can be in one of two groups: flagged and unflagged pairs. We start by placing all pairs of sites into the unflagged group.

1. Find the pair of sites in the unflagged group having the smallest distance value (i.e. that are most similar). Draw a line between these two sites, write the distance between them on top of the line, and place this pair of sites in the "flagged" group them so that you don't count these sites again.
2. Repeat step 2 until all of the sites are connected.

Figure 6.6 illustrates these steps using the Bray-Curtis dissimilarities in Table 6.1. There are six unique pairs of sites in Table 6.1. All six of these unique pairs begin as unflagged pairs. We start with our four separated sites (panel 1) and all of these sites are initially unflagged.

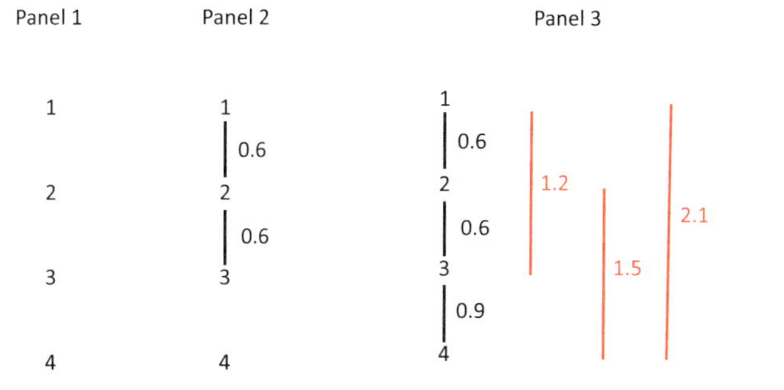

Figure 6.6. Constructing a minimum spanning tree and associated cophenetic distances.

At step 1, we look for the pair of sites in the unflagged group that have the shorted distance between them. There are actually two pairs of sites in Table 6.1 having the same shortest distance: d(1,2)=0.6 and d(2,3)=0.6.

Therefore (panel 2) we draw lines between sites 1 and 2 and also between sites 2 and 3 then place these two pairs – (1,2) and (2,3) – in the flagged group. We write the distance values beside these lines. There are now four pairs of sites that are still unflagged – (1,3), (1,4), (2,4) and (3,4) – and so we repeat step 2 of our algorithm. The pair of sites in our unflagged group having the shortest distance between them is d(3,4)=0.9. We add a line between sites 3 and 4, and write "0.9" beside it (panel 3), and then place site 4 in the flagged group. At this point, all of the sites are connected and so we stop. The resulting figure will always be a "tree" (i.e. without any closed paths looping back onto a site) and the path linking any two sites will be the minimum spanning distance, which is why it is called a "minimum spanning tree".

Now, to get the cophenetic distances (d^*) between any two sites, we simply find shortest path linking these two sites and add up the distances along each link of the path. The cophenetic distances between sites that are joined by a single line will be the same as the original distances. The cophenetic distances between pairs of sites that are separated by at least one other site will be different. There are three such pairs of sites in our example ($d^*(1,3)=0.6+0.6=1.2$, $d^*(1,4)=0.6+0.6+0.9=2.1$ and $d^*(2,4)=0.6+0.9=1.5$) and these cophenetic distances are written in red font in Table 6.1.

spantree()
cophenetic()

The vegan package has two functions that, together, produce a distance or dissimilarity matrix containing the cophenetic distances. If your distance or dissimilarity matrix is in a matrix or data frame called "D", then the function `spantree(D)` will produce the minimum spanning tree of D, `cophenetic(spantree(D))` will produce a distance object holding the cophenetic distances of D and `plot(spantree(D))` will graph the result. You might be asking yourself why can't simply always convert our dissimilarity matrix into cophenetic distances before ordinating the result? The answer is that a minimum spanning tree is always… a tree. In other words, a site will always only be connected to, at most, two other sites. However, when the underlying environmental gradient involves several environmental variables, and therefore has more than one dimension, the sites will form a network of connections, not simply a tree of connections.

The method of flexible shortest path adjustment (Bradfield and Kenkel 1987) is a variant of the cophenetic distances, as described above. Specifically, this method only replaces the actual distances or dissimilarities by the cophenetic distances when the actual values are above some threshold value; thus, the "flexible" part of the name. For instance, you can choose to replace the actual distances with cophenetic distances only when two sites have less than or equal to some minimum number (0, 1, …) species in common. (Bradfield and Kenkel 1987) suggest that this minimum number be small (0, 1 or 2).

stepacross()

The function `stepacross()` in the vegan package implements this method of flexible shortest path adjustment by using the argument `path= "shortest"` and `toolong= x`, where x is a maximum distance or dissimilarity. The path argument has two possible values: shortest or extended. The `path=extended` choice will be explained later. The output is the lower triangle of the modified dissimilarity matrix. There are actually two ways of using `stepacross()`. You can indicate the pairs of sites in the dissimilarity matrix that you want to replace by cophenetic distances by replacing them with `NA` before calling the function and then setting `toolong=-1`. The second way is to specify the dissimilarity or distance value beyond which the value will be replaced by a cophenetic distance. This second way has an inconvenience because it is not always possible to know what dissimilarity value (i.e. `toolong=`) to choose, or even what dissimilarity value represents a case in which a pair of sites have only 0, 1, 2… species in common. This is less of a problem for dissimilarity measures that have a maximum upper bound. For instance, the Bray-Curtis dissimilarity or a Euclidean distance after a Hellinger transformation both take a value of 1.0 when there are no species in common between sites. The `find.toolong()` function will allow you to get this value for any distance or dissimilarity matrix that is calculated via `vegdist()`. `D` is a distance or dissimilarity object produced by `vegdist()` or `dist()`, `siteXspecies` is the siteXspecies matrix from

find.toolong()

which the distance object was produced and `min.common.species` is the minimum number of species in common between a pair of sites.

```
find.toolong<-
function(D,siteXspecies,min.common.species=0,return.
with.NA=FALSE){
#D is an object from class dist, produced by vegdist()
or dist()
#siteXspecies is the siteXspecies matrix from which
the distance #object was produced
#min.common.species is the minimum number of species
in common #between a pair of sites for which the
threshold distance or #dissimilarity is sought.
#if return.with.NA=TRUE then the distance matrix is
returned with NA #values for those pairs of sites
having <= the minimum number of #species in common.
  logical.siteXspecies<-
t(apply(siteXspecies,1,function(x){x>0}))
  species.in.common<-
logical.siteXspecies%*%t(logical.siteXspecies)
  D.out<-D
  n.sites<-dim(siteXspecies)[1]
  kount<-0
  n<-dim(siteXspecies)[1]
  n<-attr(D,"Size")
n<-attr(D,"Size")
  d.no.species<-rep(NA,n*(n-1)/2)
  for(i in 1:(n.sites-1)){
    for(j in (i+1):n.sites){

if(species.in.common[i,j]<=min.common.species){
        kount<-kount+1
        get.pos<-n*(i-1)-i*(i-1)/2+j-i
        d.no.species[kount]<-D[get.pos]
        if(return.with.NA)D.out[get.pos]<-NA
      }
    }
  }
```

```
   out<-min(c(d.no.species,0),na.rm=TRUE)
   if(!return.with.NA)return(out)

 if(return.with.NA)return(list(threshold=out,D.out=D.
 out,N.modified.distances=kount))
 }
```

Here is how you would convert the dissimilarity matrix in Table 6.1 (saved as `D.example`) into flexible shortest path distances (the first line converts `D.example` into a distance object since it was not produced using `vegdist()`) in which all values equal to 1 are replaced by their cophenetic distances:

```
D.example<-as.dist(D.example)
stepacross(dis=D.example,path="shortest",toolong=1,tr
ace=TRUE)
Too long or NA distances: 3 out of 6 (50.0%)
Stepping across 6 dissimilarities...
    1   2   3
2 0.6
3 1.2 0.6
4 2.1 1.5 0.9
```

The output tells us that thee of the six original pairwise distances in `D.example` had values at or above the threshold value of 1 (`toolong=1`). These three original pairwise distances were then replaced by their cophenetic distances, which are the values in red in Table 6.1. We can tell which are the cophenetic distances because they are the ones with values greater than 1.

The method of "extended dissimilarities" by De'ath (1999) is very similar to the method of flexible shortest path adjustments though it does not explicitly calculate a cophenetic distance. You choose this method by specifying the argument `path="extended"` in the `stepacross()` function. For pairs of sites that do not share any species in common (say, sites A and B), the dissimilarity of this pair of sites, d(A,B), is replaced by the sum of dissimilarities with an intermediate site which has at least some species in common with both of the sites in the pair (A and B). In other words, for each site X that shares *at least some species* with both sites A and B, you calculate S(X) = d(X,A) + d(X,B). Since there are usually many such intermediate sites, you choose the intermediate site (X*) having the minimum sum of dissimilarities between it and sites A and B and then you replace the original d(A,B) with its extended dissimilarity d*(A,B) = .d(X*,A) + d(X*,B). In most cases, the result is the same as when using flexible shortest path distances but sometimes the extended dissimilarities can be a bit larger. There is no literature[45] that compares the relative ability of flexible shortest path distances versus extended dissimilarities in recovering long environmental gradients, but you are invited to try the two methods using the simulation methods in this chapter.

Let's now compare the original PCA ordination of our underlying one-dimensional environmental gradient (Figure 6.5 left) with (i) a PCA ordination using the relative (Hellinger) abundances and (ii) with a PCA

---

[45] Of which I am aware…

ordination based on flexible shortest path distances. We can no longer use the `prcomp()` function to perform the second PCA ordination since that function requires that we input a siteXspecies matrix of abundances, not a siteXsite matrix of distances. However, a principal coordinates analysis based on a Euclidean distance matrix produces the same site scores as a principal components analysis (Chapter 4). We will use this equivalence. Here is how to do it. First, we obtain the distance matrices based on Euclidean or Hellinger distances:

```
veg.dist.Euclidean<-
vegdist(community1$community,method="euclidean")
veg.dist.Hellinger<-
vegdist(decostand(community1$community,method="hellin
ger"))
```

Next[46], using the Hellinger distance matrix as input, we replace the Hellinger distances between each pair of sites having no species in common with an `NA` values, using the `find.toolong()` function and save the result in `new.D`, which is a list containing the threshold value (1) and the Hellinger distance matrix with the `NA` values.

```
new.D<-
find.toolong(veg.dist.Hellinger,community1$community,
min.common.species=0,return.with.NA=TRUE)
```

[46] Since the Hellinger distances between pairs of sites having no species in common is always 1, we could have skipped this part and used this value in the `toolong=` argument directly in the `stepacross()` function that comes next.

Next, we replace the NA values in `new.D` with the associated cophenetic distances and save this new distance matrix as veg.dist.FSPD:

```
veg.dist.FSPD<-
stepacross(new.D$D.out,path="shortest",toolong=-1)
```

Now, we can obtain the principal coordinate ordinations for the three different distance matrices. Figure 6.7 shows the result.

```
fit.pca.Euclidean<-wcmdscale(veg.dist.Euclidean,k=2)
fit.pca.Hellinger<-wcmdscale(veg.dist.Hellinger,k=2)
fit.pca.FSPD<-wcmdscale(veg.dist.FSPD,k=2,eig=T)
```

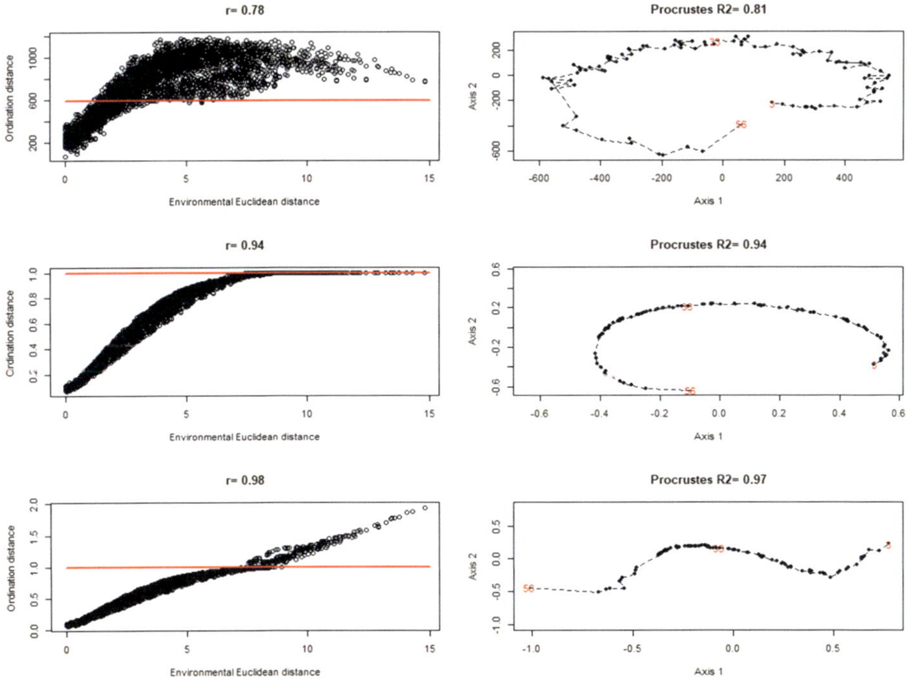

Figure 6.7. The three rows show the ordinations using the Euclidean, Hellinger and Flexible Shortest Path distances, respectively. The first column shows the relationship between the Euclidean distances between sites based on the underlying environmental variable (x axis) and the three types of distances

between sites based on the species compositions. Red numbers show the three sites having the lowest (56), highest (3) and intermediate (33) environmental values. The horizontal red line shows the threshold value for replacing the actual distances with cophenetic distances.

We again see the strong horseshoe effect when using the Euclidean distance (Figure 6.7, top row, right) and the rather poor relationship between the Euclidean distances based on species' abundances and the Euclidean distances based on the underlying environmental variable (Figure 6.7, top row, left). Converting the abundances to relative values, resulting in Hellinger distances, improves the relationship between the Euclidean distances based on species' relative abundances and the Euclidean distances based on the underlying environmental variable (Figure 6.7, middle row, left) and this reduces, but does not completely remove, the horseshoe effect (Figure 6.7, middle row, right). Using flexible shortest path distances further improves the relationship between the Euclidean distances based on species' relative abundances and the Euclidean distances based on the underlying environmental variable (Figure 6.7, bottom row, left). Notice that the first axis based on flexible shortest path distances now almost perfectly reveals the underlying environmental gradient (Figure 6.7, bottom row, right).

6.6) Recommendations for siteXspecies ordinations

We are now in a position to compare the most common distance/dissimilarity measures and the various unconstrained ordination methods that we have studied in this book. We will look at the Euclidean, Hellinger, Chi-Square, Chord and Bray-Curtis distances, since these are the ones that are most commonly recommended. In each case, we will do this with or without using flexible shortest path corrections to these distances. For each combination of distances, we will ordinate using either principal coordinates analysis (.PCoA) or non-metric multidimensional scaling (NMDS). Remember that .PCoA with Euclidean distances (with or without a Hellinger transformation) is the same as PCA and that PCoA with Chi-Square distances is the same as correspondence analysis (CA). For each simulation, we will first compare these various options in two ways. First, we will calculate the linear correlation (Pearson R^2) between the pairwise dissimilarity values between sites based on the species' abundances (or relative abundances, depending on the dissimilarity measure) and the pairwise Euclidean distances between sites based on the underlying environmental variables. R^2 values closer to 1.0 indicate a better agreement between the two. This therefore measures how well the method agrees with our fundamental assumption: *sites that are very similar in terms of their composition and abundance of species will also be very similar in terms of the values of the combination of environmental variables that control the assembly of such species. The more different the environmental conditions between sites that control the composition and abundance of species, the more different will be the communities of species found in them.* Next, we will compare these various options by calculating the Procrustes R^2 value between the site scores based on the underlying environmental variables and the site scores based on the first two ordination axes. R^2 values closer to 1.0 indicate a better agreement between the two and, therefore, a more accurate two-dimensional representation of the underlying environmental gradients. You can, of course, use more than the first two ordination axes but we will limit ourselves to two axes since this is the most common way of visualizing ordination results.

We begin by simulating the same environmental gradient as shown in Figure 6.7: a long environmental gradient consisting of species distributed along a single underlying environmental variable. The gradient is "long" because the species experience large differences in their abundances over the range of the environmental variable and this is controlled by the tolerance value (i.e. $\sqrt{5}$) relative to the range of values of the underlying environmental variables (~16) in the simulation. Here is how to generate the underlying environmental value and the resulting siteXspecies matrix of abundances:

```
set.seed(101)
env1<-gen.env(n.sites=100,r=0,n.env=1,quad=F)
```

```
community1<-
gen.community(env=env1,n.species=200,min.abund=5,max.
abund=100,tolerance=sqrt(5),random.sd=0.25)
```

The first line (`set.seed(101)`) simply initializes the random number generators so that the subsequent differences that we see between simulations will not reflect such random differences: The second line generates values for a single environmental variables (`n.env=1`) for 100 sites (`n.sites=100`). The third line calculates the abundances for 200 species (`n.species=200`) over each of the 100 sites based on the value of the single environmental variable at each site. The maximum theoretical abundance of a species is 100 (`max.abund=100`) but will randomly fluctuate between species around this value. The tolerance of all species is $\sqrt{5}$ (`tolerance=sqrt(5)`). If the simulated abundance of any species in a given site is less than five (`min.abund=5`) then its abundance is set to zero (i.e. that species is not found in that site). The last argument (`random.sd=0.25`) allows the natural logarithm of the actual abundance of a species to randomly fluctuate around the theoretical value with a standard deviation of 0.25. For instance, if the theoretical abundance is 100 then the observed abundance will be between approximately 61 – 165 about 95% of the time.

The `community1` object will now hold the observed (`community1$community`) and theoretical (`community1$predicted.community`) abundances of each of the 200 species in each of the 100 sites. We next need calculate the Euclidean distances between each pair of sites based on the underlying environmental values (`env1.dist`) and the dissimilarities between each pair of the sites based on the species' abundances using each of the dissimilarity measures (`D`). Each time we calculate a dissimilarity based on species' abundances(`D`) and then a modified dissimilarity matrix (`D.FSPD`) using flexible shortest path distances.

```
env1.dist<-vegdist(env1,method="euclidean")
D<-vegdist(community1$community,method="euclidean")
D<-vegdist(decostand(community1$community,
"hellinger"),method="euclidean")
D<-vegdist(community1$community,method="chisq")
D<-vegdist(community1$community,method="chord")
D<-vegdist(community1$community,method="bray")
new.D<-
find.toolong(D,community1$community,min.common.specie
s=0,return.with.NA=TRUE)
  D.FSPD<-
stepacross(new.D$D.out,path="shortest",toolong=-1)
```

Next, we calculate the R^2 value between the Euclidean distances based on the underlying environment and the dissimilarities based on the species (i.e. R^2(dissimilarities)):

```
r.D<-round(cor(D,env1.dist),2)
r.FSPD<-round(cor(D.FSPD,env1.dist),2)
```

Finally, we apply the principal coordinates analysis and the non-metric multidimensional scaling ordinations using each of the dissimilarity matrices and calculate the Procrustes R^2 value between two matrices: the site scores on the first two axes of the ordination and the environmental values of each site, i.e. R^2(Procrustes). Because a Procrustes analysis requires the dimensions of the two matrices to be equal, we must add extra columns of a constant (1) to the matrix having the fewer number of columns so that each matrix has the same number of rows and columns. For instance, since the above lines of code simulated a single environmental variable and so env1 is a matrix with only one column, we must create a new matrix by adding a column on 1's. Here are the relevant lines of code; you will run the last line four times, replacing `fit.PCoA` each time by each of the three other ordination objects. We can rank the results based on our two criteria by taking the average of R^2(dissimilarities) and R^2(Procrustes). The result of this series of ordinations, using this simple and strong underlying environmental gradient, is summarized in Table 6.2.

```
fit..PCoA<-wcmdscale(D,k=2,add="lingoes")
fit.nmds<-mds(D,ndim=2,type="ordinal",ties="primary")
fit.PCoA.FSPD<-wcmdscale(D.FSPD,k=2)
fit.nmds.FSPD<-
mds(D.FSPD,ndim=2,type="ordinal",ties="primary")
env.var2<-cbind(env.var,rep(1,dim(env.var)[1]))
pro.R2.PCoA<-
round(Procrustes(Y=fit.PCoA[,1:2],X=env.var2)$congcoe
f^2,2)
```

Distance measure	FSPD	Ordination	R^2 (dissimilarities)	Procrustes R^2	Score
Bray	NO	.PCoA	0.81	0.89	0.85
Bray	NO	NMDS	0.81	**0.97**	0.89
Bray	YES	.PCoA	**0.92**	**0.96**	**0.94**
Bray	YES	NMDS	**0.92**	**0.99**	**0.96**

Chi Square	NO	.PCoA	0.79	**0.93**	0.86
Chi Square	NO	NMDS	0.79	0.86	0.83
Chi Square	YES	.PCoA	0.81	**0.93**	0.87
Chi Square	YES	NMDS	0.81	**0.91**	0.86
Chord	NO	.PCoA	0.73	0.83	0.78
Chord	NO	NMDS	0.73	**0.97**	0.85
Chord	YES	.PCoA	0.88	**0.95**	**0.92**
Chord	YES	NMDS	0.88	**0.99**	**0.94**
Euclidean	NO	.PCoA	0.61	0.81	0.71
Euclidean	NO	NMDS	0.61	0.84	0.72
Euclidean	YES	.PCoA	0.76	**0.92**	0.84
Euclidean	YES	NMDS	0.76	**0.91**	0.83
Hellinger	NO	.PCoA	0.83	**0.90**	0.87
Hellinger	NO	NMDS	0.83	**0.97**	**0.90**
Hellinger	YES	.PCoA	**0.92**	**0.96**	**0.94**
Hellinger	YES	NMDS	**0.92**	**0.99**	**0.94**

Table 6.2. Summary of five different dissimilarity measures, with or without corrections for flexible shortest path distances when there are no species in common between pairs of sites, and two different ordination methods (Principal Coordinates Analysis and Non-metric Multidimensional Scaling). The flexible shortest path correction resulting in 5.1% of the community dissimilarity values being adjusted. R^2 (dissimilarities) is the R^2 value between the dissimilarities based on the underlying environmental gradient and the dissimilarities based on the species' abundances. Procrustes R^2 is the Procrustes correlation between the site scores based on the ordination and the site scores based on the underlying environmental variable.

It is clear from Table 6.2 that Euclidean distances using absolute abundances do not capture the underlying environmental gradient as well as the other dissimilarity measures. PCA using absolute abundances should be avoided when analysing siteXspecies data. This does not mean that PCA should be avoided when the data accord with its assumptions! This conclusion will be repeated for all but the shortest simulated environmental gradients. When flexible shortest path distances are not used, the ranking of the dissimilarity measures (poorest to best) in terms of the correlation between the distances based on the environmental variable and based on the community is: Euclidean < Chord < Chi Square < Bray-Curtis < Hellinger although there is virtually no difference for the last two. The ranking is essentially the same when using flexible shortest path distance corrections except that the Chord and Chi Square dissimilarities switch places. However, using flexible shortest path distance corrections always improved the result for every dissimilarity measure. I used the strictest threshold for adjusting community dissimilarity values via the `stepacross()` function since I chose to only adjust dissimilarity values between pairs of sites if they had no species in common; as a result, only 5.1% of the community dissimilarity values were adjusted by the flexible shortest path method. Using Hellinger or Bray-Curtis dissimilarities, this is the same as fixing a threshold of 1. Of course, I could choose a lower threshold for modifying the dissimilarity values. Figure 6.7 (middle panel) actually suggests that a slightly lower threshold (~0.98) would be better because the linear relationship between the Hellinger distance and the distance based on the underlying environmental gradient begins to curve before this point. The differences between PCoA and NMDS in recovering the underlying environmental gradient were not large but NMDS always performed a bit better, except when using Chi-Square dissimilarities. The general recommendation, based on this single simulated environmental gradient, is that you should use NMDS on Bray-Curtis or Hellinger distances with flexible shortest path distances. This recommendation agrees with Minchin (1987).

The above recommendation might change as we change the details of our simulated communities. For instance, what happens as we decrease the length of the underlying environmental gradient? A simple and convenient way of estimating this length is to compare the range of dissimilarity values in your data set before and after applying the flexible shortest path distance modification. For instance, the Hellinger distances in Figure 6.7 vary from 0 to 1 and a Hellinger distance of 1.0 indicates complete dissimilarity; that is, complete species turnover such that there are no species in common between two sites. The modified dissimilarities vary from 0 to ~2, meaning that the species composition completely turns over about twice along the gradient. We can simulate increasingly short environmental gradients simply by increasing the tolerance value in the `gen.community()` function without changing the range of values of the underlying environmental variable. Remember that a typical species will go from being absent, rise to its maximum abundance, and then disappear again in about 4 tolerance units in these simulations. The simulations summarized in Table 6.2 used a tolerance of $\sqrt{5}$ and so this corresponds to about 8.9 units of the underlying environmental variable. The underlying environmental variable varied over about 15 units. I will leave it to you to conduct a series of simulations like those summarized in Table 6.2 in which the tolerance is successively increased. You will see that the differences between the different dissimilarity measures, the improvement provided by the flexible shortest path distance correction, and ordination methods all decrease as the length of the environmental gradient decreases.

What happens if we increase the number of underlying environmental variables? You should definitely try this with the simulation functions. The same recommendations hold as for a single environmental variable although the overall strength of the R^2(dissimilarities) and the Procrustes R^2 will decrease as the number of environmental variables involved in the environmental gradient increases. The actual values are of little use since they depend on the many details of the simulations. The same recommendations also hold if the underlying environmental variables are correlated to various degrees although, as the strength of the correlations increase, your ability to differentiate between these environmental variables will decrease.

6.7) Implications for constrained ordination methods

The conclusions that we drew with respect to the unconstrained ordination methods also apply to the constrained ordination methods.

Since RDA includes a step in which each column of the siteXspecies matrix (i.e. the abundance of each species over the sites) is linearly regressed on the set of environmental values, and since the abundance of a species will change nonlinearly over even modestly variable environmental variables, you should not use RDA on siteXspecies data based on the absolute abundances, except over very short environmental gradients. For longer environmental gradients, you should definitely transform the siteXspecies matrix to Hellinger distances before proceeding with the RDA. However, even Hellinger distances (i.e. Euclidean distances based on relative abundances per site) will display nonlinear patterns over longer environmental gradients and so canonical correspondence analysis is probably more suited to long environmental gradients.

How can we include flexible shortest path distances with constrained ordination of siteXspecies data? We can use principal coordinate analysis using a dissimilarity matrix that has been modified using flexible shortest path distances. Distance-based redundancy analysis (db-RDA), which uses principal coordinate analysis, is the constrained principal coordinate version of RDA. In other words, you input a Hellinger or Bray-Curtis dissimilarity matrix, modified using flexible shortest path distances, into db-RDA.

6.8) Putting it all together

We have already looked at the Wisconsin understory vegetation several times in previous chapters. Here, we will put everything together and apply our recommendation: Bray-Curtis dissimilarities that are modified using flexible shortest path distances, and ordinated using PCoA and NMDS. The first line (below) calculates the Bray-Curtis dissimilarity between each pair of sites (`wis.Bray`); the siteXspecies matrix of abundances are in held in the `wisconsin[,-c(1:4)]` object. The third line uses the `stepacross()` function to replace all Bray-Curtis dissimilarities whose value is 1 (`toolong=1`) by cophenetic distances; we can do this because a value of 1.0 only occurs if the two sites have no species in common, but you could equally well use the `find.toolong()` function (second line) to do the same thing. We then perform a PCoA on the two dissimilarity matrices (`wis.Bray` and `wis.Bray.FSPD`) using the `wcmdscale()` function that performs a PCoA. Since these dissimilarity matrices will result in negative eigenvalues, we must also add a constant to correct for this via the `add="lingoes"` argument; this was explained in Chapter 4. The final two lines perform the non-metric multidimensional scaling on each of the two dissimilarity matrices via the `mds()` function of the smacofSym package.

```
wis.Bray<-vegdist(wisconsin[,-c(1:4)],method="bray")
```

```
new.D<-
find.toolong(wis.Bray,community1$community,min.common
.species=0,return.with.NA=TRUE)
wis.Bray.FSPD<-
stepacross(new.D$D.out,path="shortest",toolong=-1)
wis.Bray.FSPD<-
stepacross(wis.Bray,path="shortest",toolong=1)
fit.pcord<-wcmdscale(wis.Bray,k=2,add="lingoes")
fit.pcord.FSPD<-
wcmdscale(wis.Bray.FSPD,k=2,add="lingoes")
fit.nmds<-
mds(wis.Bray,ndim=2,type="ordinal",ties="primary")
fit.nmds.FSPD<-
mds(wis.Bray.FSPD,ndim=2,type="ordinal",ties="primary
")
```

The PCoA ordination is shown in Figure 6.8 (top, left). If we perform a multiple regression of the site scores of each axis on the environmental variables (soil organic matter content (OM), stand basal area (basal.area), soil cation exchange (cation.exchange), soil texture (soil.texture), potential evapotranspiration (pet) and the aridity index (ia)) then we can see how this combination of environmental variables relates to the site scores on each axis. These variables have been transformed as explained in the previous chapters to be approximately linearly related to each other. We display these environment variables using the `add.variable.arrows()` function that was explained in Chapter 2. This is what the arrows in Figure 6.8 are showing. Alternatively, you could use[47] the `envfit()` function in the vegan package to see how each environmental variable separately correlates with the site scores.

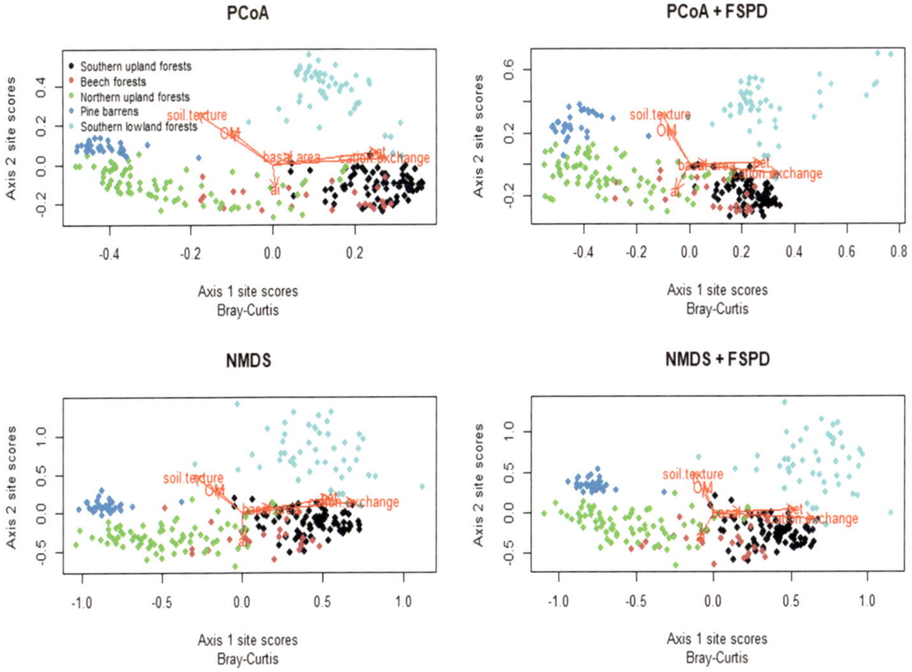

Figure 6.8. The Wisconsin understory plant communities ordinated using principal coordinates analysis (.PCoA, top row) and using non-metric multidimensional scaling (NMDS, bottom row) based on Bray-Curtis dissimilarities. Ordinations on the left row did not use flexible shortest path distances (FSPD), which replaced the Bray-Curtis dissimilarities of pairs of sites having no species in common with cophenetic distances while the ordinations on the right row did do this.

[47] `envfit()` will not work with the `mds()` function because it comes from a different package.

Differences between sites on the first axis are primarily predicted by variation in all of the environmental variables except for the aridity index. The R^2 value of the multiple regression is 0.82. Thus, sites further to the right (i.e. higher on the first axis) are ones

- with higher potential evapotranspiration and so have more heat during the growing season (i.e. above 0°C). In Wisconsin, this variable is very strongly correlated with latitude, increasing in more southerly sites;
- with higher soil cation exchange capacity and so with higher soil pH and with higher concentrations of soil Mg, Ca and K concentrations;
- with more silt and clay relative to sand and with more soil organic matter.

Differences between sites on the second axis are primarily predicted by variation in the aridity index (i.e. the ratio of mean annual precipitation to potential evapotranspiration) and, to a lesser extent, by variation in soil texture and soil organic matter content. Sites further to the top (i.e. higher on the second axis) are ones with more available water relative to temperature (i.e. less aridity) and soils with relatively more silt, clay and organic matter. The R^2 value of the multiple regression is 0.36.

The three other panels in Figure 6.8 show how replacing the Bray-Curtis dissimilarity values for pairs of sites having no species in common (i.e. flexible shortest path distances) changes the principal coordinates and non-metric multidimensional scaling ordinations with and without flexible shortest path distances. Only 1.8% (i.e. 578) of the site pairs in the

Wisconsin data had no species in common and the dissimilarities between such pairs of sites were replaced with the associated cophenetic distances. The differences between the .PCoA and NMDS ordinations are very minor. The differences between the ordination methods with and without using the flexible shortest path distances are also minor for all but the Southern lowland forest sites. Remember that the site scores in such unconstrained ordination methods are determined entirely by the patterns of abundance of these understory plants. The environmental variables play no part in the ordination; they only play a role in our *interpretation* of the resulting ordination. The two panels on the left of Figure 6.8 show the ordinations when flexible shortest path distances are not used. Notice that the Southern lowland forest sites completely overlap the Southern and Northern upland forest sites, as well as the Beech forest sites, along the first axis. The Southern lowland forest sites are only differentiated from the other three groups on the second axis. If you mentally draw straight lines from the middle of the cloud of Southern lowland forest sites to the middle of the clouds of the other three groups then such lines would have approximately the same length. This means that the understory forest plant communities in the Southern lowland forest sites are equally different from each of the other groups of plant communities using the first two axes. However, this is not the case for the ordinations using flexible shortest path distances (Figure 6.8, right). In this second set of ordinations, the Southern lowland forest sites are more similar to the Southern upland forest sites, less similar to the Beech forest sites, and less similar still to the Northern upland forest sites. The lower right panel of Figure 6.8 (NMDS + FSPD) is probably the best two-dimensional representation of these understory forest plant communities based on our simulations.

We studied three types of constrained ordinations in Chapter 4: redundancy analysis (RDA), canonical correspondence analysis (CCA) and distance-based redundancy analysis (db-RDA). As explained earlier, I do not recommend RDA of siteXspecies data based on Euclidean distances of absolute abundances because of the distortions that we have already discussed. RDA based on Hellinger distances (relative abundances per site) or CCA should be acceptable if the species are distributed along moderately long environmental gradients so long as every (or almost every) pairs of sites share some species in common. However, you should avoid these methods on longer environmental gradients that require flexible shortest path distances or if you want to use a dissimilarity measure other than Euclidean (RDA) or Chi-Square distances (CCA). In such cases, you should use distance-based redundancy analysis since this method accepts any dissimilarity measure.

As an example, we will perform a db-RDA of the Wisconsin data using Bray-Curtis dissimilarities that have been modified by flexible shortest path distances. The original vegetation classification of Wisconsin forests into the five groups (Figure 6.8) clearly describes the statewide variation in the understory plant communities and the six chosen environmental variables that have been projected onto the first two axes make biological sense but are each of these variables statistically significant? What about the variation within each vegetation group? Is the variation in understory communities within each vegetation group also significantly related to the environmental variables?

Let's begin with the first question: are all six of the chosen environmental variables statistically significant? To answer this question we first perform a db-RDA using the `capscale()` function in vegan and then test the significance of each variable using a permutation ANOVA. Since we are using a Bray-Curtis dissimilarity with cophenetic distance adjustment for pairs of sites having no species in common (in the `wis.Bray.RSPD` distance matrix that we created earlier), we must again include the `add="lingoes"`.

```
wisconsin.dbRDA<-
capscale(wis.Bray.FSPD~OM+basal.area+cation.exchange+
soil.texture+

pet+ai,data=wis.env2,sqrt.dist-FALSE,add="lingoes")
```

Next, we use the `anova()` function to perform a permutation ANOVA. Since we want to test the null hypothesis that each (transformed) environmental variable is significantly linearly associated with the variation in site scores, after taking into consideration the effects of the other environmental variables, we specify `by="margin"`. The resulting output tells us that each of the environmental variables are highly significantly related to the ordination axes.

```
anova(wisconsin.dbRDA,by="margin",model="full")
```

We next consider the second question: is the variation in understory communities within each vegetation group also significantly related to the environmental variables? This is an example of a test of conditional association. After removing the effect of (*conditional on*) the average differences between vegetation groups, does each environmental variable still explain significant variation in the community compositions between sites? We add the conditioning variable to the model, were `site.classification` is a factor variable giving the name of the vegetation type to which each site is classified. We then perform the permutation ANOVA, but now specify `model="reduced"`, since we only want to only want to test for differences between sites within each vegetation group[48]. The resulting output tells us that each of the environmental variables are highly significantly related to the ordination axes even when the test is restricted to within vegetation types.

```
wisconsin.cond.dbRDA<-
capscale(wis.Bray.FSPD~OM+basal.area+cation.exchange+
soil.texture+pet+ai+Condition(site.classification),da
ta=wis.env2, sqrt.dist=FALSE,add="lingoes")
anova(wisconsin.cond.dbRDA,by="margin",model="reduced
")
```

[48] This argument insures that the permutations of the residuals are done within each level specified by `Condition()`.

Box 6.1

```
gen.env<-function(n.sites=100,n.env=2,r=0,quad=F){
  # r is the correlation between pairs of env
variables
  #is.even tests if a number is even or odd
  is.even <- function(x) x %% 2 == 0
  L<-rnorm(n.sites,1,1)
  n<-length(L)
  env<-matrix(NA,nrow=n.sites,ncol=n.env)
  if(!quad){
    for(i in 1:n.env){
      if(r>=0)env[,i]<-r*L+rnorm(n.sites,0,sqrt(1-
r^2))
      if(r<0){
        if(is.even(i))env[,i]<--
1*r*L+rnorm(n.sites,0,sqrt(1-r^2))
        if(!is.even(i))env[,i]<-
1*r*L+rnorm(n.sites,0,sqrt(1-r^2))
        }
    }
  }
  if(quad){
    env[,1]<- (L-3)*(3*L+3)+9
    if(i>1){for(i in 2:n.env){
      env[,i]<-L+rnorm(n.sites,0,sqrt(1-r^2))
    }}
  }
#limit to ~(1,10)
  for(i in 1:n.env){
    env[,i]<-3*scale(env[,i])+5
    env[,i]<-env[,i]-min(env[,i])+1
#    env[,i]<-env[,i]-max(env[,i])+10
  }
  env
}
```

Box 6.2

```
gen.abund<-
function(env,means,tolerances,max.abund,min.abund=1,r
andom.sd){
  n.sites<-dim(env)[1]
  n.env.vars<-dim(env)[2]
#  print("Inside gen.abund")
#
print(list(tolerances=tolerances,means=means,n.env.va
rs=n.env.vars))

if(length(tolerances)!=length(means))stop("tolerances
!= means")
  if(length(tolerances)!=n.env.vars)stop("tolerances
!= n.env.vars")
  if(length(means)!=n.env.vars)stop("n.env.vars    !=
means")
  x<-(1/tolerances[1])*(env[,1]-means[1])^2
    if(n.env.vars>1){
    for(i in 2:n.env.vars){
      x<-x+(1/tolerances[i]^2)*(env[,i]-means[i])^2
    }
  }
  y<-exp(-0.5*x)
  predicted<- max.abund*y/max(y)
  y2<-exp(-0.5*x+rnorm(n.sites,mean=0,sd=random.sd))
#  observed<- max.abund*y2/max(y2)
  observed<- max.abund*y2/max(y)
  observed[observed<min.abund]<-0

data.frame(site=1:n.sites,predicted.abundance=predict
ed,observed.abundance=observed)
}
```

Box 6.3

```
gen.community<-
function(env,n.species,min.abund=5,max.abund=100,t
olerance=5,random.sd=0.25){
  #Enter the min.abundance in natural units
  #random.sd is the random variation around Gaussian
function, which is exponentiated
  #tolerance is the sd of each species around its
mean
  #min.abund is the minimum abundance at which a
species has maximum abundance at any site;
  #species less than this value are removed because
too rare
  n.sites<-dim(env)[1]
  n.env<-dim(env)[2]
#  print("Inside gen.community")
#  print(list(n.sites=n.sites,n.env=n.env))
  max.env<-max(env)
  community<-
matrix(NA,nrow=n.sites,ncol=n.species)
  predicted.community<-
matrix(NA,nrow=n.sites,ncol=n.species)
  kount<-0
  while(kount<n.species){
    #sp.mean is the mean position along each
environmental variable
    #sp.tolerances is the range around each mean
along the environmental variable
    sp.mean<-runif(n=n.env,min=1,max=max.env)
    sp.tolerances<-
rnorm(1,mean=tolerance,sd=0.2*tolerance)
    if(sp.tolerances<0)sp.tolerances<-0.1
    sp.tolerances<-rep(sp.tolerances,n.env)
    max.abund<-
rnorm(1,mean=max.abund,sd=0.2*max.abund)
    if(max.abund<1)max.abund<-1
    x<-
gen.abund(env,means=sp.mean,tolerances=sp.toleranc
es,max.abund=max.abund,

random.sd=random.sd,min.abund=min.abund)
    sp<-x$observed.abundance
```

Box 6.3 (continued)

```
#sp holds the abundance of a potential species in
each site.  Now test if this potential species
    #is above min.abund in at least one site.
    if(sum(sp>min.abund)>0){
      kount<-kount+1
      #print(kount)
      community[,kount]<-sp
      predicted.community[,kount]<-
x$predicted.abundance
    }
  }

list(site=1:n.sites,community=community,predicted.c
ommunity=predicted.community)
}
```

Chapter 7 Cheat sheet of major R functions

The major **R** functions described in this book are summarised below and are presented in alphabetical order. More information can also be obtained simply by typing help("X") in the **R** command line, where X is the name of the function, except for the few functions are new ones that are described in this book. Note that `cca()` and `rda()` are presented together under cca and that several entries include additional functions, typically for graphical display.

---

adonis (Library:vegan)

Permutational Multivariate Analysis of Variance Using Distance Matrices

Description: Analysis of variance using distance matrices for partitioning distance matrices among sources of variation and fitting linear models (e.g., factors, polynomial regression) to distance matrices; uses a permutation test with pseudo-F ratios.

Usage:
```
adonis2(formula, data, permutations = 999, method =
"bray", sqrt.dist = FALSE, add = FALSE, by = "terms",
parallel = getOption("mc.cores"), strata, ...)

adonis(formula, data, permutations = 999, method =
"bray", strata = NULL, contr.unordered = "contr.sum",
contr.ordered = "contr.poly", parallel =
getOption("mc.cores"), ...)
```

Arguments:
formula: Model formula. The LHS must be either a community data matrix or a dissimilarity matrix, e.g., from vegdist or dist. If the LHS is a data matrix, function vegdist will be used to find the dissimilarities. The RHS defines the independent variables. These can be continuous variables or factors, they can be transformed within the formula, and they can have interactions as in a typical formula. If a dissimilarity object is supplied, no species coefficients can be calculated in adonis (see Value below).
data: the data frame for the independent variables.
permutations: a list of control values for the permutations as returned by the function how, or the number of permutations required, or a permutation matrix where each row gives the permuted indices.
method: he name of any method used in vegdist to calculate pairwise distances if the left hand side of the formula was a data frame or a matrix.
sqrt.dist: Take square root of dissimilarities. This often euclidifies dissimilarities. Implemented in adonis2, ignored in adonis.
add: Add a constant to the non-diagonal dissimilarities such that all eigenvalues are non-negative in the underlying Principal Co-ordinates Analysis (see wcmdscale for details). Choice "lingoes" (or TRUE) use the recommended method of Legendre & Anderson (1999: "method 1") and "cailliez" uses their "method 2". Implemented in adonis2, ignored in adonis.
by: by = "terms" will assess significance for each term (sequentially from first to last), setting by = "margin" will assess the marginal effects of the terms (each marginal term analysed in a model with all other variables), and by = NULL will assess the overall significance of all terms together. The arguments is passed on to anova.cca. Implemented in adonis2, ignored in adonis.
strata: groups (strata) within which to constrain permutations. NB. argument strata is DEPRECATED in adonis2: you should define similar structure (blocks) in permutations. See Examples for procedures in adonis2.
contr.unordered, ***contr.ordered***: contrasts used for the design matrix (default in R is dummy or treatment contrasts for unordered factors). Ignored in adonis2.

parallel : Number of parallel processes or a predefined socket cluster. With parallel = 1 uses ordinary, non-parallel processing. The parallel processing is done with parallel package.

Value:
Function adonis2 returns an anova.cca result object with a new column for partial R-squared: This is the proportion of sum of squares from the total, and in marginal models (by = "margin") the R-squared terms do not add up to 1.

Function adonis returns an object of class "adonis" with following components:

aov.tab: Typical AOV table showing sources of variation, degrees of freedom, sequential sums of squares, mean squares, F statistics, partial R-squared and P values, based on N permutations.
coefficients: A matrix of coefficients of the linear model, with rows representing sources of variation and columns representing species; each column represents a fit of a species abundance to the linear model. These are what you get when you fit one species to your predictors. These are NOT available if you supply the distance matrix in the formula, rather than the site x species matrix
coef.sites: A matrix of coefficients of the linear model, with rows representing sources of variation and columns representing sites; each column represents a fit of a sites distances (from all other sites) to the linear model. These are what you get when you fit distances of one site to your predictors.
f.perms: An N X m matrix of the null F statistics for each source of variation based on N permutations of the data. The permutations can be inspected with permustats and its support functions.
model.matrix: The model.matrix for the right hand side of the formula.
terms: The terms component of the model.

---

boxcox (Library : MASS)

Box-Cox Transformations for Linear Models

Description : Computes and optionally plots profile log-likelihoods for the parameter of the Box-Cox power transformation.

Usage:
```
boxcox(object, lambda = seq(-2, 2, 1/10), plotit =
TRUE,  interp, eps = 1/50, xlab = expression(lambda),
ylab = "log-Likelihood", ...)
```

(Only `lm` and `aov` objects are accepted)

Arguments:
object: a formula or fitted model object. Currently only lm and aov objects are handled.
lambda: vector of values of lambda – default (-2, 2) in steps of 0.1.
plotit: logical which controls whether the result should be plotted.
interp: logical which controls whether spline interpolation is used. Default to TRUE if plotting with lambda of length less than 100.
eps: Tolerance for lambda = 0; defaults to 0.02.
xlab: efaults to "lambda".
ylab: defaults to "log-Likelihood".

Value: A list of the lambda vector and the computed profile log-likelihood vector, invisibly if the result is plotted. If plotit = TRUE plots log-likelihood vs lambda and indicates a 95% confidence interval about the maximum observed value of lambda. If interp = TRUE, spline interpolation is used to give a smoother plot.

---

capscale (Library: vegan)

[Partial] Distance-based Redundancy Analysis

Description: Distance-based redundancy analysis (dbRDA) is an ordination method similar to Redundancy Analysis (rda), but it allows non-Euclidean dissimilarity indices, such as Manhattan or Bray–Curtis distance. Despite this non-Euclidean feature, the analysis is strictly linear and metric. If called with Euclidean distance, the results are identical to rda, but dbRDA will be less efficient. Functions capscale and dbrda are constrained versions of metric scaling, a.k.a. principal coordinates analysis, which are based on the Euclidean distance but can be used, and are more useful, with other dissimilarity measures. The functions can also perform unconstrained principal coordinates analysis, optionally using extended dissimilarities.

Usage:
```
capscale(formula, data, distance = "euclidean",
sqrt.dist = FALSE,
    comm = NULL, add = FALSE,  dfun = vegdist,
metaMDSdist = FALSE,
    na.action = na.fail, subset = NULL, ...)
dbrda(formula, data, distance = "euclidean",
sqrt.dist = FALSE,
    add = FALSE, dfun = vegdist, metaMDSdist = FALSE,
    na.action = na.fail, subset = NULL, ...)
```

Arguments
formula: Model formula. The function can be called only with the formula interface. Most usual features of formula hold, especially as defined in cca and rda. The LHS must be either a community data matrix or a dissimilarity matrix, e.g., from vegdist or dist. If the LHS is a data matrix, function vegdist or function given in dfun will be used to find the dissimilarities. The RHS defines the constraints. The constraints can be continuous variables or factors, they can be transformed within the formula, and they can have interactions as in a typical formula. The RHS can have a special term Condition that defines variables to be "partialled out" before constraints, just like in rda or cca. This allows the use of partial dbRDA.
data: Data frame containing the variables on the right hand side of the model formula.
distance: The name of the dissimilarity (or distance) index if the LHS of the formula is a data frame instead of dissimilarity matrix.
sqrt.dist: ake square roots of dissimilarities.

comm: Community data frame which will be used for finding species scores when the LHS of the formula was a dissimilarity matrix. This is not used if the LHS is a data frame. If this is not supplied, the "species scores" are unavailable when dissimilarities were supplied. N.B., this is only available in capscale: dbrda does not return species scores. Function sppscores can be used to add species scores if they are missing.
add: Add a constant to the non-diagonal dissimilarities such that all eigenvalues are non-negative in the underlying Principal Co-ordinates Analysis (see wcmdscale for details). "lingoes" (or TRUE) uses the recommended method of Legendre & Anderson (1999: "method 1") and "cailliez" uses their "method 2". The latter is the only one in cmdscale.
dfun: Distance or dissimilarity function used. Any function returning standard "dist" and taking the index name as the first argument can be used.
metaMDSdist: Use metaMDSdist similarly as in metaMDS. This means automatic data transformation and using extended flexible shortest path dissimilarities (function stepacross) when there are many dissimilarities based on no shared species.
na.action: Handling of missing values in constraints or conditions. The default (na.fail) is to stop with missing values. Choices na.omit and na.exclude delete rows with missing values, but differ in representation of results. With na.omit only non-missing site scores are shown, but na.exclude gives NA for scores of missing observations. Unlike in rda, no WA scores are available for missing constraints or conditions.
subset: Subset of data rows. This can be a logical vector which is TRUE for kept observations, or a logical expression which can contain variables in the working environment, data or species names of the community data (if given in the formula or as comm argument).

Value
The functions return an object of class capscale or dbrda which inherits from rda. See cca.object for description of the result object.

---

cca and rda (Library: vegan)

[Partial] [Constrained] Correspondence Analysis and Redundancy Analysis

Description: Function `cca()` performs correspondence analysis, or optionally constrained correspondence analysis (a.k.a. canonical correspondence analysis), or optionally partial constrained correspondence analysis. Function `rda()` performs redundancy analysis, or optionally principal components analysis.

Usage:

```
cca(formula, data, na.action = na.fail, subset =
NULL, ...)
cca(X, Y, Z, ...)
rda(formula, data, scale=FALSE, na.action = na.fail,
subset = NULL, ...)
rda(X, Y, Z, scale=FALSE, ...)
```

Arguments:

formula: Model formula, where the left hand side gives the community data matrix, right hand side gives the constraining variables, and conditioning variables can be given within a special function Condition.
data: Data frame containing the variables on the right hand side of the model formula.
X: Community data matrix.
Y: Constraining matrix, typically of environmental variables. Can be missing. If this is a data.frame, it will be expanded to a model.matrix where factors are expanded to contrasts ("dummy variables"). It is better to use formula instead of this argument, and some further analyses only work when formula was used.
Z: Conditioning matrix, the effect of which is removed ("partialled out") before next step. Can be missing. If this is a data.frame, it is expanded similarly as constraining matrix.
scale: Scale species to unit variance (like correlations).
na.action: Handling of missing values in constraints or conditions. The default (na.fail) is to stop with missing value. Choice na.omit removes all rows with missing values. Choice na.exclude keeps all observations but gives NA for results that cannot be calculated. The WA scores of rows may be found also for missing values in constraints. Missing values are never allowed in dependent community data.

subset: Subset of data rows. This can be a logical vector which is TRUE for kept observations, or a logical expression which can contain variables in the working environment, data or species names of the community data.

Value :
The input data in the original ("response") form can be accessed with `fitted.cca()` and `residuals.cca()`. Function `predict.cca()` can return either working or response data, and also their lower-rank approximations. The number of observations can be accessed with `nobs.cca()`, and the residual degrees of freedom with `df.residual.cca()`. The eigenvalues of the solution can be accessed with `eigenvals.cca()`. Eigenvalues are not evaluated for partial component, and they will only be available for constrained and residual components.

The ordination scores are internally stored as (weighted) orthonormal scores matrices. These results can be accessed with `scores.cca()` and `scores.rda()` functions. The ordination scores are scaled when accessed with scores functions, but internal (weighted) orthonormal scores can be accessed by setting scaling = FALSE. Unconstrained residual component has species and site scores, and constrained component has also fitted site scores or linear combination scores for sites and biplot scores and centroids for constraint variables. The biplot scores correspond to the model.matrix, and centroids are calculated for factor variables when they were used. The scores can be selected by defining the axes, and there is no direct way of accessing all scores of a certain component. The number of dimensions can be assessed from `eigenvals()`. In addition, some other types can be derived from the results although not saved in the results. For instance, regression scores and model coefficients can be accessed with `scores()` and `coef()` functions. Partial component will have no scores.

---

cmdscale (Library: stats)

Classical (Metric) Multidimensional Scaling

Description: Classical multidimensional scaling (MDS) of a data matrix. Also known as principal coordinates analysis.

Usage: `cmdscale(d, k = 2, eig = FALSE, add = FALSE, x.ret = FALSE,list. = eig || add || x.ret)`

Arguments:
d: a distance structure such as that returned by dist or a full symmetric matrix containing the dissimilarities.
k: the maximum dimension of the space which the data are to be represented in; must be in {1, 2, …, n-1}.
eig: indicates whether eigenvalues should be returned.
add: logical indicating if an additive constant c* should be computed, and added to the non-diagonal dissimilarities such that the modified dissimilarities are Euclidean.
x.ret: indicates whether the doubly centred symmetric distance matrix should be returned.
list.: logical indicating if a list should be returned or just the n * k matrix, see 'Value:'.

Value
If .list is false (as per default), a matrix with k columns whose rows give the coordinates of the points chosen to represent the dissimilarities. Otherwise, a list containing the following components:
points: A matrix with up to k columns whose rows give the coordinates of the points chosen to represent the dissimilarities.
eig: The n eigenvalues computed during the scaling process if eig is true. NB: versions of R before 2.12.1 returned only k but were documented to return n - 1.
x: The doubly centered distance matrix if x.ret is true.
ac: The additive constant c*, 0 if add = FALSE.
GOF: A numeric vector of length 2, equal to say (g.1,g.2), where g.i = (sum{j=1..k} λ[j]) / (sum{j=1..n} T.i(λ[j])), where λ[j] are the eigenvalues (sorted in decreasing order), T.1(v) = abs(v), and T.2(v) = max(v, 0).

cophenetic (Library: stats)

Cophenetic Distances for a Hierarchical Clustering

Description: Computes the cophenetic distances for a hierarchical clustering.

Usage: `cophenetic(x)`, where x is an **R** object representing a hierarchical clustering.

Value: a list

---

corresp {MASS}   R Documentation

Simple Correspondence Analysis

Description: Find the principal canonical correlation and corresponding row- and column-scores from a correspondence analysis of a two-way contingency table.

Usage: `corresp(x, nf = 1, ...)` where x is the data (matrix or data frame) and nf is the number of dimensions (factors).

Value: An list object of class "correspondence" for which print, plot and biplot methods are supplied. The main components are the canonical correlation(s) and the row and column scores.

---

decorana (Library: vegan)

Detrended Correspondence Analysis and Basic Reciprocal Averaging

Description: Performs detrended correspondence analysis and basic reciprocal averaging or orthogonal correspondence analysis.

Usage:

```
decorana(veg, iweigh=0, iresc=4, ira=0, mk=26,
short=0,before=NULL, after=NULL)
```

```
plot(x, choices=c(1,2), origin=TRUE,
display=c("both","sites","species","none"),cex = 0.8,
cols = c(1,2), type, xlim, ylim, ...)

text(x, display = c("sites", "species"), labels,
choices = 1:2, origin = TRUE, select,  ...)

points(x, display = c("sites", "species"),
choices=1:2, origin = TRUE, select, ...)

summary(object, digits=3, origin=TRUE,
display=c("both", "species","sites","none"), ...)

print(x, head = NA, tail = head, ...)

scores(x, display=c("sites","species"), choices=1:4,
origin=TRUE, ...)
```

decostand (Library: vegan)

Standardization Methods for Community Ecology

Description: The function provides standardization methods for community ecologists.

Usage:

```
decostand(x, method, MARGIN, range.global, logbase =
2, na.rm=FALSE, ...)
```

Arguments:

x: Community data, a matrix-like object.
method: Standardization method. See Details for available options.
MARGIN: Margin, if default is not acceptable. 1 = rows, and 2 = columns of x.
range.global: Matrix from which the range is found in method = "range". This allows using same ranges across subsets of data. The dimensions of MARGIN must match with x.
logbase: The logarithm base used in method = "log".
na.rm: Ignore missing values in row or column standardizations.

The function offers following standardization methods for community data:
total: divide by margin total (default MARGIN = 1).
max: divide by margin maximum (default MARGIN = 2).
frequency: divide by margin total and multiply by the number of non-zero items, so that the average of non-zero entries is one (Oksanen 1983; default MARGIN = 2).
normalize: make margin sum of squares equal to one (default MARGIN = 1).
range: standardize values into range 0 ... 1 (default MARGIN = 2). If all values are constant, they will be transformed to 0.
rank, ***rrank***: rank replaces abundance values by their increasing ranks leaving zeros unchanged, and rrank is similar but uses relative ranks with maximum 1 (default MARGIN = 1). Average ranks are used for tied values.
standardize: scale x to zero mean and unit variance (default MARGIN = 2).
pa: scale x to presence/absence scale (0/1).
chi.square: divide by row sums and square root of column sums, and adjust for square root of matrix total (Legendre & Gallagher 2001). When used with the Euclidean distance, the distances should be similar to the Chi-square distance used in correspondence analysis. However, the results from cmdscale would still differ, since CA is a weighted ordination method (default MARGIN = 1).
hellinger: square root of method = "total" (Legendre & Gallagher 2001).
log: logarithmic transformation as suggested by Anderson et al. (2006): log_b (x) + 1 for x > 0, where b is the base of the logarithm; zeros are left as zeros. Higher bases give less weight to quantities and more to presences, and logbase = Inf gives the presence/absence scaling. Please note this is not log(x+1). Anderson et al. (2006) suggested this for their (strongly) modified Gower distance (implemented as method = "altGower" in vegdist), but the standardization can be used independently of distance indices.

All methods have a default margin. MARGIN=1 means rows (sites in a normal data set) and MARGIN=2 means columns (species in a normal data set).

---

envfit (Library: vegan)

Fits an Environmental Vector or Factor onto an Ordination

Description: The function fits environmental vectors or factors onto an ordination. The projections of points onto vectors have maximum correlation with corresponding environmental variables, and the factors show the averages of factor levels. For continuous variables this is equal to fitting a linear trend surface (plane in 2D) for a variable (see ordisurf); this trend surface can be presented by showing its gradient (direction of steepest increase) using an arrow. The environmental variables are the dependent variables that are explained by the ordination scores, and each dependent variable is analysed separately.

Usage

```
envfit(ord, env, permutations = 999, strata = NULL,
choices=c(1,2),  display = "sites", w  = weights(ord,
display), na.rm = FALSE, ...)
envfit(formula, data, ...)
plot(x, choices = c(1,2), labels, arrow.mul, at =
c(0,0), axis = FALSE, p.max = NULL, col = "blue", bg,
add = TRUE, ...)
scores(x, display, choices, ...)
vectorfit(X, P, permutations = 0, strata = NULL, w,
...)
factorfit(X, P, permutations = 0, strata = NULL, w,
...)
```

Arguments:

ord: An ordination object or other structure from which the ordination scores can be extracted (including a data frame or matrix of scores).

env: Data frame, matrix or vector of environmental variables. The variables can be of mixed type (factors, continuous variables) in data frames.

X: Matrix or data frame of ordination scores.
P: Data frame, matrix or vector of environmental variable(s). These must be continuous for vectorfit and factors or characters for factorfit.
permutations: A list of control values for the permutations as returned by the function how, or the number of permutations required, or a permutation matrix where each row gives the permuted indices. Set permutations = 0 to skip permutations.
formula, ***data***: Model formula and data.
na.rm: Remove points with missing values in ordination scores or environmental variables. The operation is casewise: the whole row of data is removed if there is a missing value and na.rm = TRUE.
x: A result object from envfit. For ordiArrowMul and ordiArrowTextXY this must be a two-column matrix (or matrix-like object) containing the coordinates of arrow heads on the two plot axes, and other methods extract such a structure from the envfit results.
choices   : Axes to plotted.
labels: Change plotting labels. The argument should be a list with elements vectors and factors which give the new plotting labels. If either of these elements is omitted, the default labels will be used. If there is only one type of elements (only vectors or only factors), the labels can be given as vector. The default labels can be displayed with labels command.
arrow.mul: Multiplier for vector lengths. The arrows are automatically scaled similarly as in plot.cca if this is not given and add = TRUE.
at: The origin of fitted arrows in the plot. If you plot arrows in other places then origin, you probably have to specify arrrow.mul.
axis: Plot axis showing the scaling of fitted arrows.
p.max: Maximum estimated P value for displayed variables. You must calculate P values with setting permutations to use this option.
col: Colour in plotting.
bg: Background colour for labels. If bg is set, the labels are displayed with ordilabel instead of text. See Examples for using semitransparent background.
add: Results added to an existing ordination plot.
strata: An integer vector or factor specifying the strata for permutation. If supplied, observations are permuted only within the specified strata.

display: In fitting functions these are ordinary site scores or linear combination scores ("lc") in constrained ordination (cca, rda, capscale). In scores function they are either "vectors" or "factors" (with synonyms "bp" or "cn", resp.).
w: Weights used in fitting (concerns mainly cca and decorana results which have nonconstant weights).

Value:
Functions vectorfit and factorfit return lists of classes vectorfit and factorfit which have a print method. The result object have the following items:
arrows: Arrow endpoints from vectorfit. The arrows are scaled to unit length.
centroids: Class centroids from factorfit.
r: Goodness of fit statistic: Squared correlation coefficient
permutations: Number of permutations.
control: A list of control values for the permutations as returned by the function how.
pvals: Empirical P-values for each variable.

---

find.toolong (this book)
Finds the tolerance level for input into `stepacross()`

Description: Finds the minimum value of a dissimilarity measure between sites that occurs when a given pair of sites have only a chosen number of species in common.

Usage:
```
find.toolong(D,siteXspecies,min.common.species=0,retu
rn.with.NA=FALSE)
```

Arguments:
`D`: Is an object from class dist, produced by `vegdist()` or `dist()`
`siteXspecies`: Is the siteXspecies matrix from which the distance object was produced

min.common.species: is the minimum number of species in common between a pair of sites for which the threshold distance or dissimilarity is sought.
return.with.NA: If TRUE then the distance matrix is returned with NA values for those pairs of sites having <= the minimum number of species in common.

Value:
The minimum dissimilarity value (threshold) between sites having the minimum common number of species in common. If return.with.NA, then, in addition to the threshold value, the dissimilarity matrix is returns with NA inserted between each pair of sites having <= the minimum number of species in common as well as the number of dissimilarity values that are to be modified.

Forkman.test.standardized.PCA (this book)
Inferential test for standardised principal components

Description: To conduct a parametric bootstrap test of significance of principal component axes obtained from a correlation matrix on centered and standardized (scaled) variables.

Usage:
```
Forkman.test.standardized.PCA(dat,alpha=0.05,N.Boot=1
000)
```

Arguments:
dat: the data matrix or data frame upon which the PCA is conducted.
alpha: the significance level
N.Boot: The number of bootstrap runs requested

Value: The null probabilities associated with each axis.

Forkman.test.unstandardized.PCA (This book)
Inferential test for unstandardised principal components

Description: To conduct a parametric bootstrap test of significance of principal component axes obtained from a covariance matrix on centered but unstandardized (unscaled) variables.

Usage:
```
Forkman.test.unstandardized.PCA(dat,alpha=0.05,N.Boot
=1000)
```

Arguments:
dat: the data matrix or data frame upon which the PCA is conducted.
alpha: the significance level
N.Boot: The number of bootstrap runs requested
Value: The null probabilities associated with each axis.

---

gen.abund (This book)
Generates species abundance values for simulations

Description: Given a matrix of environmental values per site, this function generates the abundances of a single species in each site based on its optimum and tolerance level.

Usage:
```
gen.abund(env,means,tolerances,max.abund,min.abund=1,
random.sd)
```

Arguments:
env: a matrix holding the values of each of the environmental variables (columns) in each of the sites (rows); this matrix is normally generated using `gen.env()`.
means: a numerical vector holding the optima of the species for each of the environmental variables.
tolerances: a numerical vector holding the tolerances of the species for each of the environmental variables.
max.abund: a single numerical value specifying the maximum predicted abundance of the species.

`random.sd`: a single numerical value specifying by how much the actual observed abundance of the species in a site can randomly vary from the Gaussian response curve.
Value: A matrix of abundances for each species (columns) in each site (rows).

---

gen.community (This book)

Generates the abundance of all species along an environmental gradient

Description: This function takes an input matrix of environmental values per site and generates the abundance of each of n.species species along the environmental gradient.

Usage:
```
gen.community(env,n.species,min.abund=5,max.abund=100
,tolerance=5, random.sd=0.25)
```

Arguments:
env: a matrix of environmental variables (columns) in each site (row). Usually generated from `gen.env()`.
n.species: the number of species in the simulated community.
min.abund: the minimum abundance below which the abundance in that site is set at zero.
max.abund: the average maximum predicted abundance per species.
tolerance: the tolerance of each species along the gradient
random.sd: the standard deviation of the actual ln(abundance) of species around their predicted abundances

Value: A list containing the following elements:
site: the site number
community: a matrix of species' observed abundances in each site
predicted.community: a matrix of species' predicted abundances in each site

---

gen.env (This book)

Generates environmental variables for an environmental gradient

Description: This function generates the values of a chosen number of environmental variables in each site and these variables can be correlated or nonlinearly (quadratically) related to each other.

Usage: `gen.env(n.sites=100,n.env=2,r=0,quad=F)`

Arguments :

n.sites : the number of sites to simulate

n.env: the number of environmental variables to simulate

r: the Pearson correlation coefficient specifying the correlation between the environmental variables

quad: a logical argument specifying is pairs of environmental variables should be linearly (quad=FALSE) or quadratically (quad=TRUE) related to one another

Value : a matrix of values of the environmental variables (columns) in each site (rows).

---

gowdis (Library: FD)

Gower Dissimilarity

Description: gowdis measures the Gower (1971) dissimilarity for mixed variables, including asymmetric binary variables. Variable weights can be specified. gowdis implements Podani's (1999) extension to ordinal variables.

Usage:

```
gowdis(x, w, asym.bin = NULL, ord = c("podani",
"metric", "classic"))
```

Arguments:

x: matrix or data frame containing the variables. Variables can be numeric, ordered, or factor. Symmetric or asymmetric binary variables should be numeric and only contain 0 and 1. character variables will be converted to factor. NAs are tolerated.

w: vector listing the weights for the variables in x. Can be missing, in which case all variables have equal weights.
asym.bin: vector listing the asymmetric binary variables in x.
ord: character string specifying the method to be used for ordinal variables (i.e. ordered). "podani" refers to Eqs. 2a-b of Podani (1999), while "metric" refers to his Eq. 3 (see 'details'); both options convert ordinal variables to ranks. "classic" simply treats ordinal variables as continuous variables. Can be abbreviated.

Value: An object of class dist with the following attributes: Labels, Types (the variable types, where 'C' is continuous/numeric, 'O' is ordinal, 'B' is symmetric binary, 'A' is asymmetric binary, and 'N' is nominal), Size, Metric.

---

icExplore (Library: smacof)
Exploring Initial Configurations

Description: Allows to user to explore the effect of various random starting configurations when fitting an MDS model.

Usage:
```
icExplore(delta, nrep = 100, returnfit = FALSE, ndim
= 2,type = c("ratio", "interval",
"ordinal","mspline"), weightmat = NULL, ties =
"primary",verbose = FALSE, relax = FALSE, modulus =
1, itmax = 1000, eps = 1e-6,
spline.degree = 2, spline.intKnots = 2)
```

Arguments:
delta: Either a symmetric dissimilarity matrix or an object of class "dist"
nrep: Number of initial random configurations
returnfit: If TRUE all fitted models are returned.
ndim: Number of dimensions
weightmat: Optional matrix with dissimilarity weights
type: MDS type: "interval", "ratio", "ordinal" (nonmetric MDS), or "mspline"

ties: Tie specification (ordinal MDS only): "primary", "secondary", or "tertiary"
verbose: If TRUE, replication number is printed
relax: If TRUE, block relaxation is used for majorization
modulus: Number of smacof iterations per monotone regression call
itmax: Maximum number of iterations
eps: Convergence criterion
spline.degree: Degree of the spline for "mspline" MDS type
spline.intKnots: Number of interior knots of the spline for "mspline" MDS type

---

isoMDS (Library: MASS)

Kruskal's Non-metric Multidimensional Scaling

Description: One form of non-metric multidimensional scaling

Usage:

```
isoMDS(d, y = cmdscale(d, k), k = 2, maxit = 50,
trace = TRUE,tol = 1e-3, p = 2)
Shepard(d, x, p = 2)
```

Arguments:
d: distance structure of the form returned by dist, or a full, symmetric matrix. Data are assumed to be dissimilarities or relative distances, but must be positive except for self-distance. Both missing and infinite values are allowed.
y: An initial configuration. If none is supplied, cmdscale is used to provide the classical solution, unless there are missing or infinite dissimilarities.
k: The desired dimension for the solution, passed to cmdscale.
maxit: The maximum number of iterations.
trace: Logical for tracing optimization. Default TRUE.
tol: convergence tolerance.
p: Power for Minkowski distance in the configuration space.
x: A final configuration.

Value:
points: A k-column vector of the fitted configuration.
stress: The final stress achieved (in percent).

---

isoreg (Library: stats)
Isotonic / Monotone Regression

Description: Compute the isotonic (monotonely increasing nonparametric) least squares regression which is piecewise constant.

Usage : `isoreg(x, y = NULL)`

Arguments
x, ***y***: coordinate vectors of the regression points. Alternatively a single plotting structure can be specified: see xy.coords.

Value: `isoreg()` returns an object of class isoreg which is basically a list with components
x: original (constructed) abscissa values x.
y: corresponding y values.
yf: fitted values corresponding to ordered x values.
yc: cumulative y values corresponding to ordered x values.
iKnots: integer vector giving indices where the fitted curve jumps, i.e., where the convex minorant has kinks.
isOrd: logical indicating if original x values were ordered increasingly already.
ord: if(!isOrd): integer permutation order(x) of original x.
call: the call to `isoreg()` used.

---

jackmds (Library : smacof)
SMACOF Jackknife

Description : These methods perform a SMACOF Jackknife and plot the corresponding solution.

Usage:

```
jackmds(object, eps = 1e-6, itmax = 100, verbose =
FALSE)
plot(x, plot.dim = c(1,2), hclpar = list(c = 50, l =
70),col.p, col.l, plot.lines = TRUE, legend = FALSE,
inset = c(-0.2, 0), cex.legend = 0.7,main, xlab,
ylab, xlim, ylim, asp = 1, ...)
```

Arguments:
object: Object of class "smacofB", i.e., an MDS solution from `smacofSym()`
itmax: Maximum number of iterations
eps: Convergence criterion
verbose: If TRUE, intermediate stress is printed out
x: Object of class "smacofJK"
plot.dim: Vector with dimensions to be plotted.
Hclpar: Chroma and luminance to be used for HCL colors (further details see rainbow_hcl)
col.p: Point color. If omitted, hcl colors will be used; if specified, the corresponding (single) color will be used for plotting.
col.l: Line color. If omitted, hcl colors will be used; if specified, the corresponding (single) color will be used for plotting.
plot.lines: If TRUE, the Jackknife configurations are plotted are connected with their centroid.
legend: If TRUE, the centroid labels are added as legend.
inset: Inset distance from the margins as a fraction of the plot region when legend is placed by keyword.
cex.legend: Character expansion factor for legend.
main: Plot title.
xlab: Label of x-axis.
ylab: Label of y-axis.
xlim: Scale x-axis.
ylim: Scale y-axis.
asp: Aspect ratio.

---

mds (Library : smacof)
Multidimensional Scaling

Description:Multidimensional scaling on a symmetric dissimilarity matrix using `mds()`.

Usage:
```
mds(delta, ndim = 2, type = c("ratio", "interval",
"ordinal", "mspline"), weightmat = NULL, init =
"torgerson", ties = "primary", verbose = FALSE,
 relax = FALSE, modulus = 1, itmax = 1000, eps = 1e-
06, spline.degree = 2, spline.intKnots = 2)
```

Arguments:
delta: Either a symmetric dissimilarity matrix or an object of class "dist"
ndim: Number of dimensions
weightmat: Optional matrix with dissimilarity weights
init: Either "torgerson" (classical scaling starting solution), "random" (random configuration), or a user-defined matrix
type: MDS type: "interval", "ratio", "ordinal" (nonmetric MDS), or "mspline"
ties: Tie specification (ordinal MDS only): "primary", "secondary", or "tertiary"
verbose: If TRUE, intermediate stress is printed out
relax: If TRUE, block relaxation is used for majorization
modulus: Number of smacof iterations per monotone regression call
itmax: Maximum number of iterations
eps: Convergence criterion
spline.degree: Degree of the spline for "mspline" MDS type
spline.intKnots: Number of interior knots of the spline for "mspline" MDS type

Value:
delta: Observed dissimilarities, not normalized
dhat: Disparities (transformed proximities, approximated distances, d-hats)
confdist: Configuration distances
conf: Matrix of fitted configurations
stress: Stress-1 value

spp: Stress per point (stress contribution in percentages)
resmat: Matrix with squared residuals
rss: Residual sum-of-squares
weightmat: Weight matrix
ndim: Number of dimensions
init: Starting configuration
model: Name of smacof model
niter: Number of iterations
nobj: Number of objects
Type: Type of MDS model

---

metaMDS (Library: vegan)

Nonmetric Multidimensional Scaling with Stable Solution from Random Starts, Axis Scaling and Species Scores

Description: Function metaMDS performs Nonmetric Multidimensional Scaling (NMDS), and tries to find a stable solution using several random starts. In addition, it standardizes the scaling in the result, so that the configurations are easier to interpret, and adds species scores to the site ordination. The metaMDS function does not provide actual NMDS, but it calls another function for the purpose. Currently monoMDS is the default choice, and it is also possible to call the isoMDS (MASS package).

Usage:
```
metaMDS(comm, distance = "bray", k = 2, try = 20,
trymax = 20, engine = c("monoMDS", "isoMDS"),
autotransform =TRUE, noshare = (engine == "isoMDS"),
wascores = TRUE, expand = TRUE, trace = 1, plot =
FALSE, previous.best,  ...)

plot(x, display = c("sites", "species"), choices =
c(1, 2),type = "p", shrink = FALSE,  ...)

points(x, display = c("sites", "species"),choices =
c(1,2), shrink = FALSE, select, ...)
text(x, display = c("sites", "species"), labels,
choices = c(1,2), shrink = FALSE, select, ...)
```

```
scores(x, display = c("sites", "species"), shrink =
FALSE, choices, ...)
```

Value:
Function metaMDS returns an object of class metaMDS. The final site ordination is stored in the item points, and species ordination in the item species, and the stress in item stress (NB, the scaling of the stress depends on the engine: isoMDS uses percents, and monoMDS proportions in the range 0 … 1). The other items store the information on the steps taken and the items returned by the engine function. The object has print, plot, points and text methods.

---

ord.on.env.arrows (This book)

Adds arrows using regression weights

Description: Adds arrows to 2D plots of ordination scores representing the regression weights of multiple regressions of additional variables on each ordination axis.

Usage:

```
ord.on.env.arrows(ordination.site.scores,env.matrix,c
hoices=c(1,2),arrow.scale=NA,arrow.col="red")
```

Arguments:

ordination.site.scores: A matrix or data frame of numerical ordination scores

env.matrix: A matrix or data frame of environmental variables, usually from `gen.env()`

choices: The axes for which the regressions are sought

arrow.scale: The scaling factor to fit arrows onto the ordination plot. IF NA then the function attempts to find a good scale

arrow.col: The colour of the added arrows

Value: A list giving the regression summary for each axis and the scaling value for the axes.

---

ordistep (vegan)

Choose a Model by Permutation Tests in Constrained Ordination

Description

Automatic stepwise model building for constrained ordination methods (cca, rda, capscale). The function ordistep is modelled after step and can do forward, backward and stepwise model selection using permutation tests. Function ordiR2step performs forward model choice solely on adjusted R2 and P-value, for ordination objects created by rda or capscale.

Usage

```
ordistep(object, scope, direction = c("both",
"backward", "forward"),Pin = 0.05, Pout = 0.1,
permutations = how(nperm = 199), steps = 50,trace =
TRUE, ...)
ordiR2step(object, scope, Pin = 0.05, R2scope = TRUE,
permutations = how(nperm = 499), trace = TRUE,
R2permutations = 1000, ...)
```

Arguments

object: In ordistep, an ordination object inheriting from cca or rda.
scope: Defines the range of models examined in the stepwise search. This can be a list containing components upper and lower, both formulae. If it is a single item, it is interpreted the target scope, depending on the direction. If direction is "forward", a single item is interpreted as the upper scope and the formula of the input object as the lower scope. See step for details. In ordiR2step, this defines the upper scope; it can also be an ordination object from with the model is extracted.
direction: The mode of stepwise search, can be one of "both", "backward", or "forward", with a default of "both". If the scope argument is missing, the default for direction is "backward" in ordistep (and ordiR2step does not have this argument, but only works forward).
Pin, ***Pout***: Limits of permutation P-values for adding (Pin) a term to the model, or dropping (Pout) from the model. Term is added if P <= Pin, and removed if P > Pout.

R2scope: Use adjusted R2 as the stopping criterion: only models with lower adjusted R2 than scope are accepted.
permutations: a list of control values for the permutations as returned by the function how, or the number of permutations required, or a permutation matrix where each row gives the permuted indices. This is passed to anova.cca: see there for details.
steps: Maximum number of iteration steps of dropping and adding terms.
trace: If positive, information is printed during the model building. Larger values may give more information.
R2permutations: Number of permutations used in the estimation of adjusted R2 for cca using RsquareAdj.

Value: Functions return the selected model with one additional component, anova, which contains brief information of steps taken. You can suppress voluminous output during model building by setting trace = FALSE, and find the summary of model history in the anova item.

---

prcomp (Library : stats)
Principal Components Analysis

Description: Performs a principal components analysis on the given data matrix and returns the results as an object of class prcomp.

Usage:
```
prcomp(formula, data = NULL, subset, na.action, ...)
prcomp(x, retx = TRUE, center = TRUE, scale. = FALSE,
tol = NULL, rank. = NULL, ...)
predict(object, newdata, ...)
```

Arguments:
formula: a formula with no response variable, referring only to numeric variables.
data: an optional data frame (or similar: see model.frame) containing the variables in the formula formula. By default the variables are taken from environment(formula).

subset: an optional vector used to select rows (observations) of the data matrix x.
na.action: a function which indicates what should happen when the data contain NAs. The default is set by the na.action setting of options, and is na.fail if that is unset. The 'factory-fresh' default is na.omit.
x: a numeric or complex matrix (or data frame) which provides the data for the principal components analysis.
Retx: a logical value indicating whether the rotated variables should be returned.
center: a logical value indicating whether the variables should be shifted to be zero centered. Alternately, a vector of length equal the number of columns of x can be supplied. The value is passed to scale.
scale.: a logical value indicating whether the variables should be scaled to have unit variance before the analysis takes place. The default is FALSE for consistency with S, but in general scaling is advisable. Alternatively, a vector of length equal the number of columns of x can be supplied. The value is passed to scale.
tol: a value indicating the magnitude below which components should be omitted. (Components are omitted if their standard deviations are less than or equal to tol times the standard deviation of the first component.) With the default null setting, no components are omitted (unless rank. is specified less than min(dim(x)).). Other settings for tol could be tol = 0 or tol = sqrt(.Machine$double.eps), which would omit essentially constant components.
rank.: optionally, a number specifying the maximal rank, i.e., maximal number of principal components to be used. Can be set as alternative or in addition to tol, useful notably when the desired rank is considerably smaller than the dimensions of the matrix.
object: object of class inheriting from "prcomp"
newdata: An optional data frame or matrix in which to look for variables with which to predict. If omitted, the scores are used. If the original fit used a formula or a data frame or a matrix with column names, newdata must contain columns with the same names. Otherwise it must contain the same number of columns, to be used in the same order.

Value: prcomp returns a list with class "prcomp" containing the following components:

sdev: the standard deviations of the principal components (i.e., the square roots of the eigenvalues of the covariance/correlation matrix, though the calculation is actually done with the singular values of the data matrix).
rotation: the matrix of variable loadings (i.e., a matrix whose columns contain the eigenvectors). The function princomp returns this in the element loadings.
x: if retx is true the value of the rotated data (the centred (and scaled if requested) data multiplied by the rotation matrix) is returned. Hence, cov(x) is the diagonal matrix diag(sdev^2). For the formula method, napredict() is applied to handle the treatment of values omitted by the na.action.
center, ***scale***: the centering and scaling used, or FALSE.

---

princomp (Library : stats)
Principal Components Analysis

Description: princomp performs a principal components analysis on the given numeric data matrix and returns the results as an object of class princomp.

Usage:
```
princomp(formula, data = NULL, subset, na.action,
...)
princomp(x, cor = FALSE, scores = TRUE, covmat =
NULL, subset = rep_len(TRUE, nrow(as.matrix(x))),
fix_sign = TRUE, ...)
predict(object, newdata, ...)
```

Arguments:
formula: a formula with no response variable, referring only to numeric variables.

data: an optional data frame (or similar: see model.frame) containing the variables in the formula formula. By default the variables are taken from environment(formula).
subset: an optional vector used to select rows (observations) of the data matrix x.
na.action: a function which indicates what should happen when the data contain NAs. The default is set by the na.action setting of options, and is na.fail if that is unset. The 'factory-fresh' default is na.omit.
x: a numeric matrix or data frame which provides the data for the principal components analysis.
cor: a logical value indicating whether the calculation should use the correlation matrix or the covariance matrix. (The correlation matrix can only be used if there are no constant variables.)
scores: a logical value indicating whether the score on each principal component should be calculated.
covmat: a covariance matrix, or a covariance list as returned by cov.wt (and cov.mve or cov.mcd from package MASS). If supplied, this is used rather than the covariance matrix of x.
fix_sign: Should the signs of the loadings and scores be chosen so that the first element of each loading is non-negative?
object: Object of class inheriting from "princomp".
newdata: An optional data frame or matrix in which to look for variables with which to predict. If omitted, the scores are used. If the original fit used a formula or a data frame or a matrix with column names, newdata must contain columns with the same names. Otherwise it must contain the same number of columns, to be used in the same order.

Value: princomp returns a list with class "princomp" containing the following components:
sdev: the standard deviations of the principal components.
loadings: the matrix of variable loadings (i.e., a matrix whose columns contain the eigenvectors). This is of class "loadings": see loadings for its print method.
center: the means that were subtracted.
scale: the scalings applied to each variable.
n.obs: the number of observations.

scores: if scores = TRUE, the scores of the supplied data on the principal components. These are non-null only if x was supplied, and if covmat was also supplied if it was a covariance list. For the formula method, `na.predict()` is applied to handle the treatment of values omitted by the na.action.
call: the matched call.
na.action: If relevant.

---

procrustes (Library: vegan)

Procrustes Rotation of Two Configurations and PROTEST

Description: Function procrustes rotates a configuration to maximum similarity with another configuration. Function protest tests the non-randomness (significance) between two configurations.

Usage :
```
procrustes(X, Y, scale = TRUE, symmetric = FALSE,
scores = "sites", ...)
summary(object, digits = getOption("digits"), ...)
plot(x, kind=1, choices=c(1,2), to.target = TRUE,
type = "p", xlab, ylab, main, ar.col = "blue",
len=0.05, cex = 0.7, ...)
points(x, display = c("target", "rotated"), choices =
c(1,2), truemean = FALSE, ...)
text(x, display = c("target", "rotated"), choices =
c(1,2), labels, truemean = FALSE, ...)
lines(x, type = c("segments", "arrows"), choices =
c(1, 2), truemean = FALSE, ...)
residuals(object, ...)
fitted(object, truemean = TRUE, ...)
predict(object, newdata, truemean = TRUE, ...)
protest(X, Y, scores = "sites", permutations =
how(nperm = 999), ...)
```

Arguments:
X: Target matrix
Y: Matrix to be rotated.

scale: Allow scaling of axes of Y.
symmetric: Use symmetric Procrustes statistic (the rotation will still be non-symmetric).
scores: Kind of scores used. This is the display argument used with the corresponding scores function: see scores, scores.cca and scores.cca for alternatives.
x, ***object***: An object of class procrustes.
digits: Number of digits in the output.
kind: For plot function, the kind of plot produced: kind = 1 plots shifts in two configurations, kind = 0 draws a corresponding empty plot, and kind = 2 plots an impulse diagram of residuals.
choices: Axes (dimensions) plotted.
xlab, ***ylab***: Axis labels, if defaults unacceptable.
main: Plot title, if default unacceptable.
display: Show only the "target" or "rotated" matrix as points.
to.target: Draw arrows to point to target.
type: The type of plot drawn. In plot, the type can be "points" or "text" to select the marker for the tail of the arrow, or "none" for drawing an empty plot. In lines the type selects either arrows or line segments to connect target and rotated configuration.
truemean: Use the original range of target matrix instead of centring the fitted values. Function plot.procrustes needs truemean = FALSE, and adding graphical items to the plots from the original results may need truemean = TRUE.
newdata: Matrix of coordinates to be rotated and translated to the target.
permutations: a list of control values for the permutations as returned by the function how, or the number of permutations required, or a permutation matrix where each row gives the permuted indices.
ar.col: Arrow colour.
len: Width of the arrow head.
labels: Character vector of text labels. Rownames of the result object are used as default.
cex: Character expansion for points or text.

Value: Function procrustes returns an object of class procrustes with items. Function protest inherits from procrustes, but amends that with some new items:
Yrot: Rotated matrix Y.
X: Target matrix.
ss: Sum of squared differences between X and Yrot.
rotation: Orthogonal rotation matrix.
translation: Translation of the origin.
scale: Scaling factor.
xmean: The centroid of the target.
symmetric: Type of ss statistic.
call: Function call.
t0: This and the following items are only in class protest: Procrustes correlation from non-permuted solution.
t: Procrustes correlations from permutations. The distribution of these correlations can be inspected with permustats function.
signif: Significance of t
permutations: Number of permutations.
control: A list of control values for the permutations as returned by the function how.
control: the list passed to argument control describing the permutation design.

Procrustes (Library: smacof)
Procrustean Similarity Transformations

Description: Solves the Procrustean problem of fitting one (MDS) configuration (testee) to another (target) MDS configuration.

Usage:
```
Procrustes(X, Y)
plot(x, plot.type = "jointplot", plot.dim = c(1,2),
main, xlab, ylab,  xlim, ylim, asp = 1, pch = 20,
col.X = "cadetblue", col.Y = "gray", col.Yhat =
"coral1",  label.conf = list(label = TRUE, pos = 3,
cex = 0.8),  arrows = TRUE, length = 0.10, legend =
```

```
list(plot = TRUE, labels = c("Target", "Testee"),
pos = "bottomright"), ...)
```

Arguments
X: Target configuration
Y: Testee configuration
x: Object of class procr
plot.type: Either "jointplot" or "transplot"
plot.dim: Vector with dimensions to be plotted.
main: Plot title.
xlab: Label of x-axis.
ylab: Label of y-axis.
xlim: Scale x-axis.
ylim: Scale y-axis.
pch: Plot symbol.
asp: Aspect ratio.
col.X: Color target configuration.
col.Y: Color testee configuration.
col.Yhat: Color transformed configuration.
label.conf: List with arguments for plotting the labels of the configurations in a configuration plot (logical valuc whether to plot labels or not, label position, label color).
length:length of the edges of the arrow head (in inches).
arrows: For "transplot" only, whether arrows should be plotted or not.
legend: List with arguments for plotting the legend.

Value: Returns an object of class procr with:
X: Input target configuration
Y:Input testee configuration
Yhat: Procrustes transformed (fitted) configuration
translation: Translation vector
dilation: Dilation factor
rotation: Rotation-reflection matrix
confdistX: Configuration distances X
confdistY: Configuration distances Y
confdistYhat : Configuration distances of fitted configuration

congcoef : Congruence coefficient
aliencoef: Alienation coefficient
pairdist : Pairwise object distances (sorted)

RsquareAdj (Library: vegan)

Adjusted R-square

Description: The function finds the adjusted R-square.

Usage
`RsquareAdj(x, n, m, ...)`
`RsquareAdj(x, ...)` (for rda)
`RsquareAdj(x, permutations = 1000, ...)` (for cca)

Arguments:
x: Unadjusted R-squared or an object from which the terms for evaluation or adjusted R-squared can be found.
n, ***m***: Number of observations and number of degrees of freedom in the fitted model.
permutations: Number of permutations to use when computing the adjusted R-squared for a cca. The permutations can be calculated in parallel by specifying the number of cores which is passed to permutest

Value: The functions return a list of items ***r.squared*** and ***adj.r.squared***.

Skalski.adonis (This book)

Analysis of distance (like ANOVA) for PCA groups

Description: Computes a parametric ANOVA, assuming normality, testing for equality between groups on specified PCA axis scores.

Usage: `Skalski.adonis(dat,PC.axes,Groups)`

Arguments:

dat: A matrix of data frame holding the PCA scores (output from prcomp or princomp) and an additional numeric column that gives the group to which each observational unit (row) belongs
PC.axes: a numeric vector specifying the PCA axes (i.e. columns in dat) to use
Groups: the column in dat which holds the group membership for each observation.

Value: a list containing the following:
Group.SS: the sum of squares associated with the grouping factor
Residual.SS: the residual sum of squares
F.stat: the Fisher's F value
df1: the numerator degrees of freedom
df2: the denominator degrees of freedom
null.prob: the null probability

---

spantree (Library: vegan)
Minimum Spanning Tree

Description: Function spantree finds a minimum spanning tree connecting all points, but disregarding dissimilarities that are at or above the threshold or NA.

Usage:
```
spantree(d, toolong = 0)
cophenetic(x)
plot(x, ord, cex = 0.7, type = "p", labels, dlim,FUN
= sammon,  ...)
lines(x, ord, display="sites", col = 1, ...)
```

Arguments
d: Dissimilarity data inheriting from class dist or a an object, such as a matrix, that can be converted to a dissimilarity matrix. Functions vegdist and dist are some functions producing suitable dissimilarity data.

toolong: Shortest dissimilarity regarded as NA. The function uses a fuzz factor, so that dissimilarities close to the limit will be made NA, too. If toolong = 0 (or negative), no dissimilarity is regarded as too long.
x: A spantree result object.
ord: An ordination configuration, or an ordination result known by scores.
cex: Character expansion factor.
type: Observations are plotted as points with type="p" or type="b", or as text label with type="t". The tree (lines) will always be plotted.
labels: Text used with type="t" or node names if this is missing.
dlim: A ceiling value used to highest cophenetic dissimilarity.
col: Colour of line segments. This can be a vector which is recycled for points, and the line colour will be a mixture of two joined points.

Value: Function spantree returns an object of class spantree which is a list with two vectors, each of length n-1. The number of links in a tree is one less the number of observations, and the first item is omitted. The items are:
kid: The child node of the parent, starting from parent number two. If there is no link from the parent, value will be NA and tree is disconnected at the node.
dist: Corresponding distance. If kid = NA, then dist = 0.
labels: Names of nodes as found from the input dissimilarities.
call: The function call.

---

stepacross (Library: vegan)

Stepacross as Flexible Shortest Paths or Extended Dissimilarities

Description: Function stepacross tries to replace dissimilarities with shortest paths stepping across intermediate sites while regarding dissimilarities above a threshold as missing data (NA). With path = "shortest" this is the flexible shortest path (Williamson 1978, Bradfield & Kenkel 1987), and with path = "extended" an approximation known as extended dissimilarities (De'ath 1999). The use of stepacross should improve the ordination with high beta diversity, when there are many sites with no species in common.

Usage: `stepacross(dis, path = "shortest", toolong = 1, trace = TRUE, ...)`

Arguments

dis: Dissimilarity data inheriting from class dist or a an object, such as a matrix, that can be converted to a dissimilarity matrix. Functions vegdist and dist are some functions producing suitable dissimilarity data.

path: The method of stepping across (partial match) Alternative "shortest" finds the shortest paths, and "extended" their approximation known as extended dissimilarities.

toolong: Shortest dissimilarity regarded as NA. The function uses a fuzz factor, so that dissimilarities close to the limit will be made NA, too.

trace: Trace the calculations.

Value: Function returns an object of class dist with extended dissimilarities (see functions vegdist and dist). The value of path is appended to the method attribute.

---

vegdist (Library: vegan)

Dissimilarity Indices for Community Ecologists

Description: The function computes dissimilarity indices that are useful for or popular with community ecologists. All indices use quantitative data, although they would be named by the corresponding binary index, but you can calculate the binary index using an appropriate argument. If you do not find your favourite index here, you can see if it can be implemented using designdist. Gower, Bray–Curtis, Jaccard and Kulczynski indices are good in detecting underlying ecological gradients (Faith et al. 1987). Morisita, Horn–Morisita, Binomial, Cao and Chao indices should be able to handle different sample sizes (Wolda 1981, Krebs 1999, Anderson & Millar 2004), and Mountford (1962) and Raup-Crick indices for presence–absence data should be able to handle unknown (and variable) sample sizes. Most of these indices are discussed by Krebs (1999) and Legendre & Legendre (2012), and their properties further compared by Wolda (1981) and Legendre & De Cáceres (2012).

Usage:

```
vegdist(x, method="bray", binary=FALSE, diag=FALSE,
upper=FALSE, na.rm = FALSE, ...)
```

Arguments:
x: Community data matrix.
method: Dissimilarity index, partial match to "manhattan", "euclidean", "canberra", "clark", "bray", "kulczynski", "jaccard", "gower", "altGower", "morisita", "horn", "mountford", "raup", "binomial", "chao", "cao", "mahalanobis", "chisq" or "chord".
binary: Perform presence/absence standardization before analysis using decostand.
diag: Compute diagonals.
upper: Return only the upper diagonal.
na.rm: Pairwise deletion of missing observations when computing dissimilarities.

...: Other parameters. These are ignored, except in method ="gower" which accepts range.global parameter of decostand.

Value: Should provide a drop-in replacement for dist and return a distance object of the same type.

Vif.cca goodness.cca and others (Library: vegan)

Diagnostic Tools for [Constrained] Ordination (CCA, RDA, DCA, CA, PCA, db-RDA)

Description: Functions goodness and inertcomp can be used to assess the goodness of fit for individual sites or species. Function vif.cca and alias.cca can be used to analyse linear dependencies among constraints and conditions. In addition, there are some other diagnostic tools (see 'Details').

Usage:

```
goodness(object, choices, display = c("species",
"sites"), model = c("CCA", "CA"), summarize = FALSE,
addprevious = FALSE, ...)
inertcomp(object, display = c("species", "sites"),
unity = FALSE, proportional = FALSE)
spenvcor(object)
intersetcor(object)
vif.cca(object)
alias(object, names.only = FALSE, ...)
```

Arguments:
object: A result object from cca, rda, dbrda or capscale.
display: Display "species" or "sites". Species are not available in dbrda and capscale.
choices: Axes shown. Default is to show all axes of the "model".
model: Show constrained ("CCA") or unconstrained ("CA") results.
summarize: Show only the accumulated total.
addprevious: Add the variation explained by previous components when statistic="explained". For model = "CCA" add conditioned (partialled out) variation, and for model = "CA" add both conditioned and constrained variation. This will give cumulative explanation. The argument has no effect when statistic="distance", but this will always show the residual distance after current axis and all previous components.
unity: Scale inertia components to unit sum (sum of all items is 1).
proportional: Give the inertia components as proportional for the corresponding total of the item (sum of each row is 1). This option takes precedence over unity.
names.only: Return only names of aliased variable(s) instead of defining equations.

Value: The functions return matrices or vectors as is appropriate.

wcmdscale (Library: vegan)

Weighted Classical (Metric) Multidimensional Scaling or PCoA

Description: Weighted classical multidimensional scaling, also known as weighted principal coordinates analysis.

Usage:
```
wcmdscale(d, k, eig = FALSE, add = FALSE, x.ret =
FALSE, w)
plot(x, choices = c(1, 2), type = "t", ...)
scores(x, choices = NA, ...)
```

Arguments:
d: a distance structure such as that returned by dist or a full symmetric matrix containing the dissimilarities.
k: the dimension of the space which the data are to be represented in; must be in {1,2,…,n-1}. If missing, all dimensions with above zero eigenvalue.
eig: indicates whether eigenvalues should be returned.
add: an additive constant c is added to the non-diagonal dissimilarities such that all n-1 eigenvalues are non-negative. Alternatives are "lingoes" (default, also used with TRUE) and "cailliez" (which is the only alternative in cmdscale). See Legendre & Anderson (1999).
x.ret: indicates whether the doubly centred symmetric distance matrix should be returned.
w: Weights of points.
x: The wcmdscale result object when the function was called with options eig = TRUE or x.ret = TRUE (See Details).
choices: Axes to be returned; NA returns all real axes.
type: Type of graph which may be "t"ext, "p"oints or "n"one.

Bibliography

Aitchison, J. 1983. Principal component analysis of compositional data. Biometrika **70**:57-65.

Akaike, H. 1973. Information theory and an extension of the maximum likelihood principle.*in* B. N. Petrov and F. Csaki, editors. Proceedings of the 2nd Intenational Symposium on Information Theory. Akademiai Kiado, Budapest.

Allen, R. G., L. S. Pereira, D. Raes, and Smith M. 1998. Crop evapotranspiration: Guidelines for computing crop requirements., Rome, Italy.

Benzécri, J.-P. 1973. L'Analyse des Données. Volume II. L'Analyse des Correspondances. Dunod, Paris.

Borcard, D., F. Gillet, and P. Legendre. 2018. Numerical ecology with R. Springer, Cham, Switzerland.

Borg, I., and P. Mair. 2017. The choice of initial configurations in multidimensional scaling: local minima, fit, and interpretability. Austrian Journal of Statistics **46**:19-23.

Box, G. E. P., and D. R. Cox. 1964. An analysis of transformations (with discussion). Journal of the Royal Statistical Society Series B **26**:211-252.

Bradfield, G. E., and N. C. Kenkel. 1987. Nonlinear ordination using flexible shortest path adjustment of ecological distances. Ecology **68**:750-753.

Burnham, K. P., and D. R. Anderson. 2002. Model selection and multimodel inference: A practical information-theoretic approach. Springer-Verlag, New York, NY.

Cailliez, F. 1983. The analytical solution of the additive constant problem. Psychometrika **48;**:305-308.

Crawley, M. J. 2013. The R book. Second edition. John Wiley & Sons, Chichester, UK.

Curtis, J. T. 1955. A prairie continuum in Wisconsin. Ecology **36**:558-566.

Curtis, J. T. 1959. The vegetation of Wisconsin. An ordination of plant communities. University of Wisconsin Press, Madison, WI, USA.

Daou, L., and B. Shipley. 2019. The measurement and quantification of generalized gradients of soil fertility relevant to plant community ecology. Ecology **100**.

De'ath, G. 1999. Extended dissimilarity: a method of robust estimation of ecological distances from high beta diversity data. Plant Ecology **144**:191-199.

De Leeuw, J. 2005. Monotonic regression. Pages 1260-1261 *in* B. S. Everitt and D. C. Howell, editors. Encyclopedia of Statistics and Behavioral Science. John Wiley & Sons, Chichester.

de Leeuw, J., and P. Mair. 2009. Multidimensional scaling using majorization: SMACOF in R. Journal of Statistical Software **31**:1-30.

De Leeuw, J., and J. Meulman. 1986. A special jackknife for multidimensional scaling. Journal of Classification **3**:97-112.

Escobar, D. F. E., F. A. O. Silveira, and L. P. C. Morellato. 2020. Do regeneration traits vary according to vegetation structure? A case study for savannas. Journal of Vegetation Science.

Faith, D. P., P. R. Minchin, and L. Belbin. 1987. Compositional dissimilarity as a robust measure of ecological distance. Vegetatio **69**:57-68.

Fenner, M., and K. Thompson. 2005. The ecology of seeds. Cambridge University Press, Cambridge UK.

Forkman, J., J. Josse, and H.-P. Piepho. 2019. Hypothesis tests for principal component analysis when variables are standardized. Journal of Agricultural, Biological and Environmental Sciences **24**:289-308.

Fox, J., and S. Weisberg. 2019. An R companion to applied regression. Sage, Thousand Oaks CA.

Gourlay, A. R., and G. A. Watson. 1973. Computational methods for matrix eigen problems. John Wiley, New York.

Gower, J. C. 1971. A general coefficient of similarity and some of its properties. . Biometrics **27**:857-871.

Gower, J. C. 1982. Euclidean distance geometry. Math. Scientist **7**:1-14.

Gower, J. C., and P. Legendre. 1986. Metric and Euclidian properties of dissimilarity coefficients. Journal of Classification **3**:5-48.

Grime, J. P. 1979. Plant strategies and vegetation processes. John Wiley and Sons, New York.

Hill, M. O. 1973. Reciprocal averaging: an eigenvector method of ordination. J. Ecol. **61**:237-249.
Hill, M. O. 1979. DECORANA: a FORTRAN program for detrended correspondence analysis and reciprocal averaging. Cornell University, Ithaca, NY.
Hill, M. O., and H. G. J. Gauch. 1980. Detrended correspondence analysis: an improved ordination technique. Vegetatio **42**:47-58.
Hirschfeld, H. O. 1935. A connection between correlation and contingency. Proc. Cambridge Philosophical Society **31**:520-524.
Hotelling, H. 1933. Analysis of a complex of statistical variables into principal components. Journal of Educational Psychology **24**:417–441, 498–520.
Hutchinson, G. E. 1957. Concluding remarks. Cold Spring Harbor Symposia on Quantitative Biology **22**: 415–427.
Hutchinson, G. E. 1978. An introduction to population ecology. Yale University Press, New Haven.
Ihm, P., and H. v. Groenewoud. 1975. A multivariate ordering of vegetation data based on Gaussian-type gradient response curves. Journal of Ecology **63**:767-777.
Keddy, P. A. 1981 Vegetation with Atlantic Coastal Plain affinities in Axe Lake, near Georgian Bay, Ontario. . Canadian Field - Naturalist **95**:241-248.
Keddy, P. A. 1983. Shoreline Vegetation in Axe Lake, Ontario: Effects of Exposure on Zonation Patterns. Ecology **64**:331-344.
Kooijman, S. A. L. M. 1977. Species abundance with optimum relations to environmental factors. Annals of Systems Research **6**:123-138.
Kruskal, J. B. 1964. Multidimensional scaling by optimizing goodness of fit to a nonmetric hypothesis. Psychometrika **29**:1-27.
Laliberté, E., and B. Shipley. 2009. FD: Measuring functional diversity from multiple traits, and other tools for functional ecology. R package. The Comprehensive R Archive Network (CRAN) URL http://cran.r-project. org/package=FD., Vienna, Austria.
Langsrud, O. 2003. ANOVA for unbalanced data: Use type II instead of Type III sums of squares. Statistics and Computing **13**:163-167.
Lazaridis, I., N. Patterson, A. Mittnik, G. Renaud, S. Mallick, K. Kirsanow, P. H. Sudmant, J. G. Schraiber, S. Castellano, M. Lipson, B. Berger,

C. Economou, R. Bollongino, Q. Fu, K. I. Bos, S. Nordenfelt, H. Li, C. De Filippo, K. Prüfer, S. Sawyer, C. Posth, W. Haak, F. Hallgren, E. Fornander, N. Rohland, D. Delsate, M. Francken, J. M. Guinet, J. Wahl, G. Ayodo, H. A. Babiker, G. Bailliet, E. Balanovska, O. Balanovsky, R. Barrantes, G. Bedoya, H. Ben-Ami, J. Bene, F. Berrada, C. M. Bravi, F. Brisighelli, G. B. J. Busby, F. Cali, M. Churnosov, D. E. C. Cole, D. Corach, L. Damba, G. Van Driem, S. Dryomov, J. M. Dugoujon, S. A. Fedorova, I. Gallego Romero, M. Gubina, M. Hammer, B. M. Henn, T. Hervig, U. Hodoglugil, A. R. Jha, S. Karachanak-Yankova, R. Khusainova, E. Khusnutdinova, R. Kittles, T. Kivisild, W. Klitz, V. Kučinskas, A. Kushniarevich, L. Laredj, S. Litvinov, T. Loukidis, R. W. Mahley, B. Melegh, E. Metspalu, J. Molina, J. Mountain, K. Näkkäläjärvi, D. Nesheva, T. Nyambo, L. Osipova, J. Parik, F. Platonov, O. Posukh, V. Romano, F. Rothhammer, I. Rudan, R. Ruizbakiev, H. Sahakyan, A. Sajantila, A. Salas, E. B. Starikovskaya, A. Tarekegn, D. Toncheva, S. Turdikulova, I. Uktveryte, O. Utevska, R. Vasquez, M. Villena, M. Voevoda, C. A. Winkler, L. Yepiskoposyan, P. Zalloua, T. Zemunik, A. Cooper, C. Capelli, M. G. Thomas, A. Ruiz-Linares, S. A. Tishkoff, L. Singh, K. Thangaraj, R. Villems, D. Comas, R. Sukernik, M. Metspalu, M. Meyer, E. E. Eichler, J. Burger, M. Slatkin, S. Pääbo, J. Kelso, D. Reich, and J. Krause. 2014. Ancient human genomes suggest three ancestral populations for present-day Europeans. Nature **513**:409-413.

Legendre, P., and L. Legendre. 2012. Numerical Ecology. Elsevier, Amsterdam.

Mahalanobis, P. C. 1936. On the generalised distance in statistics. Proceedings of the National Institute of Sciences of India **2**:49-55.

McCune, B., and M. J. Mefford. 1995. PC-Ord. Multivariate analysis of ecological data. MjM Software Design, Gleneden Beach, OR.

Minchin, P. R. 1987. An evaluation of the relative robustness of techniques for ecological ordination. . Vegetatio **69**:89-107.

Ndabarushimana, C. 2019. Effet des vers de terre exotiques sur les communautés des plantes de sous-bois de forêt du sud du Québec. Université de Sherbrooke, Sherbrooke, Québec.

Novembre, J., T. Johnson, K. Bryc, Z. Kutalik, A. R. Boyko, A. Auton, A. Indap, K. S. King, S. Bergmann, M. R. Nelson, M. Stephens, and C. D. Bustamante. 2008. Genes mirror geography within Europe. Nature **456**:98-101.

Økland, R. H. 1996. Are ordination and constrained ordination alternative or complementary strategies in general ecological studies? Journal of Vegetation Science **7**:289-292.

Oksanen, J., R. Kindt, P. Legendre, B. O'Hara, G. L. Simpson, P. Solynos, M. H. H. Stevens, and J. Wagner. 2010. Vegan: Community Ecology Package. The Comprehensive R Archive Network (CRAN) URLhttp://cran.r-project.org/package=vegan, Vienna, Austria.

Oksanen, J., and P. R. Minchin. 1997. Instability of ordination results under changes in input data order: explanations and remedies. Journal of Vegetation Science **8**:447-454.

Pearson, K. 1901. On lines and planes of closest fit to systems of points in space. The London, Edinburgh, and Dublin philosophical magazine and journal of science **2**:559–572.

Pielou, E. C. 1984. The interpretation of ecological data. A primer on classification and ordination. John WIley & Sons, NY.

Podani, J. 1999. Extending Gower's general coefficient of similarity to ordinal variables. Taxon **48**:331-240.

Press, W. H., B. P. Flannery, S. A. Teukolsky, and W. T. Vetterling. 1986. Numerical recipes. The art of scientific computing. Cambridge University Press, Cambridge.

R-Core-Team. 2020. R: A language and environment for statistical computing. R Foundation for Statistical Computing, Vienna, Austria.

Racimo, F., J. Woodbridge, R. M. Fyfe, M. Sikora, K. G. Sjögren, K. Kristiansen, and M. V. Linden. 2020. The spatiotemporal spread of human migrations during the European Holocene. Proceedings of the National Academy of Sciences of the United States of America **117**:8989-9000.

Rao, C. R. 1964. The use and interpretation of principal component analysis in applied research. Sankhya: The Indian Journal of Statistics **26**:329-358.

Raymond, M. 1950. Esquisse Phytogéographique du Quebec. Jardin Botanique de Montréal, Montreal (Quebec).

Rosenzweig, M. L. 1968. Net primary productivity of terrestrial communities: prediction from climatological data. The american Naturalist **102**:67-74.

Shaw, R. G., and T. Mitchell-Olds. 1993. ANOVA for unbalanced data: an overview. . Ecology **74**:1638-1645.

Shipley, B. 2010. From plant traits to vegetation structure: Chance and selection in the assembly of ecological communities. Cambridge University Press., Cambridge.

Shipley, B. 2016. Cause and correlation in biology: A user's guide to path analysis, structural equations and causal inference with R. Second edition. Cambridge University Press, Cambridge.

Skalski, J. R., S. M. Richins, and R. L. Townsend. 2018. A statistical test and sample size recommendations for comparing community composition following PCA. PLoS ONE **13**:e0206033.

Soetaert, K. 2019. plot3D: Plotting Multi-Dimensional Data. https://CRAN.R-project.org/package=plot3D.

Templ, M., K. Hron, and P. Filzmoser. 2011. robCompositions: an R-package for robust statistical analysis of compositional data. Pages 341-355 *in* V. Pawlowsky-Glahn and A. Buccianti, editors. Compositional Data Analysis. Theory and Applications. John Wiley & Sons, Chichester, UK.

Ter Braak, C. F., and N. M. Gremmen. 1987. Ecological amplitudes of plant species and the internal consistency of Ellenberg's indicator values for moisture. Vegetatio **69**:79-87.

Ter Braak, C. J. F. 1985. Correspondence analysis of incidence and abundance data: properties in terms of a unimodal response model. Biometrics **41**:859-873.

Ter Braak, C. J. F. 1987. CANOCO - a FORTRAN program for canonical community ordination by [partial] [detrended] [canonical] correspondence analysis, principal components analysis and redundancy analysis (version 2.1). TNO, Wageningen.

Ter Braak, C. J. F., and L. G. Barendregt. 1986. Weighted averaging of species indicator values: Its efficiency in environmental calibration. Mathematical Biosciences **78**:57-72.

ter Braak, C. J. F., and C. W. N. Looman. 1986. Weighted averaging, logistic regression and the Gaussian response model. Vegetatio **65**:3-11.

Ter Braak, C. J. F., and I. C. Prentice. 1988. A theory of gradient analysis. Advances in Ecological Research **18**:271-317.

van den Wollenberg, A. L. 1977. Redundancy analysis. An alternative for canonical correlations analysis. Psychometrika **42**:207-219.

Venables, W. N. 2002. Modern Applied Statistics with S. Fourth edition. Springer, New York.

Venables, W. N., and B. D. Ripley. 1994. Modern applied statistics with S-PLUS. Springer-Verlag, New York.

Vu, V. Q. 2011. ggbiplot: A ggplot2 based biplot. http://github.com/vqv/ggbiplot.

Waller, D. M., K. L. Amatangelo, S. E. S.E. Johnson, and D. Rogers. 2012. Wisconsin Vegetation Database - Plant community survey and resurvey data from the Wisconsin Plant Ecology Laboratory. Biodiversity and Ecology **4**:255-264.

Wartenberg, D., S. Ferson, and F. J. Rohlf. 1987. Putting things in order: a critique of detrended correspondence analysis. The american Naturalist **129(3)**:434-448.

Warton, D. I., I. J. Wright, D. S. Falster, and M. Westoby. 2006. Bivariate line-fitting methods for allometry. Biological Reviews of the Cambridge Philosophical Society **81**:259-291.

Whittaker, R. H. 1956. Vegetation of the Great Smoky Mountains. Ecological Monographs **26**:1-80.

Wickham, H. 2016. ggplot2: Elegant Graphics for Data Analysis. Springer-Verlag, New York.

Wright, I. J., P. B. Reich, M. Westoby, D. D. Ackerly, Z. Baruch, F. Bongers, J. Cavender-Bares, T. Chapin, J. H. C. Cornelissen, M. Diemer, J. Flexas, E. Garnier, P. K. Groom, J. Gulias, K. Hikosaka, B. B. Lamont, T. Lee, W. Lee, C. Lusk, J. J. Midgley, M. L. Navas, U. Niinemets, J. Oleksyn, N. Osada, H. Poorter, P. Poot, L. Prior, V. I. Pyankov, C. Roumet, S. C. Thomas, M. G. Tjoelker, E. J. Veneklaas, and R. Villar. 2004. The worldwide leaf economics spectrum. Nature **428**:821-827.

`adonis()` **444**
Aitchinson's centered log ratio transform 48
`anova()` 310
  by="axis" 311
  by="terms" 313
arch effect 146, 150, 402
aridity index 165
assembly by species filtering 119
Axe Lake 122, 123, 286
biplot 81
  distance preserving 83
  scaling 1 .... *See* biplot: distance preserving
  variable preserving 82
`biplot()` 84
Bonferonni correction 105
Box-Cox transformation 40
`boxcox()` 43, **447**
broken stick distribution 79
CA .... *See* Correspondence analysis
Cannonical correspondence analysis
  hypothesis testing 370
CANOCO 156
Canonical correspondence analysis 284, 348
  algorithm 349
`capscale()` 379, **448**
categorical nominal variable 190, 195
categorical ordinal variable 191
CCA .... *See* Canonical correspondence analysis
`cca()` 351, 354, **450**
classical multidimensional scaling .... *See* Principal coordinates analysis
`cmdscale()` 235, **452**
compositional data 47
cophenetic distance 412
`cophenetic()` 414, **453**
`corresp()` 143, **453**

Correspondence analysis 109
assumptions of 141
using decorana() 145
`dataEllipse()` 98
db-RDA .... *See* Distance-based redundancy analysis
DCA .... *See* Detrended correspondence analysis
DECORANA 156
`decorana()` 157, **453**
`decostand()` 212, 236, **454**
Detrended correspondence analysis 110, 146
compression effect 147
descaling 155
dimensionality 35
distance/dissimilarity measures 188
Bray-Curtis dissimilarity 214, 234, 237
chi-squared distance 214, 236
chord distance 211
definition of a "dissimilarity" 233
definition of a "distance" 232
Euclidean 90, 237
for binary variables 201
for categorical ordinal variables 204
Gower's distance 215
Hellinger distance 214, 237, 407
Jaccard's coefficient 203
Mahalanobis 92, 95
Manhattan distance 208
semimetric 233
simple matching coefficient 196
Distance-based redundancy analysis 284, 377, 431
algorithm 378
`envfit()` 167, 269, **456**
environmental gradient 383, 423
correlated environmental variables 394
European DNA variation 21
extended dissimilarities 418

`find.toolong()` **458**
flexible shortest path adjustment 414
Forkman test for PCA axes 80
`Forkman.test.standardized.PCA()` **459**
`Forkman.test.unstandardized.PCA()` **460**
Gaussian response function 125, 289, 388
Gaussian response regression .... *See* Gaussian response function
`gen.abund()` 389, 390, **460**
`gen.community()` 391, **461**
`gen.env()` 391, **462**
generalized linear model (glm) 128
`ggbiplot()` 99
`goodness()` **484**
`gowdis()` **462**
`gowdist()` 219
gradient analysis
  contrasted with environmental variables 117, 383
  direct 113, 126, 286
  indirect 109
  underlying assumption 385
Guttman's alienation coefficient 266
horseshoe effect 402, 421
Hotelling, Harold 20
`icExplore()` 252, 266, **463**
inertia 299
inferential statistics 284
interval variable 191
invasive earthworms 371
`isoMDS()` 247, **464**
`isoreg()` 245, **465**
isotonic regression .... *See* monotonic regression
`jackmds()` 259, **466**
marginal ANOVA 316
`mds()` 248, **467**
`metaMDS()` 248, **468**

Metric multidimensional scaling........244
minimum spanning tree........411
monoMDS()........248
monotonic regression........245
Multidimensional scaling........187, 241
multiple linear regression........294
niche........114
  and site history........119, 122, 386
  Hutchinson's realized niche........387
  optimum and tolerance........388
NMDS........*See* Non-metric multidimensional scaling
Non-metric multidimensional scaling........187, 237, 244
ord.on.env.arrows()........172, 269, **469**
ordination
  constrained........15, 281, 290, 382, 430
  indirect........12
  unconstrained........287, 382
ordistep()........338, **470**
Partial canonical correspondence analysis........368
Partial redundancy analysis........342
PCA........*See* Principal components analysis
PCoA........*See* Principal coordinates analysis
pcoa()........236
Pearson, Karl........20
permutation ANOVA........309
permutation test........308
plant zonation........121
potential evapotranspiration........165
prcomp()........61, 419, **471**
Principal components analysis........20
  assumptions........39
  biplots........81
  choosing which axes to consider........74
  covariances or correlations?........48
  eigenvalue decomposition vs. singular value decomposition........54
  loadings........66

predicting new values 66
scores 73
with more columns than rows 62
Principal coordinates analysis 187
algorithm 230
intuitive explanation 226
negative eigenvalues 235
relationship to CA 223
relationship to PCA 223
`princomp()` 60, **473**
Procrustes analysis 262, 275
`procrustes()` 396, **475**
`Procrustes()` 265, **477**
`protest()` 265
R
help files 17
R packages 17
ratio variable 193
RDA .... *See* Redundancy analysis
`rda()` 297, **450**
Reciprocal Averaging .... *See* Correspondence analyis
reciprocal averaging algorithm 131, 136
Redundancy analysis 284, 292
anova() 310
hypothesis testing 307
`rnorm()` 26
`RsquareAdj()` 318, **479**
`scale()` 51
scaling 2 biplots .... *See* biplot: variable preserving
scree plot 77
in PCoA 231
`screeplot()` 78
sequential ANOVA 315
`set.seed()` 26
Shepard diagram 277

similarity coefficient 189
simulating community assembly 387
Single Nucleotide Polymorphism ..... *See* European DNA variation
siteXspecies data 111
`Skalski.adonis()` 103, **479**
Skalski's analysis of distance (adonis) 103
SNP ..... *See* European DNA variation
Sorensen coefficient 214
`spantree()` 414, **480**
species optimum 114
species tolerance 114
standardized variables 51
  units of measurement 50
steepest decent 246
Steinhaus dissimilarity ... *See* distance/dissimilarity measures: Bray-Curtis
`stepacross()` 414, **482**
stepwise regression 338
stress 241
triplots 320, 322
  interpreting 334
Tucker's congruence coefficient 266
Type I ANOVA ..... *See* sequential ANOVA
Type III ANOVA ..... *See* marginal ANOVA
upper boundary problem 409
variance inflation factors 319
variance/covariance matrix 53
  obtaining via var() or cor() 50
  using standardized variables 54
variance/covarince matrix
  matrix is singular 47
`vegdist()` 208, 236, **483**
`vif.cca()` 319, 353, **484**
`wcmdscale()` 235, **485**
  GOF 268
weighted average 126
Wisconsin understory plants 159, 266, 432

worldwide leaf economic spectrum..57

Made in the USA
Las Vegas, NV
03 November 2026

57424224R00300